Mapping Across Academia

Stanley D. Brunn · Martin Dodge
Editors

Mapping Across Academia

Springer

Editors
Stanley D. Brunn
Department of Geography
University of Kentucky
Lexington, KY
USA

Martin Dodge
Department of Geography, School of Environment and Development
University of Manchester
Manchester
UK

ISBN 978-94-024-1009-9 ISBN 978-94-024-1011-2 (eBook)
DOI 10.1007/978-94-024-1011-2

Library of Congress Control Number: 2016955420

Printed on acid-free paper

This Springer imprint is published by Springer Nature
The registered company is Springer Science+Business Media B.V.
The registered company address is: Van Godewijckstraat 30, 3311 GX Dordrecht, The Netherlands

Preface

The editors of this volume share three common and overlapping interests. First, we are very interested in looking at space, place, and space/time intersections within geography and other disciplines, fields, and subfields. While space is an important component in geographical studies, whether personal, local, regional, or global spaces, we do not consider geography as the only discipline that studies space. This volume attests to our interest in seeing how other fields and disciplines, even in some very imaginative and creative ways, explore the meanings of space. Second, we are especially interested in how geography-related fields in the social and natural sciences and the humanities map spaces, places, and space/time relations. The contributors to this volume address mapping issues and topics, not only by presenting examples for wider scholarly communities, but also by issuing challenges to their own colleagues to focus new or renewed attention on space/place mapping. We also believe strongly that geography is not the only discipline with an interest in mapping, but a discipline that has much to help other fields and disciplines learn additional insights that can come from describing and analyzing map patterns and processes. Third, we are in solid agreement that there remains much more about mapping across the academy that awaits junior and senior scholars in traditional fields and disciplines and those willing to explore the margins and peripheries of their knowledge fields and specializations. It is those porous open frontiers and spaces where new ways of mapping and visualization can become realities. We anxiously await those scholars who work at these space and space/time intersections to see what kinds of maps and diagrams will emerge from studying architecture and language, art and biology, law and ecology, sustainability and green energy, genetics and brain mapping, psychology and religion, politics and climate change, music and language, gender and lifeworlds.

While assembling and editing this volume represents a shared effort, we readily admit that there are others who have influenced our career research, thinking, and ongoing creative efforts. The names would and do include our friends in geography and also those visionaries in related fields whose writings, interviews, and perspectives have challenged us to think and be comfortable outside our intellectual comfort zones. Some are ongoing inspirations; others influenced our thinking from writings thirty or forty years ago. To be sure, a complete list of all would be lengthy and probably also have some omissions. But we wish to identify the following as being especially important in our expansive views about geography and related disciplines. Stan Brunn would identify these geographers, which are not listed in any order of importance: Wilbur Zelinsky, Anne Buttimer, Yi-Fu Tuan, Gilbert White, John K. Wright, Peter Gould, Dick Morrill, Gillian Rose, Karl Butzer, Don Janelle, Bill Warntz, and David Lowenthal. Nongeographers would include Arthur C. Clarke, Carl Sagan, Jane Goddall, Stephen Gould, Lester Brown, Stephen Hawking, Isaac Asimov, Ray Bradbury, Alvin Toffler, John McHale, and John Naisbett.

Lexington, KY, USA Stanley D. Brunn
Manchester, UK Martin Dodge

Acknowledgements

This edited volume has taken a while to reach fruition, and we are very grateful to the forbearance of all the contributors for time it has taken, particularly the efficient authors who submitted their chapters at the beginning of the process.

Many thanks to Donna Gilbreath, Stan's longtime supportive managing editor on previous volumes, for her diligent help in preparing this manuscript for publication.

Thanks also to Graham Bowden, at the University of Manchester Cartography Unit, for his expertise in creating some of the illustrations.

Finally, we want to thank Stefan Einarson at Springer for shepherding the manuscript through the process. We are also grateful to earlier supporters at Springer, Evelien Bakker and Bernadette Deelen-Mans, who encouraged us to move forward with this pioneering volume.

Contents

What Is Where? The Role of Map Representations and Mapping Practices in Advancing Scholarship

Stanley D. Brunn and Martin Dodge

Abstract This chapter provides a general introduction to the Mapping Across Academia volume and gives some context regarding the development of cartography. We consider how the emergence of the impulse to map space had profoundly shaped how people understand the world around them and perform a wide array of spatial practices. Following on from this we discuss more contemporary developments in terms of the wholesale computerisation of cartography in the second half of the twentieth and on-going changes in mapping practices through the enrolment of digital technologies. The chapter switches track to think about the significance of the development of visual displays of intellectual ideas using spatial approaches and the recent growth in efforts to map out the structures of scientific knowledge.

Keywords Cartography · Map cultures · Geography of knowledge

1 The Emergence of the Mapping Impulse

> More mapping of more domains by more nations will probably occur in the next decade than has occurred at any time since Alexander von Humboldt 'rediscovered' the earth in the eighteenth century, and more terra incognita will be charted than ever before in history
>
> —Stephen S. Hall, *Mapping the Next Millennium* (1992: 22)

> How many maps, in the descriptive or geographical sense, might be needed to deal exhaustively with a given space, to code and decode all its meanings and content?
>
> —Henri Lefebvre, *The Production of Space* (1991: 85)

S.D. Brunn (✉)
Department of Geography, University of Kentucky, Lexington, KY, USA
e-mail: brunn@uky.edu

M. Dodge
Department of Geography, University of Manchester, Manchester, UK
e-mail: m.dodge@manchester.ac.uk

S.D. Brunn and M. Dodge (eds.), *Mapping Across Academia*,
DOI 10.1007/978-94-024-1011-2_1

We can state with a fair degree of certainly that map making in some form is as old as human history. That is, when we look at the world population 10,000 years ago the earliest human clans already had someone making or preparing maps in some way. Hunter gathers at that time probably numbered about ten million people who lived in forests, grasslands, and intermediate environments in multiple locations of small numbers on all continents. A major source of their well-being was concerned about basic survival questions, such as the sources of food, water, building materials for homes, places of safety and the locations of friends and adversaries. Place was a primary concern about livelihood. This "spatial" survival information was in the heads of these earliest tropical and subtropical people (Ingold 2000). It was not only information that was known, but also had to be conveyed to others in some way. Verbal communication was one way these earliest peoples could and did reveal information about "what was where." However, another way also became important in communication was by devising representations or ways of visually demarcating the location of something. In pre-literate and written language cultures some members of a clan or tribal group began to think of creative and imaginative ways to show "what was where." Here we are not talking about maps as we know them today, but other schematic ways to represent and portray vital geographical information for the community.

The earliest mapmakers would, could and did use various ways to depict the location of food sources, good drinking water and place settings that were safe from wild animals or neighboring groups competing for the same territories. They were also concerned about the navigable pathways and the best routes from point A to point B. Boundaries or territorial limits of livelihood spaces were also something always in the minds of the earliest human settlements as well. The earliest proto-"maps" were most likely crude representations of "what was where," made by drawing lines in the sand or mud or using branches of a tree with different sizes and spacings of twigs to show place features. Leafs and flowers and plant stems, all with geometric features, could be used as "maps" as well as they could and did convey information about "what was where" that could supplement orally conveyed knowledge. The arrangement of stones, with stratified by different sizes, colors and shapes, could also be used to display earth features in the surrounding landscape. (For more details on mapping through songs and stories, ritual dances and ceremonies, body art and tattoos, and rock pictographs in traditional and pre-industrial societies, see Woodward and Lewis 1998). The earliest proto-cartographers may have been village elders (women and men) who had the best cognitive maps of their territory and the most lengthy and widespread geographical experiences. Or they may have been someone groomed for the task by village leaders because of a good memory. Alternatively, perhaps they may have been someone with a physical disability who exhibited some artistic talents in making crude maps as a child or even drawing animals, rivers and food sources on a cave wall or carving place knowledge on a piece of bark or hard bone surface. These early cartographers were basically preparing representations of information on earth surface for others, not only hunters and gatherers, but also clan leaders, that is, those responsible for security and instructing others about "what was where?" We might also consider these early cartographers as having some mix of "science and art" talents, that is, not only representing something from an earth scale to some "map" scale, but also including some simple graphical sketches about the directions of a river, the types of trees, which prey animals were to be found in the grasslands, charting changing seasonal food sources and some symbols defining territorial limits. Even though examples of these earth-material maps made 5000–10,000 years ago do not exist today, we know from other artifacts (pottery, jewelry and clothing), rock art and cave wall sketches that creative artistry was a distinctive feature of many human groups in pre-history, whether they were hunters and gatherers or early coastal communities and sedentary agriculturalists (cf. Delano Smith 1987). We are also probably safe in assuming that the making of the earliest maps was not confined to one geographical or environmental location and then diffused out of a single hearth. Much more likely was that maps were independently constructed in the ways described above and were prepared by those living in multiple locations throughout the tropical and subtropical worlds of interior east and west Africa, riverine southwest Asia, mountainous southeast and northeast Asia and highland Central and South America. Indeed, research by geographer James Blaut and colleagues has been conducted to see how far mapping can be regarded as a cultural universal; they note that "many other recorded examples, from sites widely separated in space and time, suggest that it is at least plausible that mapping was and is employed by nearly all cultures, everywhere" (Blaut et al. 2003: 181).

Over time the making of maps became more important, not only as the number of clan members increased, but also knowledge about more distant geographical information was gathered which became important for ongoing struggle to survival. The same information about food, security and protection that was important for the earliest communities continued to be of paramount importance as did boundaries demarcated on the land in some way or in the minds of community leaders. The accumulation of more spatial and environmental knowledge meant that new maps and more related spatial representation had to be constructed using a variety of available surfaces: animal bones, tree bark, rocks on which one could carve distinctive spatial features (for broad regional reviews and exemplars, see Delano Smith 1987; Lopez 1986; Maggs 1998; Sutton 1998). Constructing maps was giving way to previous "maps on the ground" that displayed the location of something by the shape of a branch (which might refer to a river) or placement of stones (the size showing promising sites for edible plants and animal habitats). Early burial sites and graves themselves provide some evidence of Paleolithic humans being familiar with some geometric thinking; specific objects of the deceased were placed in a proper place and alignment, often with respect to some solar or lunar phenomena (full or half-moons or solstices) or some specific landform features that were important in the sites of early settlements. The maps of land star charts of the night skies that were now being constructed and prepared in greater detail would and could also be duplicated more easily and carried about. These characteristics are a vital aspect of Latour's (1990) conceptualization of maps as immutable mobiles. In this sense, even at this early time in human settlement and political history, the making of maps came to be associated not only with representing new knowledge about the human and environmental worlds, but as a source of power. Controlling what could be captured in a map and who possessed the map was important.

The earliest cartographers probably had a rather constrained level of power among other hunters, gatherers and herders in their clans and possibly even among the earliest agriculturalists because what they were depicting in some graphical form what was consider essential for human survival. However, as populations grew and more food sources became a priority and as the innate human instincts for knowing “what was beyond the immediate ecumene” increased, being able to illustrate that information and convey it in some form became essential. This included mapping the changing position of stars in the night sky, navigating seas and rivers and measuring important celestial phenomena, all which aided in the construction of calendars. It was not simply what was up or downstream from the encampment or beyond the forest or over the horizon that one observed daily, but where were the *terrae incognitae* to these earliest human families? It is indeed possible that gradually the clan artist or incipient map-maker achieved some sense of prestige within the community. They were an indispensable source of not only listening to what others reported or observed. In their nearby daily activities and more distant journeys, but also was able to depict that information on materials that would help the hunter, gatherer and early agriculturalist have more successes and stability. That information was also of increasingly use and value to the clan or tribal leaders. In this sense the mapmaker emerged as someone with respect, power and prestige within the group, not only being able to “know” what was where, but being able to place that information on recordable stable media, fixing territorial limits and aiding governing and even military expansion.

2 Expanding the Early Ecumenes

We can and would expect as populations increased, people migrated to nearby and distant locations and sedentary settlements with agricultural and economies based on primary resources evolved and also that the need for map representations and mapping skills would increase. Cartography became an essential part of daily life for settlements, not in the sense that maps became associated with a prehistoric or early historic popular culture, but that someone who was officially being designated as both the “recorder and keeper” of vital geographical information and was also presenting that information to someone (usually someone in power) within the community or group. Many of the early surviving maps held in archives and museums depict the crafts of these early community cartographers. Long before paper appeared as a major source, maps were drawn on materials that would be enduring as noted above, including carvings on large animal bones, slabs of rocks and drawings in caves. Some of these early graphics were intricate in detail with specific geographical details about rivers and tributaries, mountains, sacred sites and depictions of places of power and prestige within the community. Some of these were also rich in designs (plants, animals and stars and sun) and also in color, which suggests that even in early historical periods of human settlement, maps were valued as pieces of scientific information and art just like the earliest celestial calendars, which also required some collective knowledge of "what happened when" (full and partial moons, eclipses, equinoxes and solstices) and also representing that time information in some useful and meaningful way not only for the resident population, but also for those in positions of political and religious power. Map artifacts and mapping practices again became associated with territorial control and ownership, military exploits and also seeking knowledge for scientific or commercial advantage. The "keeper of knowledge" was not only the person or persons in authority, but also the individuals who knew how to make the maps.

The production of maps and the distribution of cartographic knowledge increased with the use of velum, parchment and paper to represent "what was where" (see Woodward 1975). The innovation of a paper, which came from China, greatly increased the use of maps for several reasons. Firstly, it reduced one of key costs, and, secondly, it was easier to draw information on these surfaces rather than having to etch them into hard material. Thirdly was that one could make corrections and additions easier as all that was need was some pen and ink to add that information. Fourth, one could reproduce a map easier, even if it required much hand labor, and make it available to more users. Map artefacts could themselves travel. The earliest paper maps, of course, were hand-drawn maps as were the earliest graphical depictions of royalty palaces, temples, monuments and even plants, animals and humans. These graphical artists, as we would call them today, were important members of the powerful political and commercial communities as what they were engaged in was collecting, distilling, updating and representing information that could be useful to support a ruler's ambitions.

Religious representations of the known world, in the form of T&O maps (*orbis terrarum*, a seventh century way of depicting the world with Jerusalem at the center), dominated, with the pictorial approach based on geographic descriptions in ancient teachings and the centrality of places such as Jerusalem and Mecca. Western cartography would develop with the adoption of locational grids that provided universal ways to locate the position of points on the surface of the earth and then record this location and represent it accurately on a nautical chart or territorial map. One of the origins for this ability to map space can be traced back to the work of Claudius Ptolemy, an influential 2nd century CE Greco-Roman astronomer, and the rediscovery of his text *Geographia* at the start of Renaissance period. The adoption of his grid of latitude and longitude enabled new maps of the world to be constructed with greater scientific fidelity compared to most of existing mediaeval *mappamundae* (Fig. 1). The story of the development of techniques and different forms of geographic map representation from the Renaissance period onwards is well considered in volume one of the *History of Cartography* (Harley and Woodward 1987; available free online at www.press.uchicago.edu/books/HOC/index.html).

Methods of surveying subsequently developed that were able to measure accurately large swaths of territory using triangulation techniques which were

Fig. 1 An early attempt at a world map based on Ptolemaic projection and locational data from *Geographia* as published by J. Angelius in Bologna, 1462. *Source* Bayerische StaatsBibliothek

advanced in the 16th century. These came to the fore as techniques underpinning the ambitious national topographic surveys that started in many Western counties (and some of their colonies) in the 18th century. This lineage leads directly to the detailed topographic cartography that forms the geographic framework for modern countries and the modes of government and commerce (Fig. 2).

While the sovereign, the church and even the early civil governments were the primary promoters of collecting of geospatial information (through royalties' sponsoring nearby surveys, and more distant discoveries (such as the Spanish/Portuguese 'age of exploration' in the fifteenth centuries), there were private ventures and sponsors wanting the same information. Maps were indeed prized and coveted documents that were deemed 'secret' and valued for supposed exclusive commercial, military and strategic purposes. Harley (1989) shows how the Casa de la Contración maintained the Padrón Real in the early 16th century as a secret master map to protect the key discoveries of Spanish explorers. In this environment the rivalry for funding knowledge-expeditions and finding "what was where" became a high priority for extra-continental exploits before and during the periods of exploration and discovery, notably with European sailors in the 14 and 15th centuries in and around Africa and Asia. The development of long distance sailing technologies went hand in hand with cartography advances. These emerging and expanding worlds of geographical knowledge fueled the demand for maps for those in political and military positions of power, traders and commercial colonists, scientific communities and religious elites.

The state became the major source for developing cartography and fueling the demand for professionally trained cartographers and early geographical scientists.

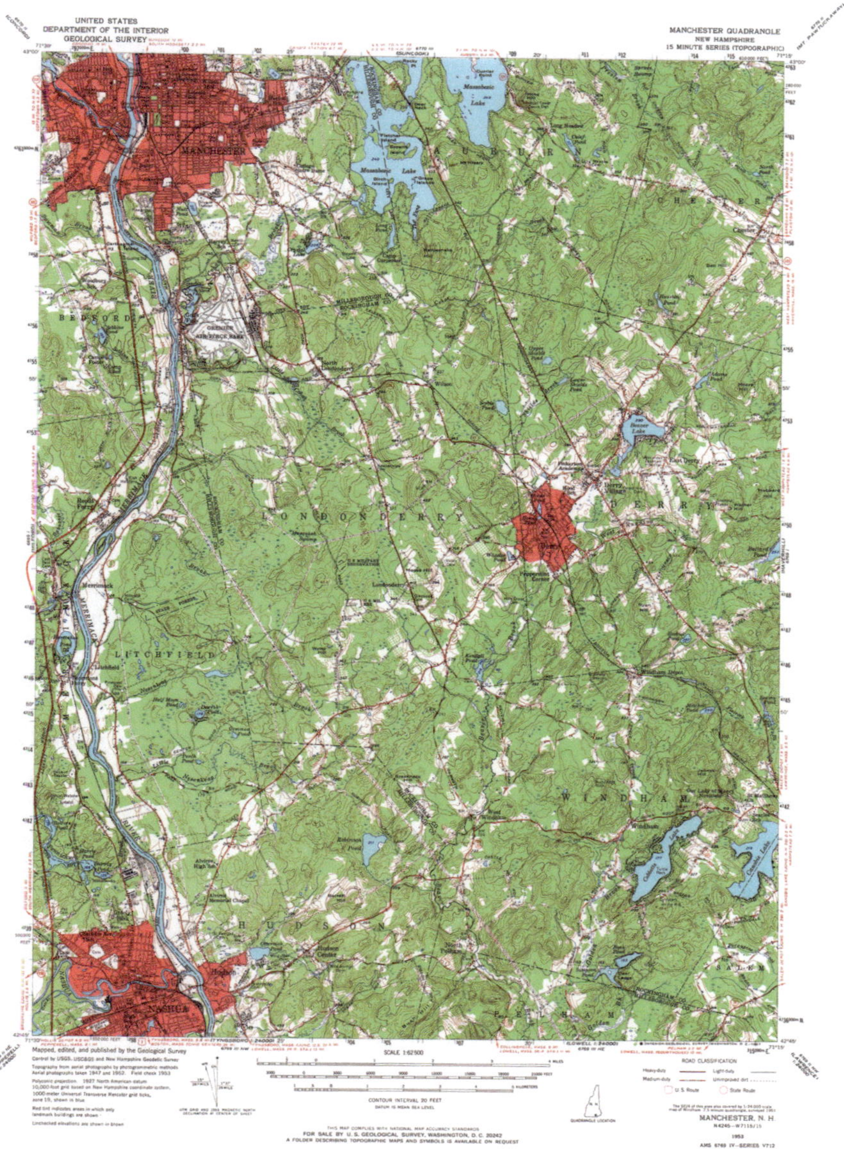

Fig. 2 An example of detail packed on a modern topographic map. This example by USGS from 1953 shows the city of Manchester NH and surrounding countryside. (*Source* USGS Historic Quad download, http://historicalmaps.arcgis.com/usgs/)

This was also the time the states were developing military institutes, which often included surveying and cartographic units, which in tandem served the state's interests of knowledge production and representation and territorial control. New and multiple single maps appeared in atlases of Africa, Asia, the Pacific, and the Western hemisphere, a far different set of *terrae incognitae* than those representing the earliest human settlement geographies and cartographies.

3 More Recent Cartographic Histories

Most of the earliest maps produced by the state or by individuals showed the "known world" or parts of the space that had been well explored and recorded, for example, the ports and coastlines shown on Portolan charts from the 13th century (Pflederer 2012). These earliest maps contained locational information about important cities and settlements as well as major physical features, including mountain ranges, rivers and road routes. Well known surviving examples, in the European context, include the Peutinger Table (circa. 13th century copy of a 4th century original) and the mid 1300s Gough map of Great Britain. However, many of these early maps also depicted areas where little was known of what was there; filling in the "empty" places with animals and strange looking peoples (to Europeans as the map producers and readers). Some places on maps delimited territories as these represented where one had or was thought to have ownership and political control.

Geographers and other spatial scientists also valued this evolving knowledge in teaching students and informing the curious public about "what was there," whether about the physical terrain, navigable rivers and port cities, new and different cultures or potential mineral possessions or valuable agricultural regions producing desired products for home countries. New world maps and atlas in the 15th century were produced by the celebrated European cartographers/publishers such as Martin Waldseemüller, Abraham Ortelius and Gerhard Mercator. Some of these new spaces also became areas for new settlements associated with colonization outside the European continent (Fig. 3).

For most of what might be considered scholarly "cartographic history," a period that would include most of the past 500 years, the words "map," "globe" or "atlas" were associated with producing the kinds of physical, environmental or exploratory maps of territories described above. Government-produced maps and commercially produced ones for the ever growing knowledge-hungry public, especially with the rapidly expanding urban-industrial populations in the 19th century and widespread primary education and promulgation of middle class sensibilities (including leisure travel and later mass tourism). These new demographic groups were curious about distant lands and changing nature of nearby places and filling in "the gaps" with information about physical environmental settings and peoples and their cultures. This cartographic and pictorial information (globes, wall hung world maps, bound atlases) became standard information for classes on geography and history. It was

Fig. 3 *The Geographer*, by Johannes Vermeer, c. 1668–1669. The scene illustrates the practices of geographical science in the 'age of discovery,' producing knowledge though topographic maps, navigation charts and terrestrial globes. A framed world nautical map hangs on the wall (Oil on canvas, 52 × 45.5 cm). (*Source* http://en.wikipedia.org/wiki/The_Geographer)

only later with more detailed information gathered on organized scientific expeditions that additional topics appeared about climates, vegetation and physical features of "newly discovered" inhabitants. (For example, consider the maps derived from the survey work of Alexander von Humboldt in central and South America at the start of the 19th century; Malcolm 1987; Fig. 5). Here we are referring to maps that would also show trade routes, heathen populations, colonial lands and economically valuable resources. Territories and boundary claims ('mine-versus-yours') and changes as a result of conflicts also entered the picture.

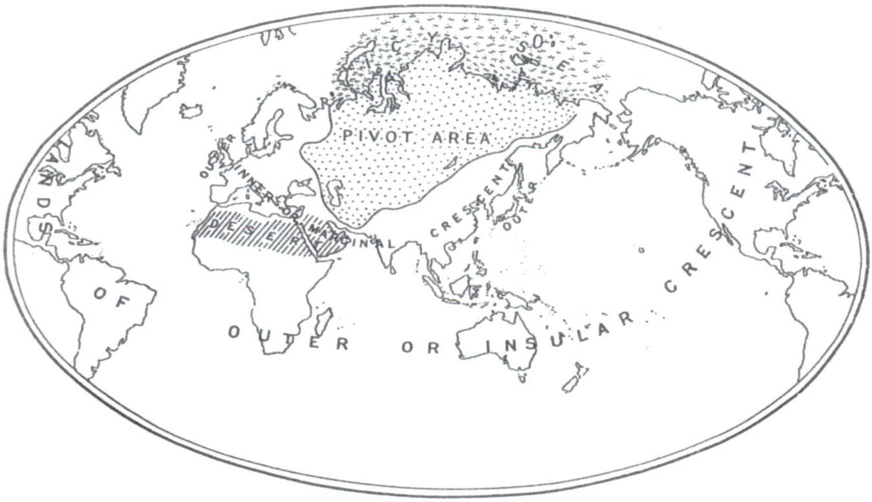

Fig. 4 The influential sketch map, entitled "The Natural Seats of Power," visually formulated the notions of zones of geopolitics. (*Source* Mackinder (1904, 435)

Fig. 5 Map of plant species on Mount Chimborazo 1803, in which Alexander von Humboldt's sought to demonstrate of the influence of altitude and climate on vegetation patterns. (*Source* R. Paselk, http://www2.humboldt.edu/scimus/AvH_HSU_Centenial Exhibit)

This is well encapsulated in geopolitical cartography, for example Halford Mackinder's famous 1904 "Geographical Pivot" of history with his 'heartlands' map (cf. Dodds and Sidaway 2004; Fig. 4). These topics not only became the standard fare for thematic maps in multiple printed public and school atlases, but also as content of elementary, high school and university geography textbooks (Cosgrove 2012). Geography to many of the teachers of the subject and students was a field of study associated with descriptive materials (graphs, globes and thematic maps) about landforms, climates and vegetation as well as about agricultural, mineral and manufacturing economies and generic cultural features (including stereotypical and racist anatomical characteristics such as skins color, facial structures, hair styles, etc. (cf. Crampton 2006), religious architecture, demographic counts, densities and trends. The development of lithographic printing techniques enabled more detailed maps to be produced more cheaply. The making of maps, or cartography, was closely associated with geography and geographers. Geographers made maps and used maps because they were considered essential elements in understanding about the planet's known and little known physical environments and cultures, and about the intersections between the natural processes and human sphere at global, regional or local scales. During the middle of the 19th century many significant national geographical societies were formed. (The French Société de Géographie was founded in 1821, the Royal Geographical Society was formed in London 1830, the American Geographical Society in New York City in 1851). Although these institutions promoted scholarship, data collection and cartography, their exploration mission was often fused to the interests of colonialism and corporate exploitation (cf. Driver 2001).

The national geographic and cartographic histories discussed above were consistent with major changes occurring in most European and North American countries throughout the first half of the 20th century. States and their scientific societies were both the major producers, consumers and disseminators of much geographical information that was collected by explorers and scientists and also reported in scholarly journals. That information not only included descriptions, as geography was still primarily a data-collecting and descriptive pursuit as were most other natural sciences at the same time, but also the production of maps with information about physical terrain, inhabited places, climates and vegetation, and probably most importantly, territorial control. That information was also the source of much of the information included in school texts at both elementary, secondary school and also university levels. Geography was considered a key discipline where the subject matter about places, economies, cultures, and environments were of central importance to the state. It also played a key role in territorial and extraterritorial conflicts and also competition for such spaces. Boundaries and borders, whether on land, in rivers or across the oceans, were important issues in resolving local, regional or global disputes. All wars in the 20th century were played out with geographical "factors" being important in possession and control, whether in Europe, Asia or places of territorial competition in South, Southeast and East Asia, the Middle East or colonies throughout the African continent. The associations of geography with national government policies and military control, including during the Cold War from the 1950s through the 1980s, are stories about map production and propaganda that only came to be topics of serious scholarly inquiry in the 1960s and 1970s, after the end of the Cold War, the emergence of the

Third and Fourth Worlds, the end of colonialism, the rise of petro economies and the later emergence of Asian 'Tiger' economies and the rise of China as a regional and world economic and political power (cf. Pickles 2000).

4 Sea Changes

While changes were occurring within the geography discipline, there were also changes emerging in other disciplines. In regards to geography the aforementioned standard "mix" of topics studied and mapped began to change from the post-war period, especially into the 1960s and 1970s with new themes and topics being studied by geographers (cf. Johnston 1996). These included economic development, environmental quality and change, and geopolitics or the intersections of power and politics. Cartography can also be added to this list. Maps and mapping were "stretching across the humanities and sciences" in ways unheard of at mid-century. Some geographers actively pursued "geography as a science" in which they sought with statistical modeling and testing to provide more "rigor" to understanding aspects of human behavior and spatial thinking. The developments in behavioral geographic and interest in the cognitive basis of spatiality broadened the understanding of the "worlds" or "environments" or "locations" studied by physical, human and human/physical geographers. Cognitive mapping or mental maps were terms that emerged and were studied by scholars with cultural, economic, political and social backgrounds (Montello 2002). A similar "scientific" disciplinary mindset was also emerging in the other social sciences. This "scientific" fabric influenced the growth in aviation along with the impact of aerial photography, both which were highly significant in changing the nature of comprehending space and envisioning place (cf. Dorrian and Pousin 2013). Sputnik, the imagery of the Earth from the Apollo missions, Skylab and Landsat and SPOT satellite remote sensing were all changing the ways scholars looked at the earth from above (the "Blue Marble" concept) and from below. Ever cheaper and more capable global telecommunications, remote sensing, computer graphics and digital cartography were integral parts of these emerging information revolutions (production, consumption, exchanges, manipulation and representations) sweeping across the sciences first and later the humanities as we discuss below. (See the debate conducted in 2000 by John Pickles and Goodchild in their papers in *Cartographic Perspective* journal.)

The emergence of transdisciplinary global development initiatives and post-colonial worlds led to focusing on a new series of topics associated with economic (the so-called 'Third World' and developing worlds as emerging fields of study), and later, social development, disasters and hazards, improved weather forecasting, income gaps and especially poverty, petro-based economies, globalization and postindustrial societies. The Cold War conflict noted above, between the Western and Eastern blocs, itself was a dominant geographical factor that was fought in part through cartography (cf. Barney 2015). It was a world with new knowledge subfields associated with transportation, trade, education, health care and diseases, tourism, social well-being, along with topics associated with race and ethnicity, language, religion, children, and gender and political/geopolitical themes related to conflicts (local and regional) including boundary changes, military power and defense agreements, political conflicts and border disputes, and developing environmental concerns related to biodiversity, endangered species, environmental protection, air and water pollution and later climate change. The spate of new topics soon began to appear as "new" maps in regional and world atlases as well as maps in primary, high school and university textbooks. This included unconventional designs, such as the use of cartograms, as maps that deliberately distort spatial scale as a way to stimulate thinking about social patterns being represented (e.g., Seager 1990, 1997; Dorling 1995; Kidron and Segal 1995). New map projections were produced and maps themselves could be produced almost instantly that are centered anywhere on the planet. Such efforts were not without controversy, a point well-illustrated by the attempt by Arno Peters (and the so-called 'Peter's Projection;) to re-frame the world in the early 1970s in the service of progressive interests with a novel equal-area projection that challenged the cartographic hegemony of the centuries' old familiar Mercator conventions (Monmonier 2005). Maps were produced not only of new and previously unmapped or little mapped places on Planet Earth, but on surfaces beyond earth: the Moon, Mars, Venus and then the diverse satellites of the outer gas giants.

In short, the field of geography was undergoing dynamic disciplinary changes, but so were the topics of maps in texts and atlases produced for the ever evolving geographically-curious public (cf. accessible review given by Hall in 1992). At the same time some of the spatial thinking and cartographic representations were starting to diffuse to other related and previously unrelated fields with common themes for the geography community, viz., surfaces, networks, landscapes, places, regions and environments.

5 Contemporary Developments

The increased social and environmental awareness that was apparent in the training of geographers in the 1970s and 1980s was increasingly evident in the atlases produced and the maps appearing in standard school and university texts. Two additional changes contributed to this evolving nature of the geography discipline and to the importance and understanding of maps and mapping. One was the introduction of new technologies associated with gathering, processing and producing representation of a "location" or "place" nature and what Michael Goodchild (2000) has characterized as the digital transition in cartography. These included computer, GIS and satellite technologies based on digital data, graphic information exchanges and powerful web-based dissemination, such as Google maps service. These innovations ushered in whole new scales of data gathering and map analysis from individual surveillance to extraterrestrial environments. Mapping went from the earth and earth surface, or from cities, countries and earth regions, to being associated with visualizing the body and the brain to studying the earth's deep interiors and outer space. Rather than distinct surveys, mapping in some domains happens continuously with results delivered in near real-time to users. No longer was cartography simply a field that was mapping earth space, but now associated with whole spectrum of spatial and temporal scales.

Coupled with these technology-related developments was a second change in what was considered geographical analysis and place knowledge in different scholarly fields. This change was evident especially across the humanities and in the natural and social sciences. There were changes in the way these subjects were considered, including some for the first time. Spatial concepts and modes of enquiry were emerging. These introductions of commonly understood concepts familiar to geographers (including map display) about place, location, networks and environment in disciplines were emerging in disciplines that had long ignored them (cf. Scholten et al. 2009). Some have gone as far as to characterize this development as a 'Spatial Turn' (Warf and Arias 2008; see also Penz et al. 2004).

For example, it is also apparent now that place has become an important concept for those studying literature, as well as for many writers, poets and lyricists. It was a theme important in understanding a society's culture or role within a larger cultural or multicultural complex (for recent reviews see Caquard and Cartwright 2014; Cooper et al. 2015; Rossetto 2014). Place itself is also an important to understanding historical and contemporary religion, language diffusion, gender and identity issues, popular culture and the human imagination. Ethnomusicologists, literary scholars, media theorists, drama, theater and museum studies began to value the importance of place, spatial relations and geographical embeddedness (e.g. in relation to human experience Van Swaaij and Klare 2000; and film, Conley 2007; Hallam and Roberts 2014). What this emergent thinking led to was a recognition that map representations and mapping practices in particular places were well worth analyzing. Particular places, regional contexts and environmental situations form unique locations where "things happened" begun to be considered by those who studied not only tourism and recreation, consumer preferences and population migration, languages and dialects, border cultures and social networks, but also those who studied transboundary economies and conflicts, international laws, regional social and economic development and planning, place promotion and preservation, infrastructure (transportation and communication) impacts and public policies. All these are examples of fields of study where maps and mapping have intersected with existing empirical techniques, modes of analysis and theoretical advancements.

The social sciences and humanities uses of map representations and mapping practices were also occurring with scholars working in various science-technology fields as well. Beyond physical geography and the earth sciences, maps are now considered fundamental to those studying the human brain and anatomy, climate modeling, natural hazard prediction and disaster planning. The "visual" world has taken on a new meaning, we would argue, with the emergence of novel maps and mapping analysis, a development that goes well beyond what has conventionally been considered as "geography." These are currently all fields where map representations and mapping practices are now considered "part and parcel" of learning about places and environments at all scales, but also making sound policies for humankind's present and future.

6 Mapping Practice Is Embedded in Academic Processes

We would argue that "mapping matters" in scholarship as a way to simplify and summarize, which is ever more crucial in science and technology, with the need to produce plausible models of reality, cope with the exponential growth in data and theoretical complexity. Mapping can help researchers work efficiently and also explore problems from new perspectives. Mapping then is part of current 'battle' for academia in utilizing the 'information explosion' and 'big data' with the hope that these developments will to lead to a better understanding and an expansion of knowledge. Concepts can be spatially envisioned and geographically mapped and also aid in our understandings and communications with other audiences. They can also be aid in efforts to convince others about the truth of the ideas (be they fellow scholars, policy-makers in government, or business investors).

Obviously, much of the above contemporary about the role for mapping can be read as part of a wider image-based working and about the primacy for the visual in scholarly practice, at least since the Renaissance. Barrow's (2008) book catalogues this well and reminds us of the long history of powerful scientific imagery that can encapsulate a complex idea and help communicate it in compelling fashion. Barrow (2008: 2) argued that:

> …some images defined our step in understanding the Universe, others have proved so effective in communicating the nature of reality that they are part of the process of thinking itself, like numbers or the letters of the alphabet. Others, equally influential, are so familiar that they appear unnoticed in the scientific process, part of the vocabulary of science that we use without thinking.

Understanding the cultural roots of visual representation as a primary means of producing believable 'truth' and convincing wider scientific communities has been the focus of much work in Science and Technology Studies. They have documented how this kind of change happens, often in mundane daily practices in labs, university classrooms and through pages of journals (e.g., Baigrie 1996; Latour 1990; Shapin and Schaffer 1985). The powerful potential of diagrams and spatial order is that they can impose provide and develop a deeper understanding to what is being studied in a spatial context, a development similar to the development of the Periodic Table and its impact on chemistry and wider material sciences (Scerri 2007). Some recent related comparative work explores these intersections, especially how different kinds of charts, diagrams, technical images, and visual practices are played out in different disciplines (Elkins 2007; Lynch and Woolgar 1990; Rose 2016).

Also, some scholars have argued that the last decade or so has seen a 'spatial turn' in a number of scholarly disciplines that have previously been a-geographical and lacking in cartographic communication (Warf and Arias 2008; Scholten et al. 2009; Penz et al. 2004). Indeed, there is much evidence of a new creation and new application of cartographic representations and map-like visual artifacts, the use of the concept of geographic location scale, spatial flows and place embeddedness (for example in explaining scientific discovery and the network of technological diffusion, Livingstone 2010), and also community mapping projects within the context of PPGIS (Public Participation GIS). Mapping as an approach and the value of GIS tools, for example, has grown in importance in the arts (e.g. scholarship around film and cinema: Conley 2007; Hallam and Roberts 2014) with some claiming the emergence of Geohumanities paradigm in recent years (Bodenhamer et al. 2010; Dear et al. 2011; Fraser 2015). This change has been defined as a "transdisciplinary and multi-methodological inquiry that begins with the human meanings of place and proceeds to reconstruct those meanings in ways that produce new knowledge and the promise of a better-informed scholarly and political practice" (Dear et al. 2011: 312). There is also active engagement in mapping practices and often the actual production of cartographic forms by a growing number of artists (see reviews in Harmon 2009; O'Rourke 2013).

Relevant to the discussion here, it is worth noting those efforts to categorize and visually map the patterns of knowledge itself. This has often been done by applying spatial structure and drawing some kind of graphical representations of the size of fields of scholarships and connections between expert practice. For example in the 1780s French philosophers of science Diderot and D'Alembert illustrated their pioneering *Encyclopedie* with hierarchical decomposition of the domains of knowledge (cf. Bender and Marrinan 2010) (Fig. 6). Subsequently many others have tried to produce genealogical trees to show change and growth in academic disciplines and fields of knowledge. Moving beyond simple hierarchy and linear branching, some have used different diagrammatic approaches and spatial arrangements to show the structure of science. There are map-like displays using spatial proximity and a visual clustering together of activities that are intellectually similar. (We intuitively understand that things close in space are likely to be similar in character) (Fig. 7). Advances in bibliographic science since the 1940s and, in particular, the creation of comprehensive citation indexes of academic literature have facilitated analysis of knowledge space through co-citation patterns and collaborative networks of researchers. The results of such analysis have most often been mapped as simple link-node diagrams, but sophisticated map-like displays have been developed in recent decades that frequently draw on ideas from information visualization (cf. Borner et al. 2003; Skupin and Fabrikant 2003). Many examples are documented in Katy Borner's recent synthesis, *Atlas of Knowledge* (Borner 2015).

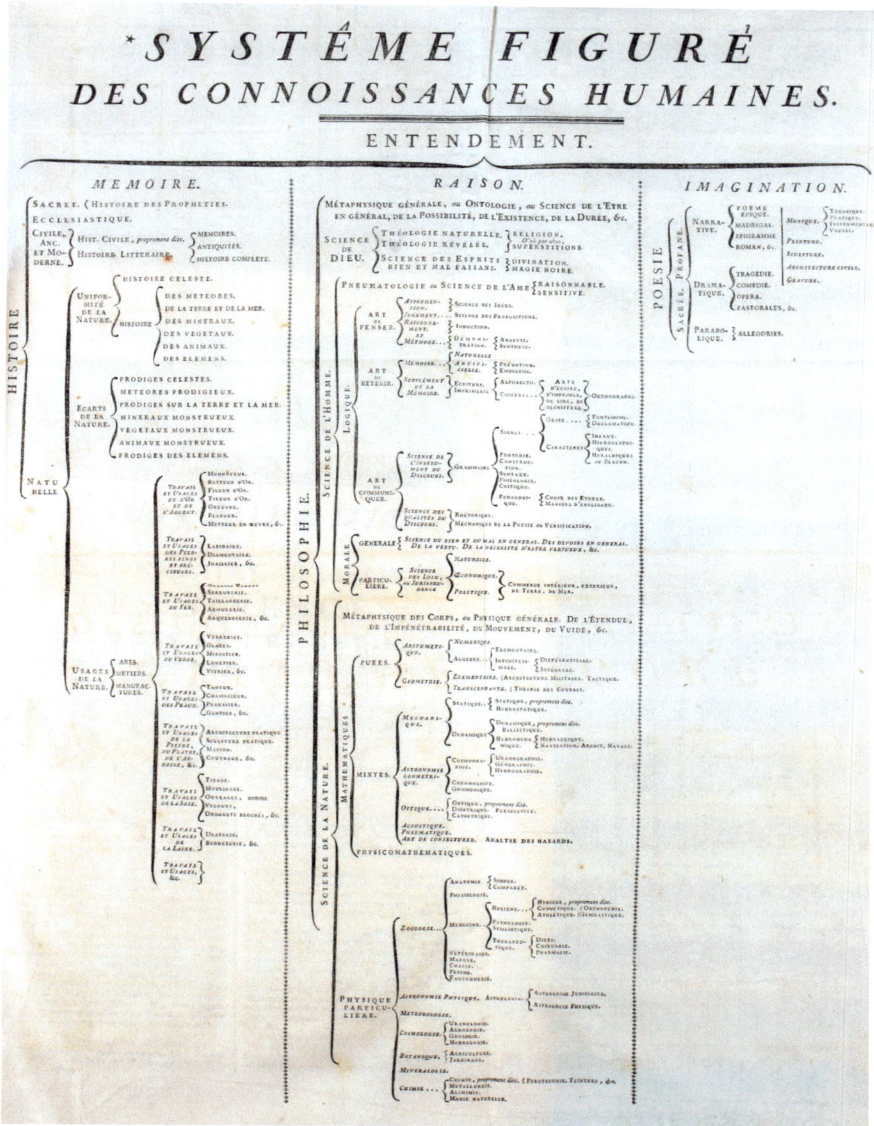

Fig. 6 An attempt to bring structure to burgeoning scale of recorded information in the European Enlightenment. This hierarchical diagram, a 'tree of human knowledge'; its taxonomy is an attempt organizes knowledge under three main branches, "memory"/history, "reason"/philosophy, and "imagination"/arts. It was deployed as an illustration in Diderot and d'Alembert's eighteenth century *Encyclopédie*. (*Source* https://en.wikipedia.org/wiki/File:ENC_SYSTEME_FIGURE.jpeg)

Fig. 7 A pioneering hand-drawn mapping of the structure of the natural sciences and technology as interpreted by Ellingham (1948), a physical chemist working in mid twentieth century. He uses the spatial metaphor of distance equating to intellectual similarity. The direction and orientation of name labeling is also used to denote commonality between fields of study. The overlaid polygons showed how scientific abstracting services at that time were covering different domains. (*Source* Ellingham (1948), Courtesy of the Royal Society)

7 Developing 'Mapping Across Academia'

This book represents an attempt, edited by two geographers but consciously drawing on a wide range of subject specialists, to delineate and discuss the importance of map representations and mapping practices across the scholarly landscape. We made a decision to edit a such volume with expert contributors from many other disciplines who both appreciate and understand the role of maps and mapping in non-traditionally geographic domains and sometimes fields of study that have been wholly a-spatial in the past. While geography, geology and other environmental sciences have long traditions of studying and using maps in their teaching, research and theory construction, and public communication and policy contributions, there are other new subjects where the making and use of maps are emerging as important perspectives and also providing valuable ingredients in what is studied. Here we would refer to map representations and mapping practices associated with the study of language, film, literature, religion, music, architecture, security, sustainability, but also with respect to studying human anatomy, and neurology. Maps are not only being employed to organize data and "show" where something is, but for providing new insights into interpreting and analyzing the problem and generating explanations and theory.

We would be the first to argue that more work needs to be done on the fields of study addressed in this volume and, of course, to those academic disciplines not covered in the context. Not included would also be the importance of maps related to what we could identify as "intersecting" fields which are emerging as the important arenas for present and future scholarly inquiry. These intersecting fields would include human consciousness and machine intelligence and faith/belief systems, preservation, climate change, environmental security, and shared futures in the Anthropocene. It is not difficult to envision dozens of maps that one could prepare that would help understand the interfacing subject matter, but also solve or resolve conflicts that might/will emerge.

We sought thoughtful and original essays that reflect upon the role of map representations, mapping practices and spatial understanding in a particular transdisciplinary disciplinary area. We tasked domain experts to consider how maps, broadly conceived, contribute to scholarship in other fields in terms of generating novels ideas, conducting empirical research, presenting and analyzing data, understanding spatial structures and patterns, and in disseminating results and communicating to different audiences. Authors could be personal, anecdotal, and opinionated.

The book is conceptually divided into two major sections—the sciences and humanities—and includes 15 chapters on different intellectual areas and related disciplinary fields (Table 1). Our selection and the division of academic endeavor was inevitably somewhat arbitrary but has provided a diverse and widely representative coverage and one that we believe will provoke an insightful range of essays. We hope this review of past and current mapping research challenges across disciplines will enhance our understanding of spatial processes and geographic visualization at scales ranging from the molecular to the galactic.

Table 1 Organization of mapping across academia

Intellectual areas	Academic experts
Section one: The Sciences	
Chap. 2: Geomorphology	Mike Smith and James Griffiths
Chap. 3: Cities and planning	Amy Hillier
Chap. 4: Ecology	David Barnett and colleagues
Chap. 5: Astronomy	Richard Gott
Chap. 6: Anatomy, human body	Jane Garb
Chap. 7: Neuroscience	Jordan Harp and Walter High
Chap. 8: Weather	James Carter
Chap. 9: Health and disease	Lance Waller
Section two: The Humanities	
Chap. 10: Language, linguistics	Roland Kehrien
Chap. 11: Landscape	Kenneth Olwig
Chap. 12: Power and geopolitics	Edoardo Boria
Chap. 13: Literature	Andrew Frayn
Chap. 14: Musicology	John Gold and colleagues
Chap. 15: Culture	Ann Kingsolver and colleagues
Chap. 16: Religion	Lillian Larsen
Chap. 17: Visual arts	Lou Cabeen
Chap. 18: Theatre	Edmund Lingan

8 Where We Go from Here

It is our hope that this book serves three purposes. First, that it informs increasingly growing academic communities and cross-disciplinary groups of scholars about the importance of map representations and mapping practices subject matter at multiple scales (from the sub-atomic, genetic social, planetary and to expanded universes). Second, that it will encourage scholars who work in the fields and subfield discussed to continue exploring, and perhaps expands the uses of maps in their instruction and research. Third, that the book will stimulate others to explore those fascinating intersections in intersecting and emerging subject matter across the sciences, humanities and public policies. In this way future scholars and students, whether in classes, seminars or workshops, will observe not only the importance of producing maps, but also recognize their basic importance in providing a sound understanding of visual representations and knowledge in "what happens where, when and why."

References

Baigrie, B. S. (1996). *Picturing knowledge*. Toronto: University of Toronto Press.
Barney, T. (2015). *Mapping the cold war*. Chapel Hill: UNC Press.

Barrow, J. D. (2008). *Cosmic imagery: Key images in the history of science*. London: The Bodley Head.

Bender, J., & Marrinan, M. (2010). *The culture of diagram*. Palo Alto, CA: Stanford University Press.

Blaut, J. M., Stea, D., Spencer, C., & Blades, M. (2003). Mapping as a cultural and cognitive universal. *Annals of the Association of American Geographers, 93*(1), 165–185.

Bodenhamer, D. J., Corrigan, J., & Harris, T. M. (Eds.). (2010). *The spatial humanities: GIS and the future of humanities scholarship*. Bloomington: Indiana University Press.

Borner, K. (2015). *Atlas of knowledge*. Cambridge, MA: MIT Press. Available at http://scimaps.org/

Borner, K., Chen, C., & Boyack, K. (2003). Visualizing knowledge domains. *Annual Review of Information Science and Technology, 37*, 179–255.

Caquard, S., & Cartwright, W. (2014). Narrative cartography: From mapping stories to the narrative of maps and mapping. *The Cartographic Journal, 51*(2), 101–106.

Conley, T. (2007). *Cartographic cinema*. Minneapolis: University of Minnesota Press.

Cooper, D., Donaldson, C., & Murrieta-Flores, P. (Eds.). (2015). *Literary mapping in the digital age*. London: Ashgate.

Cosgrove, D. (2012). *Geography and vision: Seeing, imagining and representing the world*. London: IB Tauris.

Crampton, J. W. (2006). The cartographic calculation of space: Race mapping and the Balkans at the Paris Peace Conference of 1919. *Social and Cultural Geography, 7*(5), 731–752.

Dear, M., Ketchum, J., Luria, S., & Richardson, D. (2011). *GeoHumanities: Art, history, text at the edge of place*. London: Routledge.

Delano Smith, C. (1987). Cartography in the prehistoric period in the Old World: Europe, the Middle East, and North Africa. In J. B. Harley & D. Woodward (Eds.), *Cartography in prehistoric, ancient, and medieval Europe and the Mediterranean. Volume 1 of the history of cartography*. Chicago: University of Chicago Press.

Dodds, K., & Sidaway, J. D. (2004). Halford Mackinder and the 'geographical pivot of history': A centennial retrospective. *The Geographical Journal, 170*(4), 292–297.

Dorling, D. (1995). *A new social atlas of Britain*. Chichester: Wiley.

Dorrian, M., & Pousin, F. (Eds.). (2013). *Seeing from above: The aerial view in visual culture*. London: IB Tauris.

Driver, F. (2001). *Geography militant: Cultures of exploration and empire*. Oxford: Blackwell.

Elkins, J. (2007). *Visual practices across the university*. Munich: Wilhelm Fink Verlag.

Ellingham, H. J. T. (1948). Divisions of natural science and technology. In *Report and papers of the royal society scientific information conference, July* (pp 477–484). London: The Royal Society.

Fraser, B. (2015). *Digital cities: The interdisciplinary future of the urban geo-humanities*. London: Palgrave Macmillan.

Goodchild, M. F. (2000). Cartographic futures on a digital earth. *Cartographic Perspectives, 36*, 3–11.

Hall, S. S. (1992). *Mapping the next millennium*. New York: Vintage Books.

Hallam, J., & Roberts, L. (Eds.). (2014). *Locating the moving image: New approaches to film and place*. Bloomington: Indiana University Press.

Harley, J. B. (1989). Silences and secrecy: The hidden agenda of cartography in early modern Europe. *Imago Mundi, 40*, 57–76.

Harley, J. B., & Woodward, D. (1987). *History of cartography*. Chicago: University of Chicago Press. Available free online at www.press.uchicago.edu/books/HOC/index.html

Harmon, K. (2009). *The map as art: Contemporary artists explore cartography*. New York: Princeton Architectural Press.

Ingold, T. (2000). *The perception of the environment: Essays on livelihood, dwelling and skill*. London: Psychology Press.

Johnston, R. J. (1996). *Geography and geographers: Anglo-American geography since 1945*. London: Routledge.

Kidron, M., & Segal, R. (1995). *The state of the world atlas*. London: Penguin.
Latour, B. (1990). Visualisation and cognition: Drawing things together. In M. Lynch & S. Woolgar (Eds.), *Representation in scientific practice* (pp. 19–68). Cambridge: MIT Press.
Lefebvre, H. (1991). *The production of space* (N. Donaldson-Smith, Trans). Oxford: Blackwell.
Livingstone, D. N. (2010). *Putting science in its place: Geographies of scientific knowledge*. Chicago: University of Chicago Press.
Lopez, B. (1986). *Arctic dreams: Imagination and desire in a northern landscape*. New York: Vintage Books.
Lynch, M., & Woolgar, S. (Eds.). (1990). *Representation in scientific practice*. Cambridge: MIT Press.
Mackinder, H. J. (1904). The geographical pivot of history. *The Geographical Journal, 23*(4), 421–437.
Maggs, T. (1998). Cartographic content of rock art in Southern Africa. In D. Woodward & M. Lewis (Eds.), *Cartography in the traditional African, American, Australian, and Pacific Societies. Vol: 2, Book 3, history of cartography*. Chicago: University of Chicago Press.
Malcolm, N. (1987). Alexander von Humboldt, Humboldtian science and the origins of the study of vegetation. *History of Science, 25*(2), 167–194.
Monmonier, M. (2005). *Rhumb lines and map wars: A social history of the Mercator projection*. Chicago: University of Chicago Press.
Montello, D. R. (2002). Cognitive map-design research in the twentieth century: Theoretical and empirical approaches. *Cartography and Geographic Information Science, 29*(3), 283–304.
O'Rourke, K. (2013). *Walking and mapping: Artists as cartographers*. Cambridge, MA: MIT Press.
Penz, F., Radick, G., & Howell, R. (2004). *Space in science, art and society*. Cambridge: Cambridge University Press.
Pflederer, R. (2012). *Finding their way at sea: The story of portolan charts*. Goy: Hes & De Graaf Publishers.
Pickles, J. (2000). Cartography, digital transitions, and questions of history. *Cartographic Perspectives, 37*, 4–18.
Rose, G. (2016). Rethinking the geographies of cultural 'objects' through digital technologies: Interface, network and friction. *Progress in Human Geography*, *40*(3), 334–351.
Rossetto, T. (2014). Theorizing maps with literature. *Progress in Human Geography, 38*(4), 513–530.
Scerri, E. R. (2007). *The periodic table: Its story and its significance*. Oxford: Oxford University Press.
Scholten, H. J., van de Velde, R., & van Manen, N. (Eds.). (2009). *Geospatial technology and the role of location in science*. New York: Springer.
Seager, J. (1990). *State of the Earth atlas*. New York: Routledge.
Seager, J. (1997). *State of women in the world atlas*. New York: Penguin.
Shapin, S., & Schaffer, S. (1985). *Leviathan and the air pump: Hobbes, Boyle, and the experimental life*. Princeton, NJ: Princeton University Press.
Skupin, A., & Fabrikant, S. I. (2003). Spatialization methods: A cartographic research agenda for non-geographic information visualization. *Cartography and Geographic Information Science, 30*(2), 95–115.
Sutton, P. (1998). Aboriginal maps and plans. In D. Woodward & M. Lewis (Eds.), *Cartography in the traditional African, American, Australian, and Pacific Societies. Vol. 2, Book 3, history of cartography* (pp. 387–416). Chicago: University of Chicago Press.
Van Swaaij, L., & Klare, J. (2000). *The atlas of experience*. London: Bloomsbury.
Warf, B., & Arias, S. (Eds.). (2008). *The spatial turn: Interdisciplinary perspectives*. London: Routledge.
Woodward, D. (1975). *Five centuries of map printing*. Chicago: University of Chicago Press.
Woodward, D., & Lewis, G. M. (Eds.). (1998). *Cartography in the traditional African, American, Arctic, Australian, and Pacific Societies. Vol. 2, Book 3, the history of cartography*. Chicago: University of Chicago Press.

Physical Landscapes

Mike J. Smith and James S. Griffiths

Abstract The scientific discipline of geomorphology is concerned with the processes that act upon and shape the Earth's surface to create physical landscapes. Maps have a very specific utility within this domain as they allow a spatial representation of shapes (or landforms), their material composition, age and the processes that formed them. From the creation of the very first geomorphological maps in the early 1900s, there has been continual development and increased sophistication in the representation of complex datasets. The implementation of geographical information systems, integrated with the widespread availability of satellite imagery and digital elevation models has enabled much greater application across a range of disciplines beyond geomorphology, notably in natural hazard evaluation, disaster response assessment, insurance, infrastructure planning, civil engineering and engineering geology. This chapter provides a brief outline of the development of cartographic techniques where the primary purpose is to provide maps of geomorphology that have met the requirements of different end-users. Initially this involved standard approaches of field mapping and drafting of hard-copy maps but has now developed into the use of much more sophisticated methods of digital data collection and management. This has resulted in significant growth in the use of geomorphological maps, and recognition of their wider societal significance. Whilst geomorphology, by definition, refers to mapping of the Earth's surface, there is increasing use of mapping techniques on planetary bodies across the solar system. The horizons for geomorphological mapping clearly continue to expand and this chapter concludes with a discussion on the future challenges and opportunities within the subject.

Keywords Digital elevation model (DEM) · Geomorphology · Terrain · Landscape · Legend · Symbol · Laser · Radar · Physical morphology

M.J. Smith (✉)
Department of Geography and Geology, Kingston University, Kingston upon Thames, UK
e-mail: Michael.Smith@kingston.ac.uk

J.S. Griffiths
SoGEES, University of Plymouth, Plymouth, UK
e-mail: jim.griffiths@plymouth.ac.uk

S.D. Brunn and M. Dodge (eds.), *Mapping Across Academia*,
DOI 10.1007/978-94-024-1011-2_2

1 Introduction

This chapter is concerned with the role of mapping to understand physical landscapes and more specifically the scientific discipline of *geomorphology* and its application in real-world land management tasks. The study of geomorphology deals with the form of the Earth's surface and the processes that act upon and shape it, with mapping focused upon recording the location and distribution of different landforms of interest (often including details on material composition and age) either as a specific graphical output or as data input to subsequent numerical modelling. Landscape analysis in geomorphology is inherently science based and, given the complexity of natural systems, an inductive approach to the acquisition of knowledge has often been taken. The observation of physical systems is a key tenet and geomorphologists have often used this as the basis for classifying sample sets of the environment and then generalising this complexity into a standardised theory (Chorley et al. 1985).

In geomorphology it is apparent that maps are a natural outcome of the observational process and highlight the data-driven nature of much research. This mode of knowledge construction dates back to at least the 1800s (e.g. Close 1867), although it was not until the early twentieth century that early derivatives of what we would now consider geomorphological maps first appeared (see, for example, in the work of Passarge (1912) in Germany). As geomorphology developed as a subject, more sophisticated techniques in mapping landforms began in earnest in the 1950s (e.g. Klimaszewski 1956) and 1960s (see review by Rose and Smith 2008); maps were a natural counterpart to this process as they allowed the storage, display and analysis of complex information. This inevitably led to a large number of bespoke feature classifications, symbol sets and visual designs (or more simply, legend systems), contextualised by their sub-discipline, such as glacial or country. As a result there was considerable effort put into the development of standardised systems, particularly by the International Geographical Union (1968). The richness and interdisciplinarity of geomorphology is exemplified through the diversity of application areas and close cross-over into other related disciplines such as geology, soil and natural hazard mapping.

Unfortunately, by the time academic consensus had been reached in the use of legend systems, two major changes had occurred: (1) research methods in geomorphology had changed and (2) geographic information systems (GIS) had been implemented in academic research. During the 1970s research methods in physical geography underwent a quantitative revolution and moved more towards inductive approaches to understanding physical systems; this involved detailed measurements at finer and finer scales. In short, mapping was a solution to a problem that many were no longer interested in.

Prior to this point cartographers were concerned with all aspects of data management, presentation and analysis; maps were the technological solution to harnessing the power of spatial information. The introduction of GIS and, more specifically, digital data handling was a major technological disruptor that caused a division in the cartographic community between those focused on design/communication and those on data handling. We now think of the former as "cartography," whilst the latter has become geographic information science. It is interesting to note from textbooks of the time, such as Monkhouse and Wilkinson (1971), that cartography was portrayed as an all-encompassing discipline, fundamental to the study of subjects involving the integration of two-dimensional space. Data management and analysis is conspicuous by its absence in modern volumes, with Kraak and Ormeling's (2010) *Cartography*, subtitled Visualization of Spatial Data.

Yet we would argue in the last decade or so that geomorphological mapping has undergone a period of renaissance. This is evidenced through the proliferation of maps in both the grey literature and actively being published. The establishment of the *Journal of Maps* (http://www.journalofmaps.com) in 2004 demonstrates the relevance of mapped output with a significant number of geomorphological maps (e.g. Glasser and Jansson 2008). This is underlined by the formation of the *International Association for Geomorphology (IAG)* Working Group on Applied Geomorphic Mapping (Pain et al. 2008) and publication of a new technical handbook on the topic (Smith et al. 2011).

In the next section of this chapter we detail the primary purposes of geomorphological mapping. This is followed by discussion of data collection and management issues, before the principle end-users of geomorphological maps are outlined, highlighting the breadth of contemporary inter-disciplinary involvement. In combination these sections demonstrate why geomorphological mapping has seen such recent growth and assumed a role of societal significance. The purposes of geomorphological maps are triggered by the requirements of end-users, but this is also driven by the availability of data and what can be achieved in its use. As a consequence the concluding discussion examines the future of mapping in geomorphology.

2 Purpose of Geomorphological Maps

The purpose of geomorphological maps is to systematically record the shape (or morphology), landforms types, landscape-forming processes and geological materials that constitute the surface and near-surface of the Earth. Originally described as 'physiography maps' these have a substantial history of use and development by geographers, although the earliest applications were mainly in North America (e.g. Powell 1896; Fenneman 1928). Until around 1960 geomorphological maps were used as a means of describing the landscape in fairly simply terms and typically at quite small scales. For example Hammond (1954) produced a 'geomorphic study' of part of southern California at a scale of 1:560,000, and went on to subdivide the whole of North America on 7½-minute rectangles that classified the landscape into eight types: (1) nearly flat plains; (2) rolling and irregular plains; (3) plains with

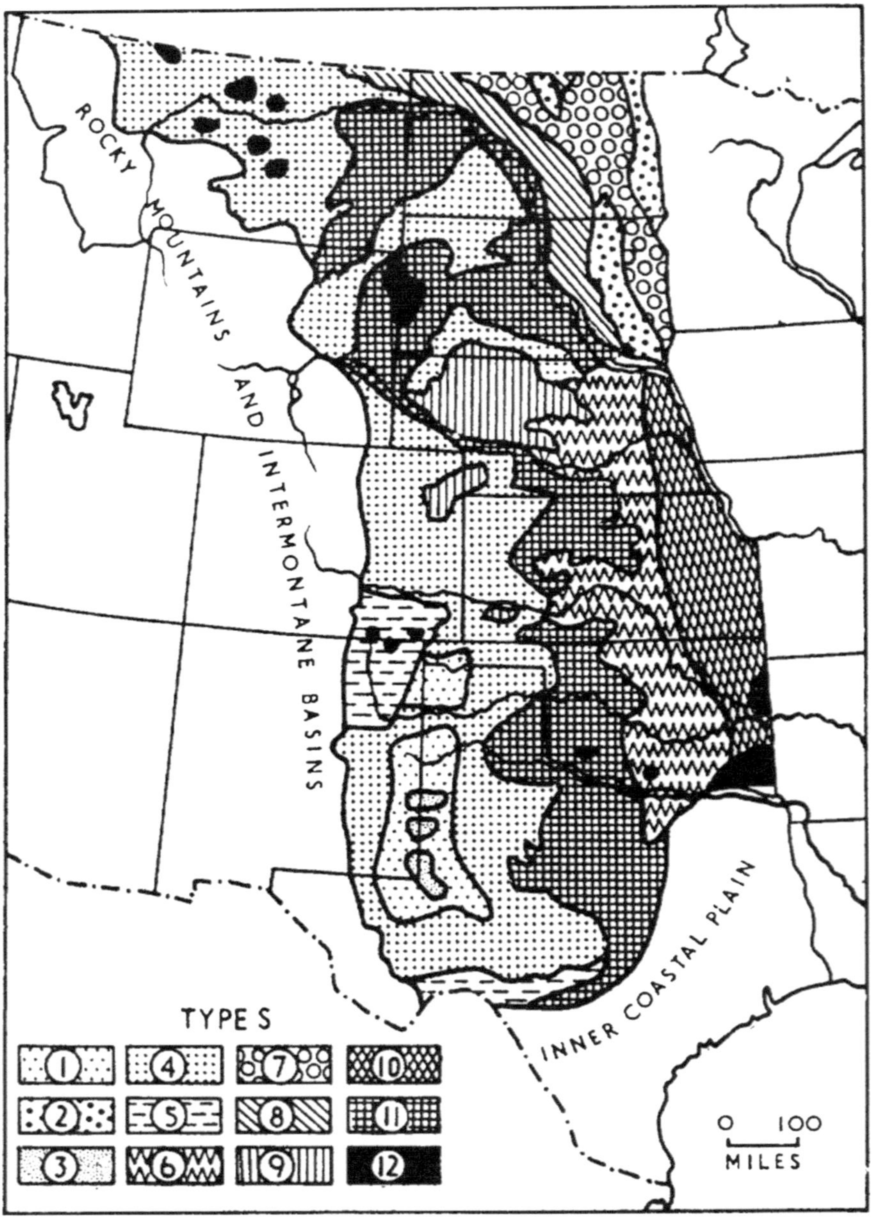

Fig. 1 Physiography of the American Great Plains (Lewis 1962)

widely spaced hills and mountains; (4) partially dissected tablelands; (5) hills; (6) low mountains; (7) high mountains; (8) ice-caps. An example of this type of map is presented in Fig. 1.

These small scale maps were useful for broad descriptions of the landscape and provided general background material but lacked the detail needed for a comprehensive understanding of the geomorphological history of an area. During the late 1950s and 60s there emerged, predominantly in Europe, an approach to geomorphological mapping that resulted in very detailed maps of the landscape and a standardised legend for this work was compiled by Demek (1972) and Demek and Embleton (1978). A summary of these developments are provided by Cooke and Doornkamp (1990) which also contains a more succinct collection of standard symbols for use in geomorphological maps. They suggested that such maps should be divided into four distinct types derived from data collected in a systematic manner using a system of morphological mapping devised by Savigear (1965):

1. Morphological maps—where the land surface is sub-divided into planar facets separated by gradual changes or sharp breaks in slope. On the maps the changes and breaks in slope are identified as either concave or convex in nature and recorded using decorated lines;
2. Morphographic maps—the distribution of a named suite of landforms and their material composition based on the boundaries shown on the morphological maps. Examples of the terms that might be used on this map are described by Griffiths and Stokes (2008);
3. Morphochronological maps—the age of formation of the landforms identified on the morphographic maps;
4. Morphogenetic maps—these show how geomorphological processes created the landforms and the overall landscape.

Otto et al. (2011) present the main symbol sets that have been developed and the contexts in which they should be used. It is important to note that geomorphological mapping seeks to capture the full complexity of the Earth's surface and represent it in a simplified form. Due to the complexity of the distinct map types noted above, symbol sets and the maps themselves can be visually complex, difficult to understand and therefore requiring considerable expertise to interpret. Much of the development and application resulted from work in Europe (Verstappen 2011) and Paron and Claessens (2011) outline the major "schools" of geomorphological mapping. Outside of Europe, it is only Australia and Brazil (e.g. DeOliveira and Vieira 2009) that have major traditions in national mapping, although the recent production of a national geomorphological atlas of China is noteworthy (Cheng et al. 2011).

Figure 2 illustrates the application and development of approaches to geomorphological mapping, with Fig. 2a presenting a morphological map for a series of landslide features in Dorset. This simply presents the *shape* of the landscape; it is not until interpretation adds information on processes and chronology that the real explanatory and communicative power of geomorphological maps are exposed.

The next stage in the development of geomorphological mapping came from an unexpected direction; it was found to be an approach that lent itself very effectively to preliminary investigations and planning for civil engineering construction

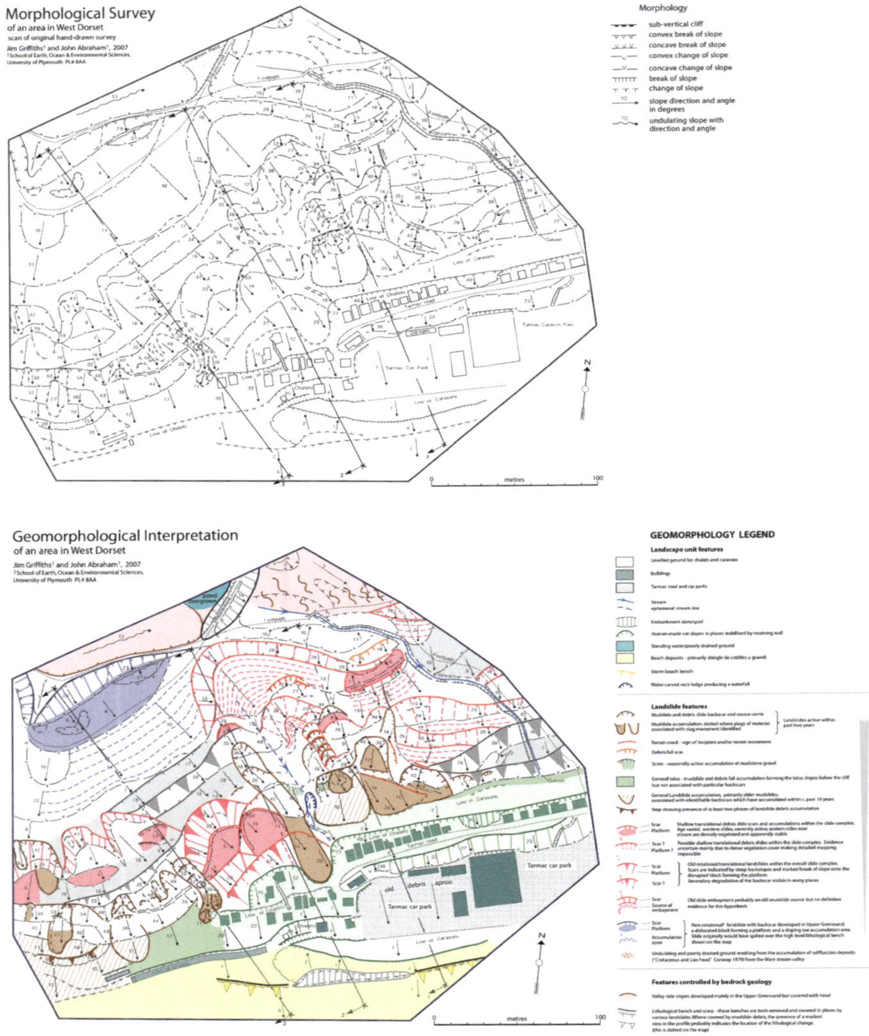

Fig. 2 Area in West Dorset, UK, illustrating active landslide features (*top*). Morphological map of the study site and (*bottom*) geomorphological map, incorporating process and chronological information (after Griffiths and Abraham 2008)

projects where an understanding of surface form, near-surface materials and surface processes is crucial. This was demonstrated by Brunsden et al. (1975) and Doornkamp et al. (1979) through a series of case studies in the UK, Nepal, and the Middle-East. As an example of this work, Fig. 3 presents the geomorphological map from the Suez New City master plan (Jones 2001). Geomorphological mapping became closely associated with developing techniques of engineering

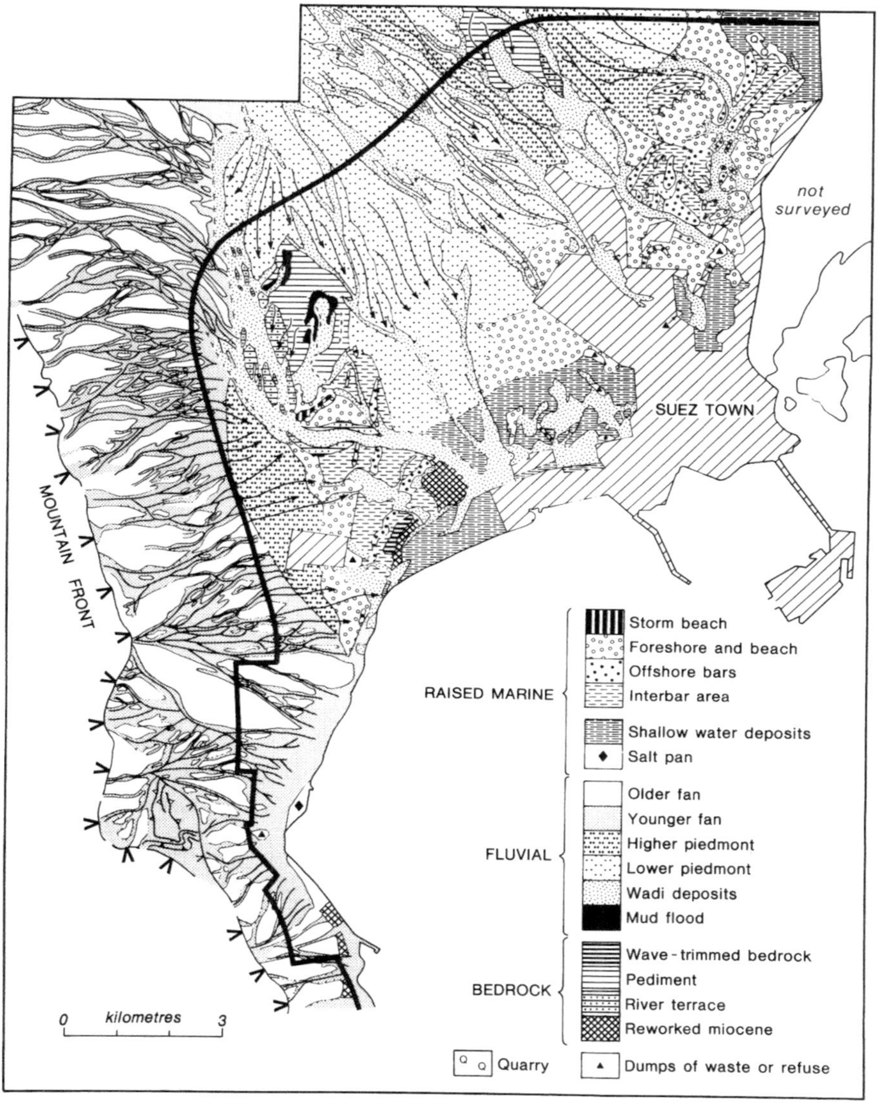

Fig. 3 Geomorphological map of Suez New City (Jones 2001). Reproduced with the kind permission of the Geological Society of London

geological mapping (Report by the Geological Society Engineering Group Working Party 1972; Dearman and Fookes 1974; International Association of Engineering Geology 1976) and in the UK recommendations were made in 1981 that both techniques should form part of future revisions to the code of practice for site investigation (BSI 1981; Griffiths and Marsh 1986). In connection with planning, it was apparent throughout many parts of the world that geomorphological data were central to more effective production of regional and local development plans (Doornkamp et al. 1987). The maps for planning that were produced were usually hybrids that incorporated a range of data including geology and geomorphology (such as the French planning maps ZERMOS; cf. Porcher and Guillope 1979), or part of a suite of maps that included specialist cartography on geomorphology (Smith and Ellison 1999; Burnett and Styles 1985, 1986; Styles et al. 1984).

By the late 1980s the on-going development of geomorphological mapping was no longer predominantly an academic pursuit but lay with applied geomorphologists and engineering geologists working in industry. In the UK examples of the application of geomorphological mapping include the Channel tunnel high speed rail link (Waller and Phipps 1996), associated with planning for the Ventnor landslide complex (Lee and Moore 1989), and as part of the Department of the Environment environmental geology mapping programme (Forster et al. 1987). In a review of these developments, Lee (2001) identified three distinct forms of geomorphological map which can still be identified in the published literature and all have distinct applications:

1. Regional surveys of terrain conditions, either for land use planning or in baseline studies for environmental impact assessment, for example the 1:25,000 scale maps of Torbay by Doornkamp (1988);
2. General assessments of resources or geohazards at scales between 1:50,000 and 1:10,000 (e.g. Bahrain Surface Materials Resources Survey; Doornkamp et al. 1980);
3. Specific-purpose large-scale surveys to delineate and characterise particular landforms (e.g. the 1:2500 scale investigations around the Parrot's Beak Ridge for the Ok Tedi copper mine in Papua New Guinea; Hearn and Blong 2001).

Whilst geomorphological mapping has now had significant resurgence as a means of data collection in academia, it still remains primarily an applied tool for engineering and planning purposes (Smith et al. 2011). An indication of the way geomorphological mapping is now part of mainstream engineering, particularly for more remote, undeveloped areas, is demonstrated by Hearn (2011) in his review of mountain road engineering. An example that illustrates an applied geomorphological map of the type recommended by Hearn (op. cit.) is presented in Fig. 4 (Griffiths 2001). This map shows a multiple rotational landslide with a distinctive backscar, a series of displaced benches, a double accumulation zone, and secondary degradational movements. The map formed the basis for subsequent ground investigations for the channel tunnel portal which was located in this landslide.

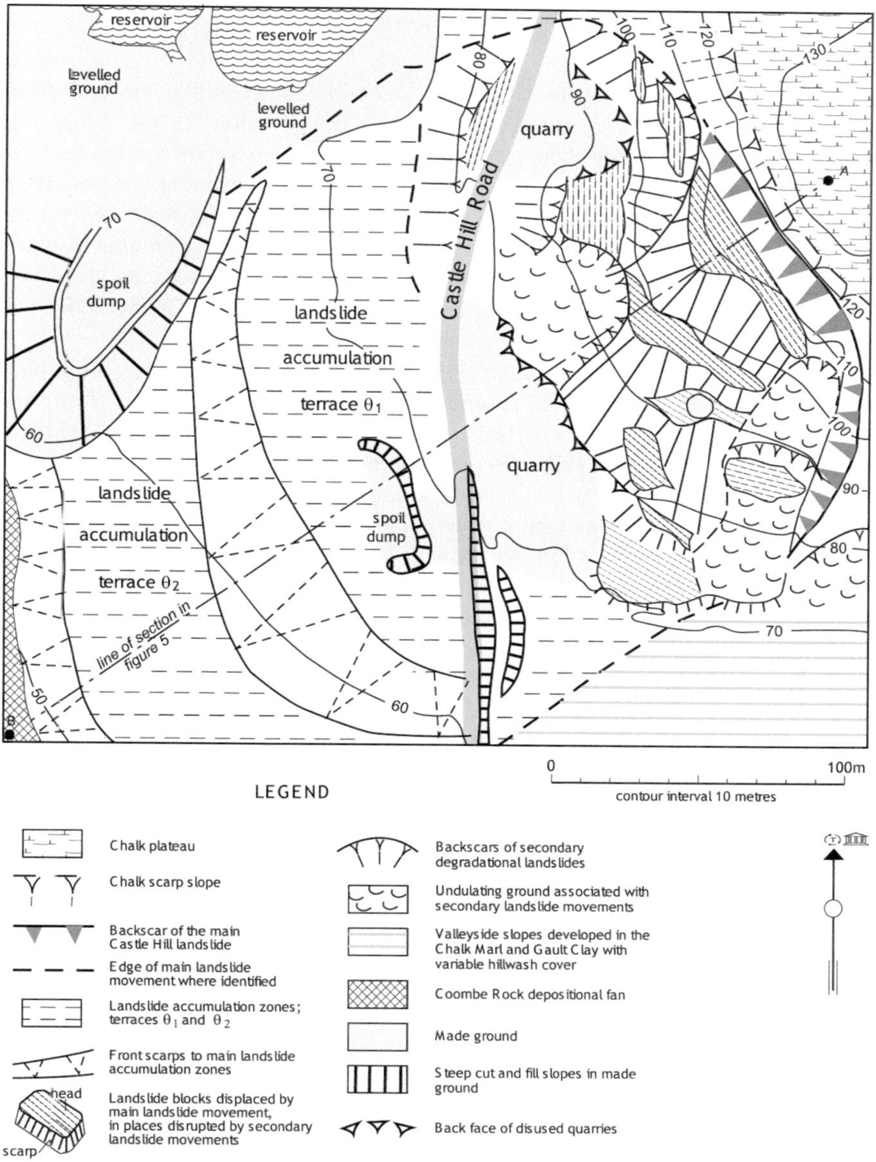

Fig. 4 Geomorphological map of the channel tunnel portal area, UK (Griffiths 2001). Reproduced with the kind permission of the Geological Society of London

3 Data Collection and Management

Cartography in its widest sense, has long been concerned with data collection, classification, management and representation. GIS is central to the future role cartography plays, however there is no need to labour its importance in the study of spatial phenomena given the extensive treatment in other chapters in this volume. It is enough to note the significance of data management in the physical sciences; the ability to reproduce the (layered) paper-based cartographic paradigm digitally through data storage, spatial analysis and map production allowed efficiency savings, the processing of very large volumes of data and the extent of analysis that had previously been impractical or impossible.

GIS therefore provides the organising framework through which geomorphological mapping takes place. As geomorphology is concerned with the form and acting processes of the Earth's surface, the spatial modelling of terrain is central to undertaking research work and effective mapping of landscapes. The digital elevation model (DEM) is a raster representation of the landscape, where each individual cell contains an elevation value. This basic framework was outlined early on (Miller and LaFlamme 1958) to allow computer processing of terrain (surface elevation), and led specifically to the development of algorithms and a theoretical underpinning through the sub-discipline of *geomorphometry* (Evans 1972; Hengl and Reuter 2008). Whilst any elevation value is inherently vector based (e.g. spot heights or contours), data processing has largely focused upon the manipulation of raster data due to the simplicity of the computation. However the relatively recent use of LiDAR (see below) has allowed the compilation of massive vector datasets that area increasingly processed and modelled within a vector framework.

Data collection is a key component in terrain modeling and visualisation and here we consider some of the important trends that have taken place. When working with historic data, conversion to a digital format is necessary. Spot heights and contours are the primary legacy datasets and these have been dealt with extensively (e.g. Gousie and Franklin 2005; Carrara et al. 1997) and are of significance because they allow the temporal analysis of landscapes. In terms of contemporary data collection, manual, field-based, techniques remain common place and are used extensively, particularly where detailed survey data is required for small areas. A theodolite with integrated distance measuring device (known as a Total Station) is often employed and will usually have a data logger allowing direct digital data capture. For many applications these have been largely replaced by global positioning system (GPS) receivers which are simpler and faster to use. Survey-grade GPS receivers allow millimetric accuracy measurements to be made. Perhaps the single greatest impact upon geomorphology over the last ten years has been the upsurge in the use of remotely sensed techniques to measure surface elevation (Smith and Pain 2009). This has allowed the collation of medium-resolution datasets over large areas thereby allowing regional-scale analyses. Three technologies have led these developments: Light Detection and Ranging (LiDAR), interferometric synthetic aperture radar (InSAR) and photogrammetry.

For terrain measurement, LiDAR typically utilises a laser pulse to calculate the distance (or range) from the scanner to the target based upon the travel time (Baltsavias 1999). With the horizontal and vertical angles of the pulse known, a three-dimensional position can be calculated; current systems can measure up to of 500,000 points per second creating a "point cloud" of data. Airborne laser scanning (ALS; e.g. Lohani and Mason 2001) has seen rapid development over this period. Output data provides a range to the *visible surface* meaning that it is often not possible to model the actual terrain surface, having to include "clutter" such as buildings and trees (although these are of significant interest for urban and ecological applications). One benefit of using laser ranging is that the pulse itself is narrow; as a result it is often able to penetrate through a tree canopy and reflect off the actual ground surface. Terrestrial laser scanners (TLS or ground-based; e.g. Hodge et al. 2009) have had far slower uptake principally due to the high unit cost and data management requirements. Heritage and Large (2009) outline the theory and application of laser scanning in the environmental sciences.

Like LiDAR, InSAR is an active remote sensing technique (Palmann et al. 2008) that has been used extensively for the collection of terrain data (Smith 2002). This operates at much longer wavelengths of "light" and, rather than using travel time, makes uses of the recorded *phase*; that is the incomplete proportion of a single wavelength received at the sensor. When the *difference* in phase between two images of the same area is calculated, the remainder is directly related to the elevation of the terrain surface; this can be extracted through a process known as *phase unwrapping*. InSAR is operated from both airborne and spaceborne platforms either in a single-pass or repeat-pass mode. In single-pass mode, *two* sensors on a single platform capture images at exactly the same moment. This produces the best terrain data, but is a more costly solution.

The final data collection technique is that of photogrammetry, which is now a fully digital process. This is concerned with quantitative measurements from photos which, when applied to landscapes, can be used to measure surface elevation (Mitchell 2007). If the exact position and orientation of a single camera is known, then it is possible to model a vector from the lens to an object of interest. If the exact position and orientation of a second camera is known for an *overlapping* photo, then trigonometry is applied to calculate the three-dimensional position of an object. When this is performed digitally, it is possible to iteratively calculate millions of points and generate point clouds in a fashion similar to laser scanners.

Traditionally photogrammetry would have used analogue (film) photography from bespoke aerial (metric) cameras acquired during dedicated surveys. However digital processing now enables a greater flexibility in approach allowing the use of off-the-shelf prosumer cameras (e.g. Chandler et al. 2005) and oblique imagery (Maas et al. 2006). The ability to mathematically model greater amounts of distortion allows the application of photogrammetry, and acquisition of surface elevation measurements, to new areas. One powerful application has been the use of historic aerial imagery (e.g. Barrand et al. 2009) to open-up this important environmental archive. A second area is *close range* imaging (within ~200 m of the Earth's surface). This scale of data collection has been relatively expensive to acquire and therefore under-utilised. However, the ability to use prosumer cameras, coupled with novel airborne platforms, is transforming the collection of imagery. Unmanned aerial vehicles (UAVs; Laliberte et al. 2010), blimps (Boike and Yoshikawa 2003) and kites (Smith et al. 2009) have all been successfully used to collect aerial imagery and subsequently processed photogrammetrically. Ultimately geomorphological mapping is reliant upon data on the form of the landscape, collected either through field-based assessment or from remotely sensed data. Where remotely sensed, there are considerable benefits to be gained from low cost, accessibility, reconnaissance and areal coverage, allowing greater utilisation of geomorphological information and therefore advancing the role of mapping in the decision making process.

4 Map Users

Robinson and Spieker (1978), researchers working at the United States Geological Survey published the book *Nature to be Commanded* which demonstrated the use and value of a range of Earth science maps including those depicting geomorphology. They make the claim that the purpose of Earth science maps is to:

> inform planners, decision makers or owners so that they can forestall or relocate new developments in areas where lives and property might be imperilled, propose appropriate design precautions in developments that cannot be placed elsewhere and/or alert inhabitants of imperilled developments to seek protection through engineering or insurance. (Robinson and Spieker 1978: 2)

On the reasonable assumption that Earth science maps will be of little value if they do not illustrate geomorphology, Table 1 identifies all the potential users of applied geomorphological maps.

In addition to geomorphological maps being used in applied settings, there will be a significant academic role by geomorphologists, geologists, environmental scientists, pedologists, foresters, cartographers, and physical geographers carrying out investigations of the landforms and processes in an area. There will also be non-specialist users, such as ramblers and farmers.

For example, Doornkamp et al. (1987) present an international review of environmental geological maps in relation to planning which demonstrates that throughout the USA and Europe geomorphological mapping is a fundamental requirement in the compilation of data. Styles et al. (1984) make a similar claim for Hong Kong in connection to the far-sighted Geotechnical Area Studies Programme and Land Use Planning. For the construction of low cost roads in mountainous regions, Hearn (1997) emphasizes the importance of understanding the geomorphology and working with the terrain for design and planning which will require the compilation of geomorphological maps. Finally, Fookes (1997) in the first Glossop Lecture makes it clear that understanding the geomorphology is fundamental to the

Table 1 Potential list of users of applied geomorphological maps

User	Application
Planners	Those responsible for local and regional development plans and also who will grant planning permission for new developments to take place
Decision makers	Local (or national) authority civil servants or politicians who have to agree to a development taking place
Owners	The clients who are paying for a development or own an existing development, whether it is a residential home, a sea wall, bridge, quarry, waste dump, tunnel, sea outfall, nuclear power station, sea defences or reservoir with a 100 m high dam
Architects	Who will design any new development
Engineers	Who will investigate sites, design and build new developments or remediate existing developments that are "imperilled"
Resource exploitation	Water authorities, exploration companies or national geological surveys seeking resources
Insurers	Those responsible for insuring or reinsuring existing, planned, under-construction or built developments
Banks	Who provide the financial package that allows a development to proceed
Disaster response	Those who will prepare disaster response plans for 'imperilled' areas (e.g. areas of flooding, landsliding, earthquake, tsunami, or volcanic hazard), including NGOs

creation of a ground model necessary for the safe and economic design of all engineering works.

Whilst the references cited above mainly refer to traditionally compiled two-dimensional maps, GIS has allowed the development of three-dimensional (surface and sub-surface) and four-dimensional (temporal) geomorphological maps which will only increase their importance and value to end-users.

5 Planetary Geomorphological Mapping

Whilst the importance of mapping terrestrial landscapes is well-established, new opportunities for scientific investigation by geomorphologists were identified in the physical landscapes of other planets in the solar system (Sharp 1967). Realistically this is limited to the Moon and Mars, where the solid surface can be clearly seen and analogues with terrestrial processes can be established, plus more speculative work on asteroids, comets and some of the larger moons, notably those orbiting the gas giants Jupiter and Saturn, Mercury and Venus.

The International Association for Geomorphology has a working group (IAG PGWG) investigating planetary geomorphology (http://planetarygeomorphology.wordpress.com). Their work has identified a range of active and relict geomorphological activities including meteorite impact cratering, volcanism, plus processes identified as aeolian, fluvial, lacustrine, deltaic, mass wasting, rock disintegration, glacial and periglacial. The IAG PGWG notes that "Whilst the landforms appear similar to those on Earth, there are issues of equifinality in addition to important differences in denudation rates, landform scale and indeed geomorphic processes." The IAG PGWG web site hosts a collection of images displaying some of these landforms coupled with a description of the features, comparisons with possible terrestrial equivalents and further reading.

With the exception of meteorites landing on Earth and samples collected from the Moon landings 40 years ago, all the mapping and interpretation of the geomorphology is based on terrestrial telescope observations and remote sensing data. Whilst the Moon is dominated by impact craters (Ronca 1972), the Martian landscape presents a much more complex geomorphological picture (Balme et al. 2011). Craddock and Howard (2002) make the case for rainfall on a warm, wet early Mars and it is now accepted that flowing water has created many landforms on the Martian surface (Carr 1983), with Towner et al. (2011) suggesting there is evidence of water flow as recent as 100 Ma. Nekum et al. (2004) suggest there is evidence for recent volcanic (within the past 2 million years) and glacial activity (within the past 4 million years) on the planet. Studies such as these coupled with work notably at NASA (i.e. http://marsoweb.nas.nasa.gov/globalData/) have led to the creation of maps of many facets of the physical landscape of Mars which are now available interactively through Google Maps (http://www.google.com/mars/).

6 Future Challenges

It is perhaps strange in a book about mapping to spend so much time on data acquisition and management. Yet representing terrain is fundamental to understanding the environment and the recent renaissance in geomorphological mapping has been driven by the development of new techniques for data collection. Data are now available for larger areas, in greater detail and over short time periods. There has now been nearly 20 years of topographic data collection from space and this has been transformative for geomorphology and the physical sciences more widely. Whilst geomorphologists are concerned with terrain morphology and surface/subsurface material properties, it was really only surface material properties that were available prior to using remote sensing. The complementary addition of elevation datasets has greatly extended the capability of geomorphological investigation and provided the context for the expansion in mapping. This has been an evolutionary period for the discipline, yet the outputs and outcomes from research that have benefited from these data has been truly revolutionary. The scope and capability of such work has been significantly extended and this is explored by Smith and Pain (2009) where they show the temporal/spatial resolution constraints on current work (Fig. 5). In particular, there is a growing trend for more detailed imagery available over shorter timescales; this considerably expands the scope and scale of geomorphological investigation open to researchers. Whilst the technology, data and applications are available, not all researchers will have the knowledge or skills to take best advantage of them and one key challenge will be the

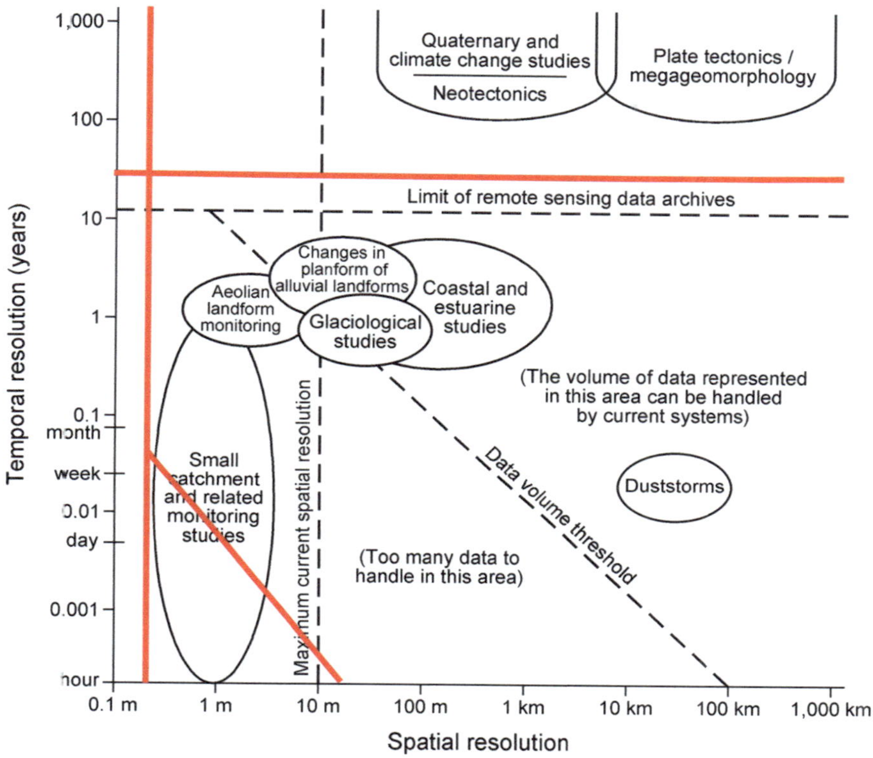

Fig. 5 Temporal and spatial resolution constraints of satellite sensors on geomorphological research (Smith and Pain 2009: 569)

dissemination of best practise. In this sense academics are self-regulating and the growth in published outputs such as the *Journal of Maps* and key textbooks including Smith et al. (2011) and Hengl and Reuter (2008) play their part.

Future technological changes in geomorphology can best be seen in current *airborne* technology, as sensor development and deployment is commensurately shorter than for space. Passive systems, such as radiometrics (composition of the upper 50 cm of the surface) and aeromagnetics (imaging subsurface features), and active systems, such as airborne electromagnetics (three-dimensional detail on conductivity down to 100 m), have great utility and are beginning to allow the development of a true three-dimensional understanding of surface and near-surface landforms and their material properties. An area currently in transition is that of hyperspectral remote sensing; rather than collecting images at single wavelengths, these sensors collect images at tens or hundreds of different wavelengths. This provides a far richer 'data cube' that enables more sophisticated investigation. At the opposite end of the spectrum, developments in close-range remote sensing will drive the acquisition of low-cost field-scale data. Already remotely controlled model aircraft or quad-copters can be pre-programmed to a set flight plan to acquire imagery using a prosumer camera. Subsequent imagery can then be processed photogrammetrically or as a series of mosaics. Costs will be driven down in this area, particularly with the use of *Arduino* (http://www.arduino.cc/), an open source prototyping hardware platform (e.g. http://ardupilot.org). Also expect to see far greater interest in the development of multispectral prosumer cameras that can be used on these platforms. These technological developments do not necessarily expand upon current capabilities, but significantly reduce the cost to implement meaning that the collection and application of these data will become widespread and ubiquitous, transforming the spatial and temporal scales that geomorphologists can investigate landforms and the processes that act upon them.

Not surprisingly the wealth of data is becoming overwhelming to manage, whilst finding relevant information appropriate to the task at hand is increasingly difficult. Data management has become a vital component in any project and GIS is ideally suited to fill this role, yet the increasing requirements for large datasets makes this problematic and current storage and processing paradigms lag behind the needs of the end user. In early 1980s 'cutting edge' remotely sensed data volumes involved the manipulation of $\sim$240 Mb per dataset, while current systems can easily produce $\sim$1.2 Gb of data for *one tenth* of the area. Ynnerman (2010) describes data explosion in the medical sciences where volumes from computed tomography (CT) scanners have dramatically increased; a single scan now generates 24,000 images and $\sim$20 Gb of data. The whole area of spatial data visualisation and analysis requires a step-change in capabilities moving towards immersive three-dimensional environments, making use of multi-modal applications and greater leveraging of haptic interaction (Lundin et al. 2008). Much of the current investment and development is coming from the computer games industry (e.g. Microsoft Kinect) and the following years should see greater penetration into scientific visualisation.

Searching for data is an increasingly difficult problem and the development of data warehouses, such as the *Global Land Cover Facility* (http://glcf.umd.edu), are an attempt to solve this. It remains a fragmented area, however, with warehouses segregated between governments, research agencies, universities, subject specialisms, commercial aggregators, professional bodies and industry sectors. A key aspect is the provision and utilisation of metadata; often simple elements such as the producer, owner, copyright and spatial extent are missing. Details such as processing algorithms, precision, accuracy and currency are limited, particularly where secondary data is concerned. Many academic and research funding bodies now require the submission of data upon the completion of projects and, with the mandating of metadata, this is helping to improve their downstream discovery and use.

Access, use and re-use of some of the data highlighted above is mixed. Federally collected data in the US remains in the public domain and therefore entirely free from restriction. Yet much of the wealth in data and development of new applications come from recent commercial initiatives (e.g. high spatial resolution satellites such as *GeoEye-1*). These fall under copyright and their use protected.

Access to other datasets is often variable; professional bodies, funding agencies and universities may require some kind of affiliation before they grant access (although the data will remain under copyright). Governments are increasingly seeing the benefit and power of releasing data and there has been a movement to place these in the public domain or under a Creative Commons license (http://creativecommons.org/). Discoverability and reuse is of fundamental importance to researchers; geomorphological information can often be limited and it is therefore important that it is easy to find, access and reuse.

With all this technological change, another aspect that continually needs revising is that of *communication*. Researchers are increasingly aware of requirements by funding agencies to disseminate their work, maximise the impacts upon society and make their outputs widely available. Yet communication is often limited to conferences and papers, with spatial outputs ignored. Maps are a key part of this strategy, yet effective visual communication is often limited. With a renaissance in mapping comes a requirement to better communicate; dissemination of cartographic best practise is an important part of this. Likewise, researchers are finding new ways to communicate and disseminate; digital globes such as *Google Earth* and *NASA Worldwind* provide a platform through which they can provide online interaction with the general public. Likewise, the value and need for open spatial data standards are paramount and the support and traction gained by the Open Geospatial Consortium (http://www.opengeospatial.org) is of significance in easing this process.

Finally, given the wealth of digital data that can now be collected, collated and compiled, there remains the question of how to produce an actual accurate 'map' of the physical landscape. The ability to interpret the information to provide a genuine appreciation of the landforms and processes is dependent of the skill and experience of the geomorphologist. Such skills and experience can only be gained in the field and no matter how accurately the data can be represented on a map or in a GIS there will always be the requirement for the end product to be 'ground-truthed.'

7 Conclusions

Geomorphological mapping has a long history in the use of applied cartography. A range of specialised maps have been a fundamental 'enabler' in studying the landscape as both a method for visualisation and a paradigm through which data storage and analysis could be performed. The move of physical geography (and geomorphology) to quantitative, field-scale, approaches of study in the 1970s limited the application of maps. And, ironically, GIS led to the demise of maps in the published scholarly literature with almost a generation of spatial outputs largely absent from this permanent archive (Smith 2005). Yet the mantle of mapping was taken up in the 1980s by engineering geologists who saw the power of integrating morphology with an understanding of surface and subsurface materials and processes. Geomorphology represents one of the key linkages in understanding and managing the environment and can therefore be seen as a societal necessity. The real power of geomorphological mapping is unleashed through this interdisciplinary approach, between practitioners in, for example, engineering, geology, geomorphology, planning and land management. The full utility of this realisation has become apparent through a move in physical geography to combine process level understanding at the landscape scale; changes in data availability have now made such studies technologically feasible. Maps and mapping are an important part of this process and new demands are being placed upon visualisation and dissemination. Techniques for the automated production of maps directly from GIS, the presentation of complex multi-dimensional variables and the integration into electronic workflows are important and remain a developing area. It is with this backdrop that geomorphological mapping can look forward to the next decade with anticipation.

References

Balme, M. R., Bargery, A. S., Gallagher, C. J., & Gupta, S. (Eds.). (2011). *Martian geomorphology* (p. 356). London: Geological Society, Special Publications.

Baltsavias, E. P. (1999). Airborne laser scanning: basic relations and formulas. *ISPRS Journal of Photogrammetry and Remote Sensing, 54*, 199–214.

Barrand, N. E., Murray, T., James, T. D., Barr, S. L., & Mills, J. P. (2009). Optimising glacier DEMs for volume change assessment using laser-scanning derived ground control points. *Journal of Glaciology, 55*(189), 106–116.

Boike, J., & Yoshikawa, K. (2003). Mapping of periglacial geomorphology using kite/balloon aerial photography. *Permafrost and Periglacial Processes, 14*, 81–85.

Brunsden, D., Doornkamp, J. C., Fookes, P. G., Jones, D. K. C., & Kelly, J. M. N. (1975). Large scale geomorphological mapping and highway engineering design. *Quarterly Journal of Engineering Geology, 8*, 227–253.

BSI. (1981). *BS 5930: Code of practice for site investigations*. London: British Standards Institution.

Burnett, A. D., & Styles, K. A. (1985). Geotechnical area studies programme and land use planning in Hong Kong. *Planning and Development, 1*(2), 321–340.

Burnett, A. D., & Styles, K. A. (1986). A review of the reliability of an engineering geological mapping system used for the delineation of landslide prone terrain in Hong Kong. In S. Wang (Ed.), *Engineering geological problems in Asia* (pp. 321–340). Beijing: Science Press.

Carr, M. H. (1983). Stability of streams and lakes on Mars. *Icarus, 56*, 476–495.

Carrara, A., Bitelli, G., & Carla, R. (1997). Comparison of techniques for generating digital terrain models from contour lines. *International Journal of Geographical Information Science, 11*, 451–473.

Chandler, J. H., Fryer, J. G., & Jack, A. (2005). Metric capabilities of low-cost digital cameras for close range surface measurement. *Photogrammetric Record, 20*(109), 12–26.

Cheng, W., Zhou, C., Chai, H., Zhao, S., Liu, H., & Zhou, Z. (2011). Research and compilation of the Geomorphologic Atlas of the People's Republic of China (1:1,000,000). *Journal of Geographical Sciences, 21*(1), 89–100.

Chorley, R. J., Schumm, S. A., & Sugden, D. E. (1985). *Geomorphology*. Routledge: Kegan and Paul.

Close, M. H. (1867). Notes on the general glaciation of Ireland. *Journal of the Royal Geographical Society of London, 1*, 207–242.

Cooke, R. U., & Doornkamp, J. C. (1990). *Geomorphology in environmental management*. Oxford: Oxford University Press.

Craddock, R. A., & Howard, A. D. (2002). The case for rainfall on a warm, wet early Mars. *Journal of Geophysical Research (Planets), 107*, 5111.

Dearman, W. R., & Fookes, P. G. (1974). Engineering geological mapping for civil engineering practice in the United Kingdom. *Quarterly Journal of Engineering Geology, 7*, 223–256.

Demek, J. (Ed.). (1972). *Manual of detailed geomorphological mapping*. Prague: Academia.

Demek, J., & Embleton, C. (Eds.). (1978). *Guide to medium-scale geomorphological mapping*. Stuttgart: International Geographical Union.

DeOliveira, F. A., & Vieira, C. V. (2009). Semi-detailed geomorphological map of northeastern Santa Catarina State, Brazil—The Garuva sheet. *Journal of Maps, 5*, 66–74.

Doornkamp, J. C. (Ed.). (1988). *Planning and development: Applied earth science background—Torbay*. Nottingham: M1 Press.

Doornkamp, J. C., Brunsden, D., Cooke, R. U., Jones, D. K. C., & Griffiths, J. S. (1987). Environmental geology mapping—An international review. In M. G. Culshaw (Ed.), *Planning and engineering geology* (pp. 215–220). Engineering Geology Special Publication. London: Geological Society.

Doornkamp, J. C., Brunsden, D., Jones, D. K. C., & Cooke, R. U. (1980). *Geology, geomorphology and pedology of Bahrain*. Norwich: GeoBooks.

Doornkamp, J. C., Brunsden, D., Jones, D. K. C., Cooke, R. U., & Bush, P. R. (1979). Rapid geomorphological assessments for engineering. *Quarterly Journal of Engineering Geology, 12*, 189–204.

Evans, I. S. (1972). General geomorphometry, derivatives of altitude and descriptive statistics. In R. J. Chorley (Ed.), *Spatial analysis in geomorphology* (pp. 17–90). New York: Harper and Row.

Fenneman, N. M. (1928). Physiographic divisions of the United States. *Annals of the Association of American Geographers*, 18: end folding map.

Fookes, P. G. (1997). Geology for engineers: The geological model, prediction and performance. *Quarterly Journal of Engineering Geology, 30*, 293–424.

Forster, A., Hobbs, P. R. N., Wyatt, R. J., & Entwisle, D. C. (1987). Environmental geology maps of Bath and the surrounding area for engineers and planners. *Geological Society Engineering Geology Special Publication, 4*, 221–235.

Glasser, N. F., & Jansson, K. N. (2008). The glacial map of southern South America. *Journal of Maps, 4*, 175–196.

Gousie, M. B., & Franklin, W. R. (2005). Augmenting grid-based contours to improve thin-plate DEM generation. *Photogrammetric Engineering and Remote Sensing, 71*(1), 67–79.

Griffiths, J. S. (2001). Development of a ground model for the UK Channel Tunnel portal. In J. S. Griffiths (Ed.), *Land surface evaluation for engineering practice* (Vol. 18, pp. 129–133). Geological Society Engineering Geology Special Publication.

Griffiths, J. S., & Marsh, A. (1986). BS5930: The role of geomorphological and geological techniques in a preliminary site investigation. In A. B. Hawkins (Ed.), *Site investigation practice: Assessing BS5930* (pp. 261–267). Engineering Geology Special Publication. London: Geological Society.

Griffiths, J. S., & Stokes, M. (2008). Engineering geomorphological input to ground models: A possible approach based on earth systems. *Quarterly Journal of Engineering Geology and Hydrogeology, 41*, 1–19.

Griffiths, J.S., & Abraham, J. (2008). Factors affecting the use of applied geomorphological maps to communicate to different end-users. *Journal of Maps, 4*, 201–210.

Hammond, E. H. (1954). A geomorphic study of the cape region of Baja California. *University of California Publications in Geography, 10*, 50.

Hearn, G. J. (1997). *Overseas road note 16: Principles of low cost road engineering in mountainous roads* (p. 149). Berkshire: Transport Research Laboratory.

Hearn, G. J. (2011). *Slope engineering for mountain roads*. Geological Society of London.

Hearn, G. J., & Blong, R. (2001). Landslide hazard for the Ok Tedi Mine in Papua New Guinea. In: J. S. Griffiths (Ed.), *Land surface evaluation for engineering practice* (Vol. 18, pp. 143–150). Geological Society Engineering Geology Special Publication.
Hengl, T., & Reuter, H. I. (2008). *Geomorphometry: Concepts, software, applications*. Amsterdam: Elsevier.
Heritage, G., & Large, A. (2009). *Laser scanning for the environmental sciences*. Chichester: Wiley-Blackwell.
Hodge, R., Brasington, J., & Richards, K. S. (2009). In situ characterization of grain-scale fluvial morphology using terrestrial laser scanning. *Earth Surface Processes and Landforms, 34*, 954–968.
International Association of Engineering Geology. (1976). *Engineering geological maps: A guide to their preparation*. Paris: The UNESCO Press.
International Geographical Union. (1968). The unified key to the detailed geomorphological map of the World 1:25,000–1:50,000. Part I. Problems of the detailed geomorphological map, by Klimaszewski, M. Part II. Project of the unified key to the detailed geomorphological map of the World, by Bashenina, N. V., Gellert, J. Jolly, F. et al. Commission on Applied Geomorphology, Subcommission on Geomorphological Mapping, Folia Geographica, Series Geographica-Physica 2, Krakow.
Jones, D. K. C. (2001). Ground conditions and hazards: Suez City development, Egypt. In: J. S. Griffiths (Ed.), *Land surface evaluation for engineering practice* (Vol. 18, pp. 159–170). Geological Society Engineering Geology Special Publication.
Klimaszewski, M. (1956). The principles of the geomorphological map of Poland. *Przeglad Geograficzny, 28*(suppl), 32–40.
Kraak, M.-J., & Ormeling, F. (2010). *Cartography: Visualisation of spatial data* (3rd ed.). Harlow, Essex: Prentice Hall.
Laliberte, A. S., Herrick, J. E., Rango, A., & Winters, C. (2010). Acquisition, orthorectification, and object-based classification of unmanned aerial vehicle (UAV) imagery for rangeland monitoring. *Photogrammetric Engineering and Remote Sensing, 76*(6), 661–672.
Lee, E. M. (2001). Geomorphological mapping. In J. S. Griffiths (Ed.), *Land surface evaluation for engineering practice* (Vol. 18, pp. 53–56). Geological Society Engineering Geology Special Publication.
Lee, E. M., & Moore, R. (1989). *Report on the study of landsliding in and around Luccombe Village*. Department of the Environment, HMSO.
Lewis, G. M. (1962). Changing emphases in the description of the natural environment of the American Great Plains area. *Transactions of the Institute of British Geographers, 30*, 75–90.
Lohani, B., & Mason, D. C. (2001). Application of airborne scanning laser altimetry to the study of tidal channel geomorphology. *ISPRS Journal of Photogrammetry and Remote Sensing, 56*, 100–120.
Lundin, K., Cooper, M., & Ynnerman, A. (2008). Haptic interaction with dynamic volumetric data. *IEEE Transactions on Visualization and Computer Graphics, 14*(2), 263–276.
Maas, H.-G., Dietrich, D., Schwalbe, E., Bäßler, M., & Westfeld, P. (2006). Analysis of the motion behaviour of Jakobshavn Isbræ glacier in Greenland by monocular image sequence analysis. *International Archives of Photogrammetry, Remote Sensing and Spatial Information Sciences, XXXVI*(5), 179–183.
Miller, C. L., & LaFlamme, R. A. (1958). The digital terrain model—Theory and application. *Photogrammetric Engineering, 24*, 433–442.
Mitchell, H. (2007). Fundamentals of photogrametry. In J. Fryer, H. Mitchell, & J. Chandler (Eds.), *Applications of 3D measurement from images* (pp. 9–40). Dunbeath: Whittles Publishing.
Monkhouse, F. J., & Wilkinson, H. R. (1971). *Maps and diagrams*. London: Routledge.
Nekum, G., Jaumann, R., the HRSC Co-investigator Team, et al. (2004). Recent and episodic volcanic and glacial activity on Mars revealed by the high resolution stereo camera. *Nature, 432*, 971–979.

Otto, J.-C., Gustavsson, M., & Geilhausen, M. (2011). Cartography: Design, symbolisation and visualisation of geomorphological maps. In M. J. Smith, P. Paron, & J. Griffiths (Eds.), *Geomorphological mapping: Methods and applications*. London: Elsevier.

Pain, C. F., Paron, P., & Smith, M. J. (2008). Applied geomorphological mapping (AppGeMa): A working group of the International Association of Geomorphologists. *Geophysical Research Abstracts, 10*, EGU2008-A-05888.

Palmann, C., Mavromatis, M., Sequeira, J., & Brisco, B. (2008). Earth observation using radar data: An overview of applications and challenges. *International Journal of Digital Earth, 1*, 171–195.

Paron, P., & Claessens, L. (2011). Makers and users of geomorphological maps. In M. J. Smith, P. Paron, & J. Griffiths (Eds.), *Geomorphological mapping: Methods and applications*. London: Elsevier.

Passarge, S. (1912). Physiologische Morphologie. *Mitteilungen der Geographischen Gesellschaft Hamburg, 26*, 135–337.

Porcher, M., & Guillope, P. (1979). Cartography des risqué ZERMOS appliqués a des plans d'occupation des sols en Normandie. *Bulletin Mason Laboratoire des Ponts et Chaussees, 99*, 43–54.

Powell, J. W. (1896). *The physiography of the United States*. New York: National Geographic Society.

Report by the Geological Society Engineering Group Working Party. (1972). The preparation of maps and plans in terms of engineering geology. *Quarterly Journal of Engineering Geology and Hydrogeology, 5*, 293–382.

Robinson, G. D., & Spieker, A. M. (1978). *Nature to be commanded*. USGS Professional Paper 950.

Ronca, L. B. (1972). The geomorphic evolution of the lunar surface. In S. K. Runcorn & H. C. Urey (Eds.), *The Moon, International Astronomical Union* (pp. 43–54). Dordrecht: Reidel.

Rose, J., & Smith, M. J. (2008). Glacial geomorphological maps of the Glasgow region, western central Scotland. *Journal of Maps, 4*, 399–416.

Savigear, R. A. G. (1965). A technique of morphological mapping. *Annals of the Association of American Geographers, 53*, 514–538.

Sharp, R. P. (1967). Geomorphology in the space age. *Engineering and Science, 31*, 61–63.

Smith, A., & Ellison, R. A. (1999). Applied geological maps for planning and development: A review of examples from England and Wales, 1983 to 1996. *Quarterly Journal of Engineering Geology, 32*, S1–S44.

Smith, L. C. (2002). Emerging applications of interferometric synthetic aperture radar (InSAR) in geomorphology and hydrology. *Annals of the Association of American Geographers, 92*, 385–398.

Smith, M. J. (2005). The Journal of Maps: An electronic journal for the presentation and dissemination of map based data. *Journal of Maps, 1*, 1–6.

Smith, M. J., Chandler, J., & Rose, J. (2009). High spatial resolution data acquisition for the geosciences: Kite aerial photography. *Earth Surface Processes and Landforms, 34*, 155–161.

Smith, M. J., Griffiths, J., & Paron, P. (Eds.). (2011). *Geomorphological mapping: Methods and applications. Developments in earth surface processes*. London: Elsevier.

Smith, M. J., & Pain, C. (2009). Applications of remote sensing in geomorphology. *Progress in Physical Geography, 33*(4), 568–582.

Styles, K. A., Hansen, A., Dale, M. J., & Burnett, A. D. (1984). Terrain classification methods for development planning in and geotechnical appraisal: A Hong Kong case study. In *Proceedings of the 4th international symposium on landslides*, pp. 561–569.

Towner, N. C., Eakin, C., Conway, S. J., & Harrison, S. (2011). Geologically recent flow inferred in channel systems in the NE Sulci Gorddii region, Mars. In M. R. Balme, A. S. Bargery, C. J. Gallagher, & S. Gupta (Eds.), *Martian geomorphology* (Vol. 356, pp. 229–256). Geological Society, London, Special Publications.

Verstappen, H. T. (2011). Old and new trends in geomorphological and landform mapping. In M. J. Smith, P. Paron, & J. Griffiths (Eds.), *Geomorphological mapping: Methods and applications*. London: Elsevier.

Waller, A., & Phipps, P. (1996). Terrain systems mapping and geomorphological studies for the Channel Tunnel rail link. In C. Craig (Ed.), *Advances in site investigation practice* (pp. 25–38). London: Thomas Telford.

Ynnerman, A. (2010). *Visualizing the medical data explosion [Online video]*. http://www.ted.com/talks/anders_ynnerman_visualizing_the_medical_data_explosion.html. Last accessed: Aug 27, 2015.

Making Sense of Cities: The Role of Maps in the Past, Present, and Future of Urban Planning

Amy Hillier

Abstract Cities are by definition complex and dynamic physical and social systems. Professionals charged with planning for and building our constantly-changing cities, including urban planners and designers, depend upon a wide range of maps and spatial technologies to make decisions. Planning is essentially a visual and spatial discipline where ideas, plans, arguments and empirical analyses are all expressed through paper and digital maps, including GIS (geographic information systems). This chapter critically explores the way professional and academic planners who work in transportation, community and economic development, international development, environmental conservation, real estate and urban use maps at all scales. Maps will continue to be critical in planners' efforts to understand the spatial form of cities and how residents conceive of places, make and analyze plans, model social and environmental impacts, engage citizens and hold government officials accountable. The interdisciplinary fields of public health, energy and sustainability and civic engagement are examined as opportunities for future collaboration.

Keywords Maps and planning · Spatial technologies · Participation · Empowerment · Policy

Cities are by definition complex and dynamic physical and social systems. We who live in them depend upon street signs, GPS gadgets, "smart" phones, subway and bus maps, online routing systems, fellow residents, and that "sense of direction" gene simply to find our way around. Visitors who are more easily intimidated by the tall buildings, massive street networks, and unfamiliar social landscape are forced to rely even more upon such devices to feel safe as they travel into the city and then home again.

Professionals charged with planning for and building our constantly-changing cities, including urban planners and designers, depend upon a wider range of maps and spatial technologies to make decisions as do scholars who strive to make

A. Hillier (✉)
School of Design, University of Pennsylvania, Philadelphia, PA, USA
e-mail: ahillier@design.upenn.edu

S.D. Brunn and M. Dodge (eds.), *Mapping Across Academia*,
DOI 10.1007/978-94-024-1011-2_3

intellectual sense of cities and the decisions of planners, developers, and politicians. Academic planners rely on maps and spatial analysis to assess sustainability and equity—values central to the profession—as well as the distribution of power, historical and future trends in population and real estate, and to train the next generation of planners, developers, and academics.

Planning is essentially a visual and spatial discipline where ideas, plans, arguments, and empirical analyses are all expressed through paper and digital maps. Mapping has become nearly synonymous with geographic information systems (GIS) in recent years, and masters-level urban planners are now expected to have at least basic skills in GIS. However, while GIS and related geospatial technologies such as GPS are relatively new and rapidly developing, the reliance upon spatial thinking and spatial models is a constant in this field.

This chapter critically explores the way professional and academic planners who work in the subfields of transportation, community and economic development, international development, environmental conservation, real estate, and urban design use maps and the scales—from intimate neighborhood to the entire globe—at which they focus their attention. It considers maps as ideals in the tradition of Ernest Burgess, Ebenezer Howard, and Clarence Perry, mental images in the tradition of Kevin Lynch, plans in the tradition of urban renewal and expressway construction, pre- and post-natural disaster and terrorism planning, and modeling in the form of three dimensional mapping, inferential statistics, overlay analysis, and cartographic modeling techniques. The role of maps in civic engagement and the process of planning are also explored.

In addition to celebrating some of the major contributions of planning to mapping—cadastral maps, zoning maps, and population forecasting—this chapter considers some of the failures and limitations of the use of maps by planners. Finally, it considers new areas ripe for inter-disciplinary collaboration where planners' technical expertise and knowledge of how cities work can complement what specialists in international development, energy efficiency, and public health bring to the study of cities.

1 Maps as Ideals

Before urban planners and scholars can aspire to alter the shape of the city, they must understand its form—or at least make efforts to understand it. Toward this end, scholars have used maps to model the ideal urban structure. Rather than empirically representing the reality of a specific city or casting a detailed vision for a specific place, these types of maps are used to convey the basic spatial form—the morphology—of the city. Sociologist Ernest Burgess, who helped establish the Chicago School as a leader in urban studies starting in the 1920s, proposed a theory of urban ecology that understood cities as places where different groups competed for space, leading to the development of distinct zones (Park et al. 1925). He famously drew the city as a series of concentric rings around the central business

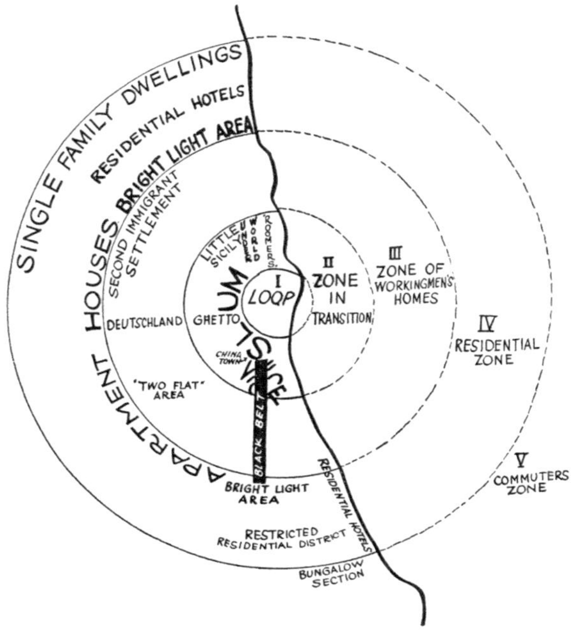

Fig. 1 Ernest Burgess' simplified Concentric Zone city model

district (Fig. 1). The first "zone" was synonymous with the slums or ghetto; the second was home to the working class; and the two outlying zones were for the middle and upper classes who could afford single family homes and the commute into the city.

Over subsequent decades, urban scholars critiqued the simplicity of the concentric zone model, offering new ways of thinking about city growth through sector (Hoyt 1939), polycentric, and multiple nuclei (Harris and Ullman 1945) ideals. But despite its clear limitations, Burgess' diagram is still among the most frequently-referenced maps of the city because scholars still find some value in his diagram. For example, Beijing's spatial form has been described as four metropolitan regions, including inner and outer centers and inner and outer peripheries or five zones, following the six ring-roads around the city. Never mind that the composition of Beijing's zones bears little resemblance to Chicago and Burgess' model of American urbanism; his rings are emblazoned on the imagination of planners and scholars around the world.

Ebenezer Howard and Clarence Perry used maps in a similar way to model ideal land use within an urban community. Inspired by Edward Bellamy's utopian *Looking Backward*, Howard outlined his idea of garden cities in his self-published 1898 book, *To-Morrow: A Peaceful Path to Real Reform*. Intended to stem the migration of rural residents to urban areas and the consequent over-crowding, garden cities were to be planned towns of a limited size surrounded by agricultural land. They took the form of a series of connected satellite communities across the country-side, connected by train lines and roads.

Also concerned about the spatial patterns of urban living, Perry (1929) conceived of a neighborhood unit as a way to protect children from the new but rapidly increasing phenomena of vehicle traffic in cities in the 1920s. The basic principles of the neighborhood unit included: (1) children should live within a ¼ mile walk of school; (2) schools should be used for community events beyond educating children; (3) arterial streets around the perimeter help define the neighborhood and reduce through-traffic; and (4) 10 % of land should be dedicated to parks and open space (Howard 1898, 1902) (Fig. 2). Though criticized for reinforcing racial segregation and inspiring gated communities, Howard's spatial scheme and the design principles of Clarence Stein and Henry Wright, including their ideas of superblocks and road hierarchy, are directly linked to the contemporary field of neighborhood planning.

None of these ideals matches perfectly the spatial form of cities, the neighborhoods within them, or the towns surrounding them, but the concepts of zones or districts, garden cities, and neighborhood units have a permanent place in urban planning. Beyond historical texts about the profession, the idea that cities have a certain spatial organization that separates land uses and can be deliberately influenced for the well-being of residents lives on in much of contemporary urban planning practice and research. Scholars point to New Urbanism, regional growth boundaries, and transit-oriented development as evidence of the enduring legacy of Howard's garden city and Perry's neighborhood unit (Fishman 1998) although like the original idea, these strategies have not been consistently effective in preventing suburban sprawl, soulless suburbs, ugly strip malls or automobile-centered development or resolving the persistent inequity within cities and regions.

2 Maps as Mental Images

In *The Image of the City* (1960), a classic urban planning text, Kevin Lynch used map in a less abstract way than the ideals of Burgess, Howard, and Perry. His maps reflected the way that individual residents experienced cities based on 4 years of observations in Boston, Los Angeles and Jersey City (Fig. 3). He introduced the concepts of "legibility" and "way-finding" and argued that beyond maps and street signs, individuals relied upon mental map images to orient themselves. "The need to recognize and pattern our surroundings is so crucial, and has such long roots in the past, that this image has wide practical and emotional significance to the individual" (Lynch 1960: 4). Furthermore, understanding the process by which individuals navigated through a "complex, shifting environment" was critical to efforts to improving the design of cities.

Scholars continue to look to Lynch's work for its conceptual and utilitarian value, including serving as a framework for considering how marginalized communities—such as post-Katrina New Orleans and the Lower Ninth Ward—might

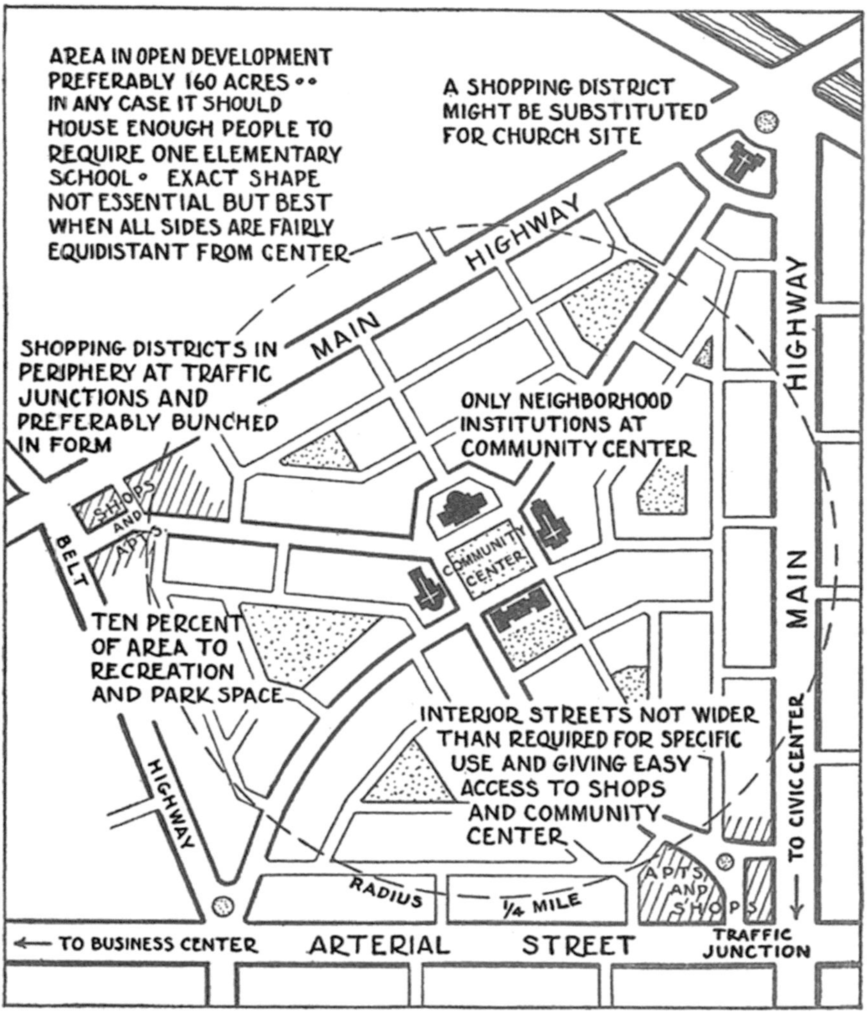

Fig. 2 Clarence Perry's design for a Neighborhood Unit

become valued or help identify distinctive elements of cities to inform their marketing strategies (Hospers 2010). Subsequent professional and scholarly work has continued to explore the nature of wayfinding and "cognitive maps," (Golledge 1999) to understand resident perceptions of neighborhood boundaries (Guest and Lee 1984), children's experience of the city (Halseth and Doddridge 2000), and older adults' perceptions of space (Evans et al. 1984).

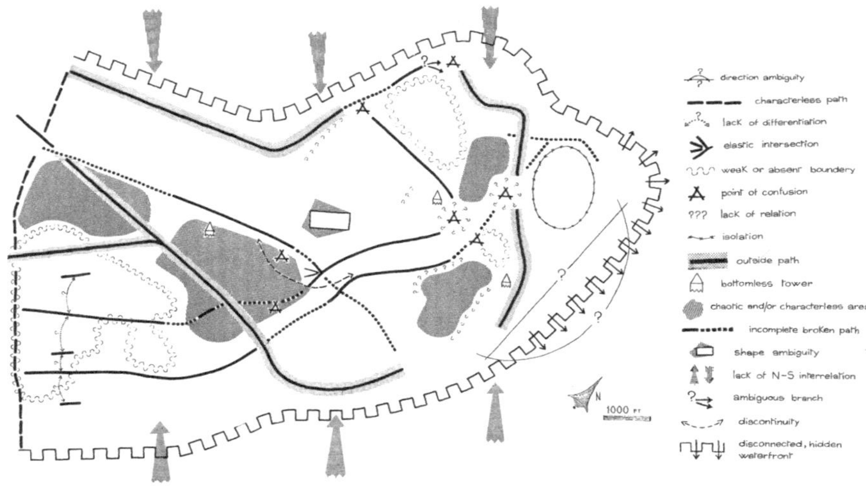

Fig. 3 Example of a mental map created by Kevin Lynch

3 Maps as Plans

At the heart of urban planning is the process of developing and implementing schemes for reconfiguring the city. More than as ideals, urban planning professionals and scholars have used maps in the process of making and communicating short-term and long-term plans to adapt cities to changing physical, environmental, and social conditions. Some plans are created with the hope they will never be used, such as disaster plans, while others carry forward visions and dreams that public officials and private developers scheme to realize—new expressways, Olympic parks, desert cities, and 400 m skyscrapers.

Cadastral and zoning maps form the foundation of many of these plans. Cadastral maps show lines for all real properties along with measurement and ownership information for tax purposes while zoning maps outline allowable land uses. Together, they distinguish the village or city from the countryside, carving it into small, mostly regular-shaped pieces, and assigning them to technically precise categories with colors, numbers, and letters. While zoning maps are nominally intended to serve the needs of the general public—by establishing legal boundaries and promoting public health and safety by establishing standards for air and light and keeping incompatible land uses separate—they represent the city as person-less and rational. Zoning maps and the plans to reinforce, adapt or replace land use decisions are the basis for endless heated civic meetings, legal action, and political wrangling because they are recognized as the tools for controlling cities, creating jobs, winning votes, establishing legacies, and making money.

The Urban Renewal era in the U.S. provides numerous examples of how planners used maps to cast their visions for removing blight, reducing crime, and increasing property values. Expressways connecting suburban residents with central cities, public housing developments segregating poor whites and blacks from middle-class neighbors, and the development of racially exclusive tract housing in the suburbs all found their first expression in maps. A 1960 map of the extensive network of expressways planned for Philadelphia points to the speculative nature of planning, as fewer than half of the planned expressways were ultimately built, but also the power of printed maps. The Crosstown expressway was never built, but maps like Fig. 4 convinced residents—many of them African American who had lived in the area for several generations—to move, effectively emptying out downtown and making it ripe for gentrification.

Comprehensive plans for cities are intended to guide the long-term physical development of cities. New York City's PlaNYC 2030 Comprehensive Plan used maps extensively to show, for example, the future demands on the city and potential solutions to problems relating to energy consumption and commuting (PlaNYC 2009). Academics intersect with the comprehensive planning process as direct participants, advising through public service stints and teaching courses and studios, and critics, mercilessly reviewing plans as if they were new movies or restaurants.

While comprehensive plans frequently focus on meeting future demands for housing, transportation, and other infrastructure created by population growth, they are also used to limit growth and development. The establishment of growth boundaries in Canada, the United Kingdom, and Australia and U.S. cities in Washington, Oregon, Tennessee was intended to reduce sprawl establish minimum residential densities within and maximum residential densities outside. They follow in the tradition of London's greenbelt, a post-war effort to secure England's self-image as pastoral and idyllic by restricting the development of rural land, and the greenbelts that surround many of Britain's other cities.

Maps serve as the basis for plans to avoid, respond to, and recover from natural and man-made disasters including flooding and terrorism. In order to be eligible for federal flood insurance in the U.S., local communities must create flood hazard maps identifying 100-year and 500-year flood plains (FEMA 2011). The Federal Emergency Management Administration (FEMA) provides resources for communities about using desktop GIS to create the requisite maps. FEMA also developed HAZUS-MP, GIS software for estimating potential population and building losses from natural disasters—including earthquakes, hurricane winds, and floods (FEMA 2010)—that communities can download for free. HAZUS-MP is designed to help local communities with emergency preparedness, mitigation plans and policies, and disaster response and recovery planning. Events such as Hurricane Katrina underscore the uncertainty that characterizes the process of assessing risk and anticipating impacts while they also serve to highlight the political and economic forces that influence emergency preparedness and response.

Post 9/11, maps are increasingly used to prepare for possible acts of terrorism. Cities around the globe are investing hundreds of thousands of dollars to develop three-dimensional maps of buildings and below-surface areas such as subway concourses (Hadhazy 2010). Planners, engineers, and architects have relied on three-dimensional physical models for decades, using blocks and voids to convey

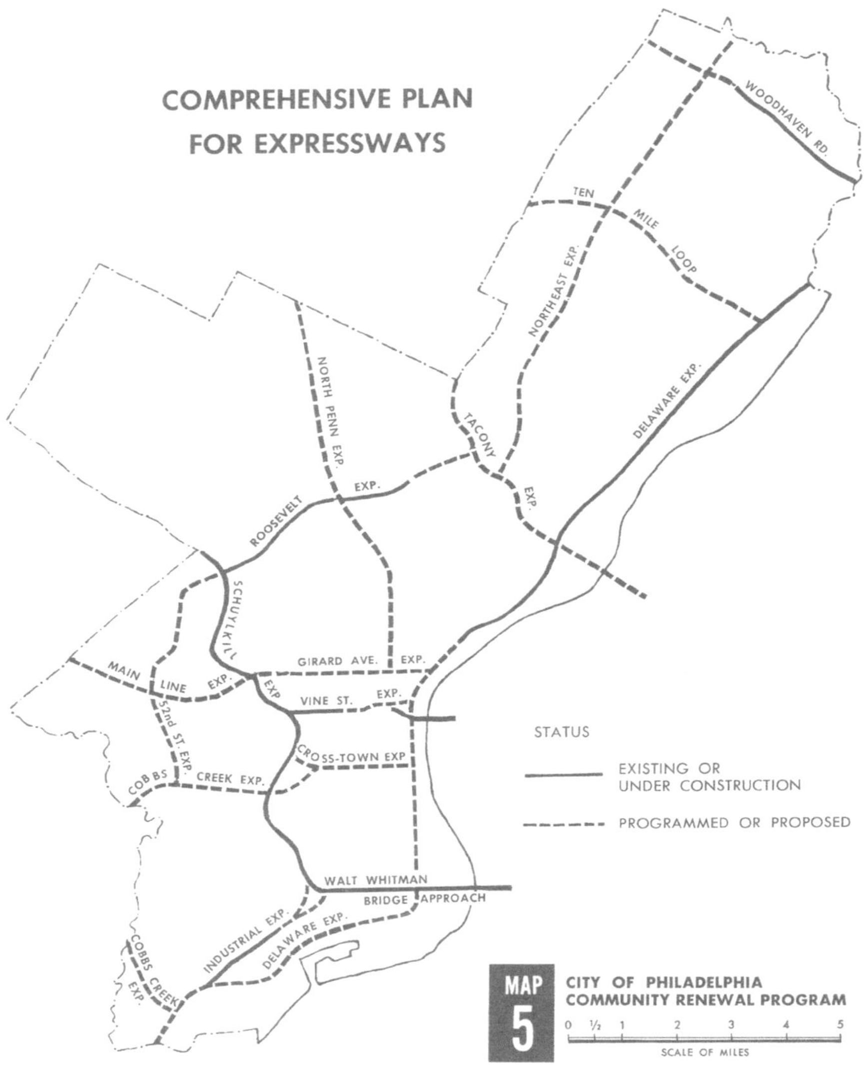

Fig. 4 Schematic display of proposed Philadelphia expressways, 1960

volume, sightlines, and skylines. The three-dimensional maps of today are developed with Light Detection And Ranging (LIDAR) technology, using low-flying aircraft equipped with lasers that measure the precise distance to the ground or buildings (Fig. 5), and are capable of supporting real-time disaster management (Kwan and Lee 2005). For example, emergency management officials in New York City used LIDAR-generated images of Ground Zero in the wake of 9/11 to monitor

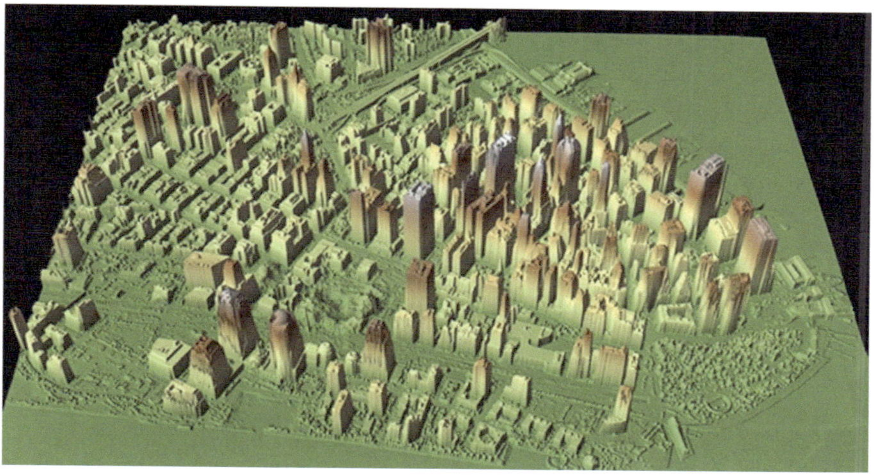

Fig. 5 Downtown area of Philadelphia captured as a LIDAR model

shifting debris as well as to generate models for PlaNYC, to estimate future energy consumption and greenhouse gas emission (Hadhazy 2010).

4 Maps as Models

As urban planning concerns itself so much with the built form, the ability to visualize different design scenarios and estimate their social, visual, and environmental impact of new development is essential to the planning process. For several decades, computer-aided design packages such as AutoDesk's AutoCad and Bentley's Microstation have given high-skilled design professionals the tools to construct precise three-dimensional models, albeit without the benefit of any context. Today, software products such as Google Sketch-up make it relatively simple for planners to create basic three-dimensional models of future developments and visualize them in their geographic context through products like Google Earth (Fig. 6). The integration of computer-aided design (CAD) software with GIS and packages such as Rhino, 3D Studio Max, and Maya allow for mapping detailed buildings in their geographic context that, when animated, can recreate fly-overs, showing proposed development from all angles or change over time.

For more than a decade, scholars have been anticipating what web-based mapping and virtual reality technologies will contribute to modeling urban form. GeoDesign is the new name for this idea of blending planning and design methods with simulations in a geographic context. More specifically, GeoDesign is a set of techniques and technologies for conceptualizing, planning, analyzing, and designing environments in an integrated process. In theory, the process can include significant stakeholder participation and collaboration and consideration of social

Fig. 6 Google Sketch-up model in Google Earth

needs. One example of how integrating technologies and professional disciplines through GeoDesign can generate innovative designs is the 40-story Solar Power Tower in Seville, Spain, the world's first commercial solar power plant designed to produce no greenhouse emissions while generating enough electricity for most of Seville's residents (ESRI 2010; Zwick 2010).

Planners also use agent-based modeling and simulation (ABMS) to try to recreate and predict how "agents," generally individuals or groups, will interact. CommunityViz, an extension for ArcGIS software that allows planners to model new developments and estimate the impact on energy, environmental management, schools, infrastructure, integrates ABMS through its Policy Simulator (Walker and Daniels 2011). Academic planners have used the CommunityViz Policy Simulator to compare future development scenarios for one Washington state municipality with comprehensive growth plans, considering the potential impact of new protections for endangered salmon, population in-migration from retirees, and visitors seeking recreation on development patterns (Ligmann-Zielinska and Jankowski 2007).

Academic planners utilize statistical models more than visualization, especially in the subfields of transportation modeling and population forecasting. Transportation planners and engineers use a variety of software packages that incorporate statistical algorithms, transportation networks, and visualization tools to conduct travel demand forecasting, operate public transit systems, route freight vehicles, and to assess the efficiency and equity of transportation systems.

Professional and academic planners rely upon current and projected population estimates in much of their work, giving rise to for-profit vendors that sell population data as well as new methods that combine GIS and statistical modeling (Landis and Zhang 2000). GIS is also integrated with statistical algorithms in location-allocation modeling, used for site selection of particular types of facilities, and multi-criteria analyses, an alternative to cost-benefit analysis that allows for consideration of quality, time, and environmental and social impacts in addition to monetary costs (Dodgson 2009).

Geostatistical analysis and cartographic modeling borrow concepts and measures from the earth sciences and landscape architecture. In contrast to representations of space that rely on discrete parcels, streets, administrative and political boundaries, and census geographies to carve up and make sense of space, the idea of "landscape" encourages holistic and continuous representations. Scottish-born Ian McHarg developed a system for using transparent acetate map overlays in his landmark book, *Design with Nature*, to model landscapes. Deeply disturbed by the implications of massive highway building on the natural environment, he advocated for ecological planning that incorporated social costs and benefits. Each map layer represented a different social value, helping to identify areas of greatest social value that should be protected from development.

McHarg's method of synthesizing multiple map layers into a single end map took on new form with the advent of GIS. Fellow landscape architect C. Dana Tomlin developed the concept and algorithms of "map algebra" with his 1983 dissertation and 1990 book, *Geographic Information Systems and Cartographic Modeling*. The process of cartographic modeling uses a combination of simple operations—local (on individual raster cells), focal (on individual cells and their neighbors), global (on the entire map layer) and zonal (on cells that share the same value)—on input layers to perform complex tasks and generate a new output layer. Map algebra has been incorporated into a number of GIS software packages including the Spatial Analyst Extension for ESRI's ArcGIS (Tomlin 1990). Cartographic modeling has been used to model urban heat islands, risk of crime, and watersheds. Geostatistical analysis relies upon the same spatial data format, using statistical modeling including Kriging to interpolate a continuous surface of values—such as air quality or precipitation—from a sample of points at different locations.

5 Maps as Participation

Planning is a process as much as a product, and stakeholder participation in all phases of planning—infamously avoided at times, historically—has become a key ingredient in good planning. Managing public participation requires a set of skills distinct from design such as conflict negotiation and group facilitation (Bollens 2004). The field of Public Participation GIS (PPGIS) emerged partly in response to concerns from Critical GIS scholars that GIS technology was being used by "experts" to further exacerbate power differentials among professionals and the communities they professed to serve (Craig et al. 2002; Pickles 1995). Conferences and books dedicated to PPGIS proposed using GIS technology to engage people through methods such as cognitive temporal mapping that uses in-depth interviews, drawing of cognitive temporal maps, and interviews between planners and residents. Relatively new tools such as CommunityViz and Google Sketchup offer additional technologies to facilitate such participation, but the promise of greater civic engagement through GIS is still largely unrealized as expert technical knowledge is still prejudiced and essential.

Web 2.0 tools, focused on building collaboration rather than merely content, are beginning to be integrated into planning processes (Bugs et al. 2010). Tools for promoting participation allow participants to visualize plans, email feedback, post comments directly on a digital map, and interact with spatial data through applications like WikiMapia, which combines features of Google Maps and Wikipedia, and Second Life (Foth et al. 2009). Programmers and practitioners are leading the development and application of these new technologies in planning with academic planners watching from the sidelines, ready to evaluate, rather than initiate, such innovation.

Citizen stakeholders have not waited for GIS to come to them; they, along with universities and non-profit organizations, have developed web-based GIS systems for collecting and sharing spatial data that have practical and political implications for residents. The Brooklyn Food Coalition, for example used volunteers to conduct a food census, stating: "We believe that a positive first step toward a healthy food environment is understanding the current terrain and that maps are one way to gain that understanding" (Diep 2011). A link from the organization's website takes visitors to a simple interactive map where they can view or download data about the location of food stores by category. The map offers an accessible tool for locating food resources in the community while also highlighting the spatial nature of food access and inequality.

6 Maps as Accountability

Beyond using maps to have citizen voices heard in the planning process, planning organizations use maps to hold local, state, and federal governments accountable. Community information systems (CIS) provide public access to administrative data that individuals and community organizations can use to research property ownership, tax arrearages, home sales, and zoning regulations, among other interests (Hillier et al. 2005). The National Neighborhood Indicators Partnership (NNIP), launched by The Urban Institute, provides networking and support for dozens of local organizations that maintain web-based CIS. Foreclosure has been a popular focus among partners, and through NNIP, organizations in three cities—Baltimore, Washington DC, and New York City—studied the impact of the foreclosure crisis on school-aged children (NNIP 2011).

Building web-based mapping systems that provide access to basic housing, population, and health, and crime data has proven easier than sustaining them as keeping geographic boundaries, attribute data, and software up-to-date often demand dedicated professional staff and long-term funding. Similarly, promoting their use beyond generating thematic maps from census data and looking up ownership information for nuisance properties that would facilitate transformation of disadvantaged communities has proven challenging. Finally, web-based mapping systems depend upon municipal, state, and federal agencies to make data available (Hillier and Culhane 2012; Hillier et al. 2005). Arguments about the proprietary nature of data and the legal ramifications of making them public often mask practical concerns about data quality and exposing government failures.

Access to mortgage loans is another topic of interest. In compliance with the 1975 Home Mortgage Disclosure Act (HMDA), the Federal Financial Institutions Examination Council (FFIEC) requires certain types of lending institutions to provide data about loan applications and dispositions. FFIEC then makes public this HMDA data in the form of individual loan records and aggregations at the census-tract level that can be analyzed to look for patterns of redlining (Hernandez 2009). The term "redlining" was first used in the 1960s to describe a much older process by which lenders refused to make loans to residents based on the perceived credit-worthiness of their neighborhood. Historically, redlining had the greatest negative impact on African American neighborhoods in cities (Squires 1992). The US federal government, under the aegis of the Home Owners' Loan Corporation and Federal Housing Administration, made maps during the 1930s indicating the credit-worthiness of neighborhoods for real estate investment (Hillier 2003). HMDA data can be viewed through digital maps, including The Reinvestment Fund's online PolicyMap (www.policymap.org), a subscription-based service aimed at policy decision-makers, and CenTrax (Marquis 2011), software designed to help lenders meet their fair housing responsibilities.

In an effort to uphold the principles of Title VI of the 1964 Civil Rights Act, the Department of Transportation (DOT) requires local Metropolitan Planning Organizations (MPOs) to analyze the impact of Transportation Improvement Projects (TIP) funded by DOT on its region. Specifically, the MPOs must produce a series of "environmental justice" maps showing the location of projects relative to the "degrees of disadvantage," including areas with high poverty, ethnic/racial minority residents, older adults, limited English proficiency, and car-less households (DVRP 2011).

7 Opportunities for Interdisciplinary Work

Urban planning has historically borrowed ideas and techniques from many other fields. The use of mapping is no exception; the mapping practices described in this chapter reflect those in use in ecology, architecture and landscape architecture, business and real estate, environmental studies and others. Looking to the future, the greatest opportunities for developing new conceptual and technical ways of mapping will rely upon interdisciplinary partnerships. Some of these opportunities include expanding public participation and civic engagement, examining the role of the built environment on public health, reducing energy usage and promoting sustainability, and supporting equitable development. Professional planners and urban planning scholars, alike, have substantial experience with mapping that can help address some of these major local and global societal challenges.

While planners have already found ways to use mapping technology to engage the public in discussions about the future of cities, the emerging Web 2.0 tools and mobile devices provide significant and exciting possibilities for greater civic engagement. "Crowd-sourcing" is already being used to support disaster relief efforts (Bessis et al. 2010), and designers have used "Volunteered Geographic Information" (VGI) systems to collect detailed spatial data about a site. Together with computer scientists and video game developers, a new generation of planners and designers will find ways to expand and integrate these technologies into planning processes. For example, graduate students with landscape architecture and computer science training proposed to develop a multi-player Augmented Reality Game (ARG), combing a real-world environment and real-time social interactions with virtual challenges and rewards, for the City of Annaheim, California to "spark conversation, action, and activism" relating to environmental issues (Lee and Niu 2011).

The qualitative GIS methods developed by sociologists and feminist geographers also hold great potential as methods for incorporating resident perspectives and feedback into the planning process (Cope and Elwood 2009). Geo-ethnography, which combines ethnography and GIS (Matthews et al. 2005), places value on the everyday lived experience of individuals—particularly low-income families—and incorporates data gathered qualitatively about where people spend time into GIS. Kwan (2002) has shown how, for example, the experience of fear among Muslim women completing daily activities can be modeled using a three-dimensional space-time aquarium. These types of techniques would allow planners to better incorporate information about how residents actually experience their communities and anticipate how they might experience new places.

Planning has its biggest role to play in developing countries around the world, places that are experiencing tremendous population growth but have neither the physical nor data infrastructure to support conventional planning processes. China has demonstrated how quickly a nation can move to the forefront of capital development and technological innovation. The challenge for Western-trained planners will be to apply their knowledge to help other countries develop GIS infrastructure, tools for engaging citizens and holding government accountable, and professional schools to train their own planners. Similarly, while alternative-energy solutions and sustainability are critical issues facing the US and most Western nations, the real challenges lie abroad and will rely upon cross-disciplinary collaborations that bring together new ways of collecting and modeling spatial data.

A final area of opportunity for urban planning is in the area of public health. The development of health impact assessment (HIA) practices is reviving old professional ties between planning and public health (Northridge and Sclar 2003; Sloane 2006; Forsyth et al. 2010). While conceptually-well developed, the tools for conducting HIAs efficiently and successfully are just emerging. The San Francisco Bay

Area Health Impact Assessment Collaborative is among the leaders, offering standards, a toolkit, training, and case studies (http://www.hiacollaborative.org/). The number of municipalities that are incorporating HIAs into their planning activities is limited, but web-based mapping tools that allow citizens to participate in collecting data about current conditions, provide feedback about proposed changes, and evaluate the potential and actual impact of those changes on population health—including obesity, access to healthful foods, walkability, air quality, open space, and mental well-being—could transform planning processes. Only by working with colleagues in public health, itself an interdisciplinary field including medicine, epidemiology, nursing, social work, and government administration, will planners succeed in using conventional physical planning to promote the health and well-being of citizens.

8 Conclusion

Could the new and emerging geospatial tools of today finally allow us to move beyond the conventional approaches to mapping the city—concentric zones, parcels, administrative units—that have constricted new ways of seeing the city? The answer is still unclear because the success of these tools depends more upon the conception of what planning is and aspire to than technology. The role of cities as population, cultural, and economic centers is not diminishing in this twenty-first century, nor the need for planners to help insure that cities are the locus of opportunity rather than inequity for future generations. Maps will continue to be critical to planners' efforts to understand the spatial form of cities and how residents conceive of places, make and analyze plans, model social and environmental impacts, engage citizens and hold government officials accountable. With more tools at our disposal, the future is promising but still uncertain.

Suggested Readings

Drummond, W. J., & French, S. P. (2008). The future of GIS in planning: Converging technologies and diverging interests. *Journal of the American Planning Association, 74*(2), 161–174.

Lynch, K. (1960). *The image of the city*. Cambridge: M.I.T. Press.

Walker, D., & Daniels, T. (2011). *The planners' guide to communityviz*. APA Planners Press and the Orton Family Foundation.

References

Bessis, N., Asimakopoulou, E., French, T., Norrington, P., & Xhafa, F. (2010). The big picture, from grids and clouds to crowds: A data collective computational intelligence case proposal for

managing disasters. In *P2P, parallel, grid, cloud, and Internet computing, international conference*, pp. 351–356.
Bollens, S. A. (2004). Urban planning and intergroup conflict: Confronting a fractured public interest. In B. Stiftel & V. Watson (Eds.), *Dialogues in urban & regional planning*. New York: Routledge.
Bugs, G., Garnell, C., Fonts, O., Huerta, J., & Painho, M. (2010). An assessment of public participation GIS and Web 2.0 technologies in urban planning practice in Canela, Brazil. *Cities, 27*, 172–181.
Cope, M., & Elwood, S. (Eds.). (2009). *Qualitative GIS: A mixed-methods approach*. Thousand Oaks, CA: Sage.
Craig, W. J., Harris, T. M., & Weiner, D. (Eds.). (2002). *Community participation and geographical information systems*. New York: Taylor & Francis Inc.
Diep, F. (2011). Amateur cartographers map local 'food desert'. *Scientific American,* February 28. Available at http://www.scientificamerican.com/article.cfm?id=mapping-the-food-desert
Dodgson, J. S. (2009). *Multi-criteria analysis: A manual*. London: Department for Communities and Local Government.
DVRP (Delaware Valley Regional Planning Association). (2011). http://www.dvrpc.org/GetInvolved/TitleVI/
ESRI. (2010). *Changing geography by design: Selected readings in geodesign*. Redlands, CA: ESRI.
Evans, G. W., Brennan, P. L., Skorpanich, M. A., & Held, D. (1984). Cognitive mapping and elderly adults: Verbal and location memory for urban landmarks. *Journal of Gerontology, 39* (4), 452–457.
FEMA (Federal Emergency Management Agency). (2010 and 2011). *HAZUS-MP*. http://www.fema.gov/plan/prevent/hazus/hz_overview.shtm
Fishman, R. (1998). Beyond utopia: Urbanism after the end of cities. In *Ciutat real, ciutat ideal. Significat i funció a l'espai urbà modern [Real city, ideal city. Signification and function in modern space.]* Barcelona: Centre de Cultura Contemporània de Barcelona, 1998 (Urbanitats; 7). Available at http://www.cccb.org/rcs_gene/beyond_utopia.pdf
Forsyth, A., Slotterback, C. S., & Krizek, K. (2010). Health Impact Assessment (HIA) for planners: What tools are useful? *Journal of Planning Literature, 24*, 231.
Foth, M., Bajracharya, B., Brown, R., & Hearn, G. (2009). The second life of urban planning? using NeoGeography tools for community engagement. *Journal of Location Based Services, 3* (2), 97–117.
Golledge, R. G. (Ed.). (1999). *Wayfinding behavior. Cognitive mapping and other spatial processes*. Baltimore: Johns Hopkins University Press.
Guest, A. M., & Lee, B. A. (1984). How urbanites define their neighbourhoods. *Population and Environment, 7*(1), 32–56.
Hadhazy, A. (2010). Lasers used to create 3-D model of New York City. *TechNewsDaily,* May 12. Available at http://www.technewsdaily.com/lasers-used-to-create-3-d-model-of-new-york-city-0548/
Halseth, G., & Doddridge, J. (2000). Children's cognitive mapping: A potential tool for neighbourhood planning. *Environment and Planning B: Planning and Design, 27*(4), 565–582.
Harris, C. D., & Ullman, E. L. (1945). The nature of cities. *Annals of the American Academy of Political and Social Science, 242*, 7–17.
Hernandez, J. (2009). Redlining revisited: Mortgage lending patterns in Sacramento 1930–2004. *International Journal of Urban and Regional Research, 33*(2), 291–313.
Hillier, A. (2003). Redlining and the home owners' loan corporation. *Journal of Urban History, 29* (4), 394–420.
Hillier, A., & Culhane, D. (2012). GIS applications and administrative data to support community change. In M. Weil (Ed.), *Handbook of community practice* (2nd ed.). Thousand Oaks, CA: Sage.
Hillier, A., Wernecke, M., & McKelvey, H. (2005). Removing barriers to the use of community information systems. *Journal of Community Practice, 13*(1), 121–139.

Hospers, G. (2010). Lynch's *The Image of the City* after 50 years: City marketing lessons from an urban planning classic. *European Planning Studies, 18*(12), 339–349.
Howard, E. (1898). *To-morrow: A peaceful path to real reform*. London: Swan Sonnenschein.
Howard, E. (1902). *Garden cities of to-morrow*. London: Swan Sonnenschein.
Hoyt, H. (1939). *The structure and growth of residential neighborhoods in American Cities*. Washington, DC: Federal Housing Administration.
Kwan, M. (2002). Feminist visualization: Re-envisioning GIS as a method in feminist geographic research. *Annals of the Association of American Geographers, 92*(4), 645–661.
Kwan, M., & Lee, J. (2005). Emergency response after 9/11: The potential of real-time 3D GIS for quick emergency response in micro-spatial environments. *Computers, Environment and Urban Systems, 29*, 93–113.
Landis, J. D., & Zhang, M. (2000). Using GIS to improve urban activity and forecasting models: Three examples. In S. A. Fotheringham & M. Wegener (Eds.), *Spatial models and GIS: New potential and new methods*. London: Taylor and Francis.
Lee, J., & Niu, M. (2011). *Multiplayer gaming as a tool to build better communities*. Proposal to the Graduate and Professional Student Association, University of Pennsylvania.
Ligmann-Zielinska, A., & Jankowski, P. (2007). Agent-based models as laboratories for spatially explicit planning policies. *Environment and Planning B: Planning and Design, 34*(2), 316–335.
Marquis. (2011). *CenTrax*. Available at http://www.gomarquis.com/fair-lending-hmda-and-cra-software.aspx
Matthews, S. A., Detwiler, J. E., & Burton, L. M. (2005). Geo-ethnography: Coupling geographic information analysis techniques with ethnographic methods in urban research. *Cartographica, 40*(4), 75–90.
NNIP (National Neighborhood Indicators Partnership). (2011). http://www2.urban.org/nnip/foreclosures.html
Northridge, M. E., & Sclar, E. (2003). A joint urban planning and public health framework: Contributions to health impact assessment. *American Journal of Public Health, 93*(1), 118–121.
Park, R., Burgess, E. W., & McKenzie, R. D. (1925). *The city*. Chicago: University of Chicago Press.
Perry, C. (1929). The neighbourhood unit as seen by Clarence A. Perry. *New York Regional Survey*, Vol. 7.
Pickles, J. (1995). *Ground truth: The social implications of geographic information systems*. New York: The Guilford Press.
PlaNYC. (2009). *A greener, greater New York*. New York City: Planning Commission.
Sloane, D. (2006). From congestion to sprawl: Planning and health in historical context. *Journal of the American Planning Association, 72*(1), 10–18.
Squires, G. (Ed.). (1992). *From redlining to reinvestment: Community responses to urban disinvestment*. Philadelphia: Temple University Press.
Tomlin, C. D. (1990). *Geographic information systems and cartographic modeling*. New Jersey: Prentice Hall College Division.
Walker, D., & Daniels, T. (2011b). *The planners guide to communityviz: The essential tool for a new generation of planning*. Chicago: APA Planners Press.
Zwick, P. (2010). The world beyond GIS. *Planning, 76*(6), 20–23.

Ecology and Space: A Case Study in Mapping Harmful Invasive Species

David T. Barnett, Catherine Jarnevich, Geneva W. Chong, Thomas J. Stohlgren, Sunil Kumar and Tracy Holcombe

Abstract The establishment and invasion of non-native plant species have the ability to alter the composition of native species and functioning of ecological systems with financial costs resulting from mitigation and loss of ecological services. Spatially documenting invasions has applications for management and theory, but the utility of maps is challenged by availability and uncertainty of data, and the reliability of extrapolating mapped data in time and space. The extent and resolution of projections also impact the ability to inform invasive species science and management. Early invasive species maps were coarse-grained representations that underscored the phenomena, but had limited capacity to direct management aside from development of watch lists for priorities for prevention and containment. Integrating mapped data sets with fine-resolution environmental variables in the context of species-distribution models allows a description of species-environment relationships and an understanding of how, why, and where invasions may occur. As with maps, the extent and resolution of models impact the resulting insight. Models of cheatgrass (*Bromus tectorum*) across a variety of spatial scales and grain result in divergent species-environment relationships. New data can improve

D.T. Barnett (✉)
Battelle Ecology, Boulder, CO, USA
e-mail: dbarnett@battelleecology.org

D.T. Barnett · T.J. Stohlgren · S. Kumar
Natural Resource Ecology Lab, Colorado State University, Fort Collins, CO, USA
e-mail: thomas.stohlgren@colostate.edu

S. Kumar
e-mail: sunil@nrel.colostate.edu

G.W. Chong
U.S. Geological Survey, Northern Rocky Mountain Science Center, Bozeman, MT, USA
e-mail: Geneva_Chong@usgs.gov

C. Jarnevich · T. Holcombe
U.S. Geological Survey, Fort Collins Science Center, Fort Collins, CO, USA
e-mail: jarnevichc@usgs.gov

T. Holcombe
e-mail: holcombet@usgs.gov

S.D. Brunn and M. Dodge (eds.), *Mapping Across Academia*,
DOI 10.1007/978-94-024-1011-2_4

models and efficiently direct further inventories. Mapping can target areas of greater model uncertainty or the bounds of modeled distribution to efficiently refine models and maps. This iterative process results in dynamic, living maps capable of describing the ongoing process of species invasions.

Keywords Invasive species · Species distribution models · Cheatgrass · *Bromus tectorum*

1 Introduction

An invasive species is defined as a species (or genotype) from another ecosystem (usually another country) that causes harm to the economy, the environment, or human health (Executive Order 13112; Feb 3, 1999). The rapid spread of invasive species in North America has prompted the urgent need to map their current and potential distributions to initiate containment programs (Stohlgren and Schnase 2006). Mapping a species sounds easy. An ecologist might locate the coordinates of several individuals, assess the environmental factors associated with the presence or absence of the species (Guisan and Thuiller 2005; Thuiller et al. 2005), and then create a range map of suitable habitat. However, there are three significant challenges to overcome. First, occurrence and abundance data for many species are often limited, or they are collected in a biased manner (e.g., close to roads, in flat areas, or in backyards), or collected in only a small subset of the actual species range. Second, assessing species-environment relationships is made difficult when a species is "present" temporarily in poor habitat, or "absent" in suitable habitat due to dispersal limitations or control—it's just not there yet. Third, there are significant theoretical and ecological limitations associated with predicting species distributions and abundances in space and time. In this chapter, we: (1) provide a brief review of species distribution mapping; (2) explore several of the many ways ecologists are using maps of invasive species; and (3) discuss the theoretical and ecological limitations of the maps and models currently used by invasion ecologists.

2 Observations Across Ecological Gradients

Observing patterns in species distributions is fundamental to ecological mapping and crucial to science. In the middle of the nineteenth century, Charles Darwin placed observations in a spatially explicit context to fundamentally change the understanding and theory of species-environment and evolutionary relationships (Darwin 1859). The belief in the importance of species-specific responses to environmental gradients fluctuated through time as the turnover of ideas, theory, and evidence moved the discipline of ecology forward (Barbour et al. 1987). Early in the twentieth century, Henry Gleason championed an individualistic understanding of ecological patterns (Gleason 1926) by maintaining that the distribution of species was more likely controlled by environmental factors than associations with other species (Gleason 1920). Robert Whittaker later developed quantitative tools that both described and furthered Gleason's work. Species response curves, pervasive in current *species mapping* and distribution modeling, typically describe occurrence, abundance, or fitness as a function of an environmental variable (Whittaker 1967, 1970). These curves are a graphical representation of niche theory that both influenced Whittaker's thinking and was furthered by his research contributions.

While the concept of a niche has evolved through time, most research incorporates the role of environmental gradients on species distributions. Early references used the niche to describe the functional role of a species (Elton 1927, 1958) and those resources required to support a species (Grinnell 1917). A contemporary and often cited treatment defined the niche as that space at the intersection of multiple environmental variables where a species can survive and reproduce (Hutchinson 1957). Hutchinson separated this "hypervolume" into two components: the fundamental niche described the environmental space suitable for persistence, and the realized niche described the environmental space in which a species actually survives and reproduces. The realized niche, moderated by ecological processes such as dispersal and biotic interactions, was thought to occupy a subset of the environmental space and geographical extent of the fundamental niche. Departures from this scenario occur in instances when a species occupies "sink" habitat (areas where a species can persist but not successfully reproduce; Pulliam 2000), by chance, or through facilitation (for example, a symbiotic interaction that facilitates species establishment; Pulliam 2000; Soberon 2007). Conversely, species frequently do not occur in suitable habitat or portions of the realized niche. The species might have been extirpated by disturbance, natural life span, or through biotic interactions. Dispersal may also play a role in occupation of the realized niche through time and space.

3 Species Distribution Mapping

Invasive species distribution mapping aims to describe areas at risk to the invasion of non-native plant species by identifying environmental gradients at the intersection of habitat suitability, establishment and persistence. Charles Elton, regarded as the father of invasive species biology, collected examples of early efforts to map invasive species (Elton 1958). The maps, dating to early in the twentieth century, generally exist at the global, continental, or regional (for example, many US states) scale; the species distributions are described by coarsely shaded zones or polygons. Presumably expert opinions influenced the lines drawn around observations defining the extent of invasion. Elton's collection effectively supported communication of the magnitude of the invasion problem and launched a vital discipline of ecology (Levine and D'Antonio 1999; Richardson and Pysek 2008). However, the utility of these early mapping efforts, challenged by technology and lacking hindsight was limited. The failure to show variability (absent areas) within the very general shaded polygons prevents specific assessments of risk, the ability to document spread within coarsely defined areas, containment and eradication efforts (Beaumont et al. 2009). Additionally, coarse records of occurrence cannot be linked to environmental gradients at scales that allow definition of vulnerable habitat in space and facilitate forecasts of future distributions as the environment changes (Gallien et al. 2010; Thuiller et al. 2005).

Mapping of invasive species has improved with time. County level maps of species presence are popular with government agencies (PLANTS Database, http://plants.usda.gov/java/) and organizations (Biota of North America Program, http://www.bonap.org/) in the United States. The county resolution mapping tracks distributions at relatively fine scales, easily aggregates location specific (and less specific) records of occurrence, and is congruent with the work and lists maintained by county weed managers (Jarnevich et al. 2010). Rigorous collection efforts in the northwestern part of the United States documented the temporal progression of species-specific invasion at the county scale (INVADERS database; http://invader.dbs.umt.edu/). Maps at this scale have proven useful for describing spatial and temporal patterns of invasion, developing lists of species of concern for managers (Jarnevich et al. 2010), and contributing to theory guiding invasion ecology and ecosystem stability (Stohlgren et al. 2003, 2006). The subsequent insights from these efforts contribute to general strategies for containment and resource allocation, but the county scale is too coarse for calibrating species-environment relationships, and management implemented at the local scale.

Location specific (or point distribution) maps have improved in recent decades. Technological advancements in geography allow (and reduce the cost of) global positioning systems (GPS) to document the occurrence of species at sub-meter accuracy. Repeated field surveys and rigorous monitoring efforts can document change through time and target control efforts (Barnett et al. 2007). Now-common geographic information systems (GIS) display points and polygons in space, can be easily updated to reflect temporal invasion or efficacy of control, and display data at multiple spatial extents. These spatially explicit representations of information allow relation of invasion to other landscape features, resources, and descriptions of environmental variables that may control the distribution or niche of target invasive species.

4 Modeling Species Distributions

Improvements in the ability to document species occurrence greatly increased understanding of the extent of the invasive species problem, but coarse range maps and even highly accurate dots on a map were incapable of providing solutions for the management of landscapes facing invasion by multiple species and pathways that threatened a variety of native species and natural resources. Management of risks to natural resources associated with invasive species requires answering questions at relevant spatial and temporal scales: Which particular species should be managed in which areas? How far will an invader spread? How fast will it spread? How will control efforts impact native species? How will changes in climate and land use, at a variety of spatial and temporal scales, impact future patterns of invasion? Operating with limited budget and time, managers might confront these questions in a risk assessment context that requires insight beyond simple maps of invasion occurrence.

Managers need to understand extant patterns of invasion and forecast the impact of environmental change, land use, or management on vulnerability at multiple spatial and temporal scales. Tools must interpolate and extrapolate from point observations to continuous, 'wall-to-wall' surfaces that can be subjected to a variety of future scenarios, convey levels of uncertainty, and describe where more data are needed to reduce uncertainty. Untangling the relationship of a species to the biotic and abiotic environment has long been a goal of ecologists that can also provide a crucial tool for confronting invasive species questions and challenges.

Realizing a need to extrapolate information from sampled to un-sampled areas to gain a more complete picture of suitable habitat or areas that are at risk of invasion, ecologists have invested heavily in species distribution models (also called species-environmental matching models, niche models, and other names, Franklin 2009). The concept is to overlay the locations of species with several environmental

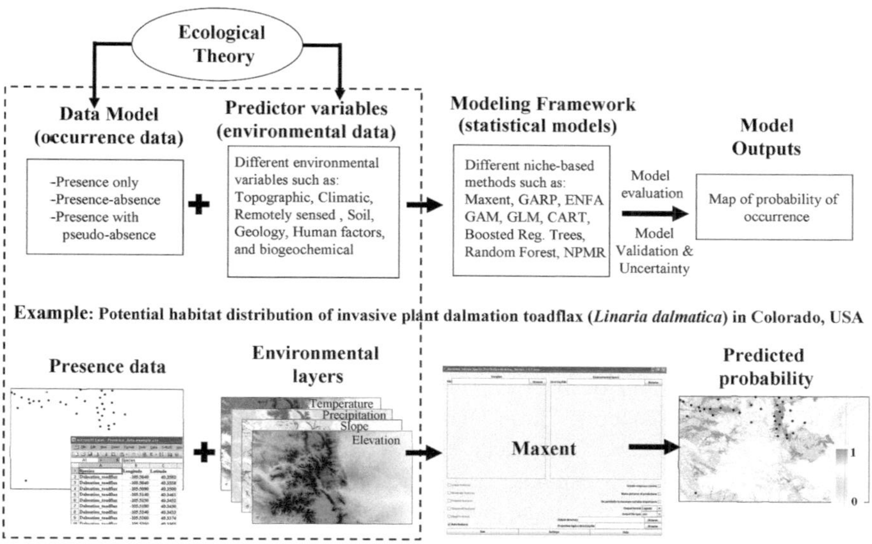

Fig. 1 Generalized schematic of species distribution modeling from points to probabilities. Future climate scenarios can be swapped with current climate layers to get a first approximation of future species "habitat suitability" (but see the discussion)

factors (typically held as data layers in a GIS) to interpolate and extrapolate the probable species distribution from the sample points (Fig. 1). Species distribution models are now commonly used to predict the distribution of invasive species (Kumar et al. 2009), explore potential impacts of climate change (Thuiller et al. 2008; Jarnevich and Stohlgren 2009), establish conservation priority areas and select reserve designs (Fuller et al. 2008; Pawar et al. 2007), identify new populations of rare and endangered species (de Siqueira et al. 2009; Guisan et al. 2006), predict distributions of known and unknown species (Raxworthy et al. 2003; Pearson et al. 2007), in public health (Neerinckx et al. 2008) and in phylogeographic studies (Miller and Knouft 2006; Moritz et al. 2009).

5 Examples of Mapping and Modeling Invasive Species

5.1 Species Distribution Modeling Techniques

Here we describe applications of species distribution modeling to invasive species risk assessments through modeling invasive species at a variety of spatial and temporal scales. Models relate the documented species occurrence data to environmental layers. Development of these modeling techniques is an active and rapidly evolving field and a significant body of literature compares and explains competing approaches (Elith et al. 2006). In an effort to cover applications of species distribution modeling in the context of managing the risk of invasive plant species instead of another comparison of model alternatives (Elith et al. 2006), this chapter demonstrates a single modeling approach.

Maxent is a machine learning method (version 3.1; http://www.cs.princeton.edu/~schapire/maxent/) and is based on the maximum entropy principle. It estimates the probability distribution of a species by estimating the probability distribution of maximum entropy (Phillips et al. 2006). Model comparison ranked Maxent as the best performing model algorithm among 16 different modeling methods (Elith et al. 2006). Maxent compares mapped presence locations of a species to the available environment as described by a set of background points to create a potential habitat suitability surface by projecting the relationships it finds across the entire landscape. When applying the model across the landscape, Maxent 'clamps' environmental values outside the range of those used in training the model (that is, locations with climate outside the range of that covered by the presence and background locations used to develop the model). These 'clamped' areas can be used to access where novel environmental conditions occur by generating a surface highlighting where clamping occurs. In each example described below, we ran the model 25 times, withholding a different 25 % of the presence locations from each model run as a test dataset for model evaluation; the final surface is an average of the 25 model runs. The models can be assessed by the Area Under the Receiver Operating Characteristic Curve (AUC), which is a threshold independent indication of model performance (Phillips et al. 2006). In examples where future species distributions were modeled, the climatic model derived from current climatic conditions was applied to future climate scenarios in 2020 and/or 2050 as described by the CCCMA a2a climate scenario (http://www.worldclim.org/futdown.htm).

5.2 *Assessing Vulnerability at Multiple Spatial and Temporal Scales*

Documenting the progress of invasive species establishment at a variety of spatial and temporal scales facilitates an understanding of how specific environmental factors dictate species distributions, how those distributions might change in time and space, and how models can be used to iteratively define species distributions. The following examples focus on assessing risk of cheatgrass (*Bromus tectorum*) establishment at multiple spatial and temporal scales.

5.3 *Developing Continental Frameworks for Invasive Plant Species Management*

Cheatgrass (*Bromus tectorum*), a Eurasian, cool-season, annual grass (Thill et al. 1984), was first introduced to the United States from Europe late in the nineteenth century (Mack and Pyke 1983). The subsequent rapid spread across the west was facilitated by human movement and settlement, grazing and agriculture (Mack and Pyke 1984). Cheatgrass modified plant community composition, decreased forage for livestock and wildlife (Thill et al. 1984), and altered fire frequency (D'Antonio and Vitousek 1992) on the way to dominating more than 20 % of sagebrush steppe in the Great Basin (Knapp 1996) and other parts of the United States.

To better understand areas of the continental United States vulnerable to cheatgrass invasion, two databases were queried (National Institute of Invasive Species Science, NIISS.org; EDDMapS, eddmaps.org) for specific presence locations that served as dependent variables for use in Maxent models. The resulting model is likely compromised by the quality of the dependent data. Despite the inclusion of many mapped locations (7915 for training and 2638 for model testing) the majority of the data reflect presence records in California, Arizona, Nevada, and Colorado; very few points from heavily invaded states, such as Idaho and Oregon, have been assimilated into regional online databases. Maxent was able to model areas vulnerable to invasion by cheatgrass in unsampled areas (Fig. 2a, AUC = 0.75), but the possibility that cheatgrass may have invaded novel environments in these areas suggests that Maxent may not have adequately described the climatic envelope or fundamental niche of cheatgrass across the Western United States.

Understanding the leading edge of an invasion by a non-native plant species is crucial to early detection and establishing priorities at national, regional and local

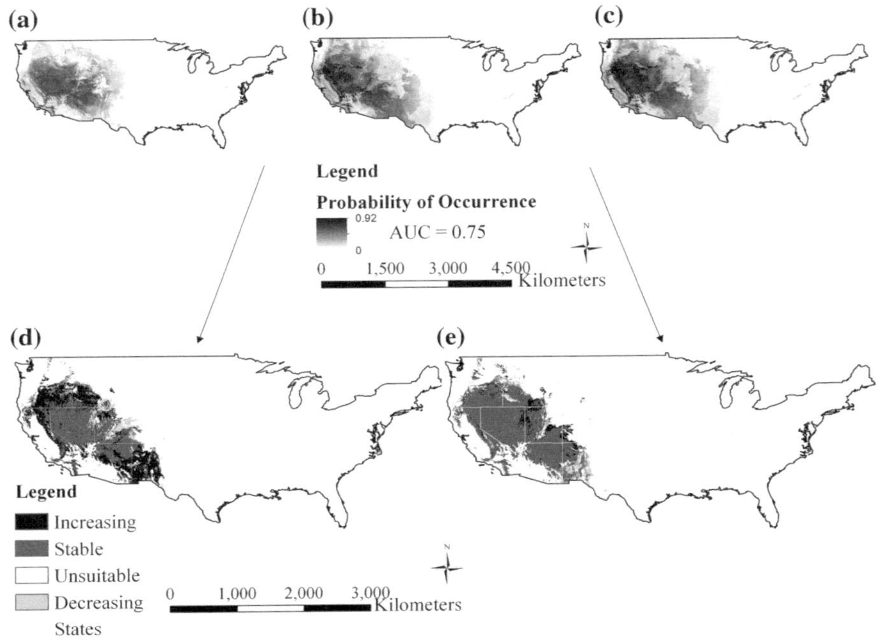

Fig. 2 An example of habitat suitability maps for cheatgrass (*Bromus tectorum*) at the continental scale. Temporal forecasts include suitability based on current (**a**), and future climate scenarios in 2020 (**b**) and 2050 (**c**). The leading edge of invasion ("increasing") through time was described by creating binary maps of suitability according to Maxent-derived thresholds and evaluating change in suitability between two successive time steps: the current and 2020 models (**d**), and the 2020 and 2050 models (**e**)

scales. To better understand how future environmental conditions might alter the regions susceptible to invasion by cheatgrass, the Maxent-generated distribution of habitat suitable to cheatgrass based on current environmental conditions was applied to climate scenarios for 2020 (Fig. 2b) and 2050 (Fig. 2c). The Maxent-generated ten percentile logistic threshold (Phillips et al. 2006) transformed the continuous probability surface into the binary condition 'suitable' and 'unsuitable' habitat. Comparison of suitability maps identified the leading edge of invasion (e.g. unsuitable habitat in the current model that becomes suitable in the 2020 model) and trailing edge of invasion for example, suitable habitat in the 2020 model that becomes unsuitable in the 2050 model), and areas that remained vulnerable to invasion across time steps (Holcombe et al. 2010); Fig. 2c, d). Elevation made the primary contribution to the model (Table 1). While topography will remain static through time, future changes in climatic variables associated with cheatgrass distribution and correlated with elevation. The amount of precipitation during the warmest quarter (Table 1), will likely extend the areas vulnerable to cheatgrass invasion into more northern latitudes and higher elevations (see Fig. 4).

Table 1 The relative contribution of independent environmental variables to models of cheatgrass (*Bromus tectorum*) distribution at multiple spatial extents and grain sizes

Environmental variables	Continental US	Greater Yellowstone area	Northern range	Jackson Hole
Temperature seasonality	0.69	10.29	–	–
Max temperature of warmest month	1.36	6.94	–	–
Temperature annual range	0.78	13.01	–	–
Mean temperature of driest quarter	1.35	4.16	–	–
Mean temperature of coldest quarter	3.89	0.0	–	–
Precipitation of wettest quarter	0.17	10.21	–	–
Precipitation of driest quarter	3.07	6.91	–	–
Precipitation of warmest quarter	24.23	19.13	–	–
Elevation	55.06	4.41	44.07	42.34
Slope	0.07	1.64	3.44	3.06
Distance to road	–	10.65	7.00	10.21
Brightness	–	–	10.62	13.21
Enhanced vegetation index	–	–	5.63	0.49
Greenness	–	–	0.55	15.15
Landform[a]	–	–	4.17	–
Vegetation type[a]	–	–	5.11	5.46
Wetness	–	–	0.66	5.69

Variables with less than 3.0 contribution across all models not shown
[a]Categorical variable

5.4 *Understanding Regional Vulnerability Across Space and Time*

The Great Basin and the Snake River Plain may be the region most invaded by cheatgrass in the US. The idea that certain combinations of climatic, disturbance, soil, and vegetation characteristics make the sage steppe especially vulnerable to invasion is reflected by both the impacts and the abundance of literature (Mack 1981; Thill et al. 1984; Mack and Pyke 1984). The bulk of the attention from the management and science community suggested that forest canopy, colder temperatures, and a shorter growing season at higher elevations prevent cheatgrass from becoming dominant across the Mountain West and into the Greater Yellowstone Area (GYA) comprised of Yellowstone and Grand Teton National Parks, multiple National Forests and National Wildlife Refuges, and tribal and private lands (Fig. 3).

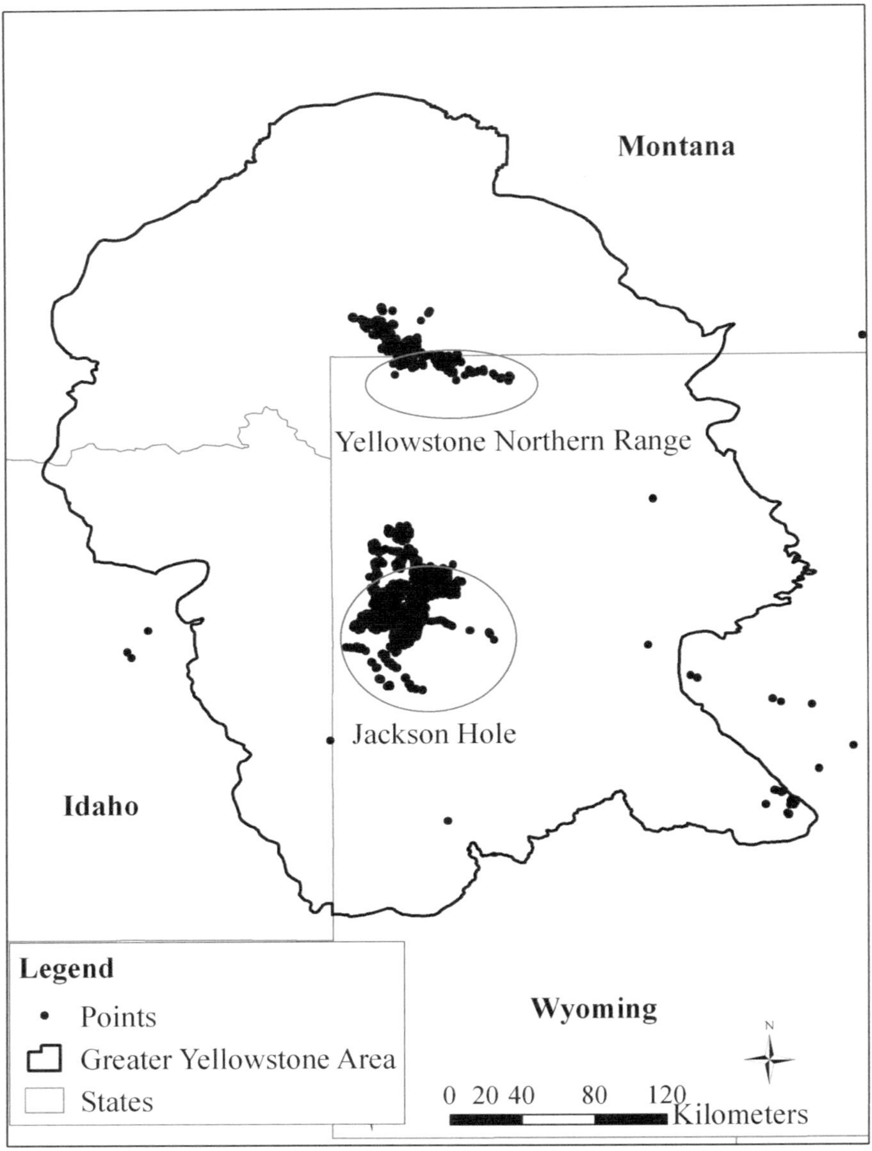

Fig. 3 The Greater Yellowstone Area (GYA) includes a matrix of public and private land in Wyoming, Montana, and Idaho. Temporal and spatial forecasts of landscape-scale vulnerability to the invasion of cheatgrass relied on a regional mapping database supported by the Greater Yellowstone Coordinating Committee (GYCC)

The continental model of areas vulnerable to invasion suggests that conditions exclude the GYA from the current climatic envelope occupied by cheatgrass (Fig. 2a, d). The results even imply that changes in climate might allow the leading edge of potential invasion or suitable habitat to reach only the lowest latitudes of the GYA by 2050 (Fig. 2e). However, largely overlooked records of cheatgrass in the GYA date back to 1927 (INVADERS database; http://invader.dbs.umt.edu/), and observations suggest it might be increasing in specific areas. Populations of cheatgrass along roads and disturbed areas, especially in lower elevations where establishment displaces native forage in winter range important to the maintenance of large ungulate herds, prompted localized mapping efforts that facilitate an assessment of extant distribution and the potential spread of cheatgrass in the GYA.

Species distribution models at regional scales provide the chance to evaluate the risks associated with occurrence of cheatgrass at the scale of the entire GYA. The Greater Yellowstone Coordinating Committee (GYCC) was formed to allow representatives from the National Park Service, US Forest Service, and the US Fish and Wildlife Service to pursue opportunities to cooperate and coordinate in the management of federal lands in the GYA. The GYCC initiative to collect and organize mapped locations of invasive plant species across the GYA provided the opportunity to map and model current cheatgrass distribution and model potential future distribution using a variety of environmental descriptors of the landscape (Table 1). The actual location records of occurrence (Fig. 3) are a static picture of where favorable conditions and process for establishment collided, and the model (Fig. 4a) describes the likelihood of these playing out across the landscape: a potential range or environmental envelope for cheatgrass in the Greater Yellowstone Area.

Forecasting future non-native species distributions is a substantial part of understanding risk and threats to a landscape. However, substantial, and

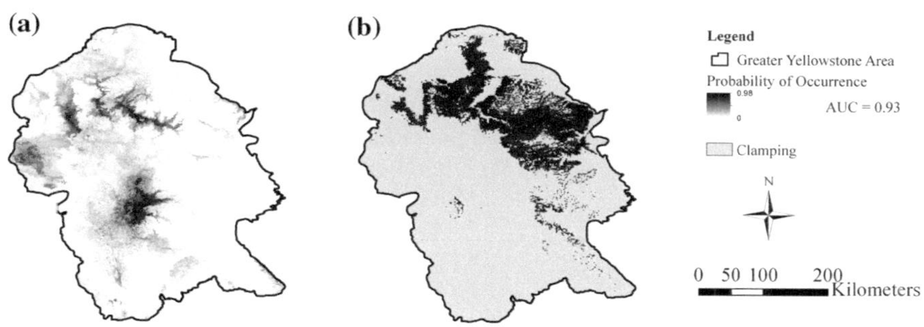

Fig. 4 Regional cheatgrass distribution models across the Greater Yellowstone Area (GYA) were capable of describing habitat suitability based on current climatic conditions (**a**), but the application of the model to future climate scenarios was limited by the significant change in climate predicted in 2050 shown by the areas where 'clamping' occurred (**b**)

unpredictable climate shifts at local to global scales can hinder temporal projections of habitat vulnerability. In the case of the GYA, locally novel climatic conditions might dominate much of the regional landscape by 2050. Lacking the opportunity to train the model according to these conditions, Maxent is unable to make reliable forecasts at this regional scale. As such, areas where the model is extended beyond climatic conditions on which it was trained are removed from consideration in a process termed 'clamping.' Clamping applies a limit to how the model will predict occurrence in novel conditions, which results in controlling the shape of the response curve so that it can only reach a certain limit in novel conditions, and those areas are displayed as novel surfaces (Fig. 4b).

5.5 *Assessing Risk and Directing Management at Local Spatial Scales*

The understanding that species distribution models generated at regional scales may not be appropriate for management at local scales, and that the processes that define patterns are scale dependent, directed mapping efforts towards understanding distributions at local scales. The one km^2 grain (pixel size) of the GYA regional model (appropriate for the inclusion of climate variables at the regional scale), is too coarse for managers of the Northern Range of Yellowstone National Park who hope to understand landscape vulnerability and the impacts on native forage, and perhaps control specific patches of cheatgrass according to distribution models. A targeted search of the equivalent of a 30-m^2 pixel (or smaller) is far more realistic and efficient than a one km^2 area (Fig. 5).

Species distribution models at finer grain size provide more detail. In complex terrain, significant small-scale environmental heterogeneity can be lost in coarse landscape descriptions (Stohlgren et al. 1997). Cheatgrass might respond favorably to disturbances detected by remotely sensed imagery at 30-m^2 resolution. Similarly, small patches of dry grassland associated with bits of south-facing slopes in an area otherwise dominated by densely forested vegetation would likely be suitable for cheatgrass establishment. Ignoring these potential sources, locations where cheatgrass can persist on the landscape, might otherwise result in an under-representation of vulnerable habitat.

5.6 *Sampling with Species Distribution Models*

Species distribution models direct efficient sampling. Rare species inventory and monitoring efforts have followed Maxent probability surfaces to otherwise difficult to detect individuals (Kumar et al. 2009), and invasive species inventories must

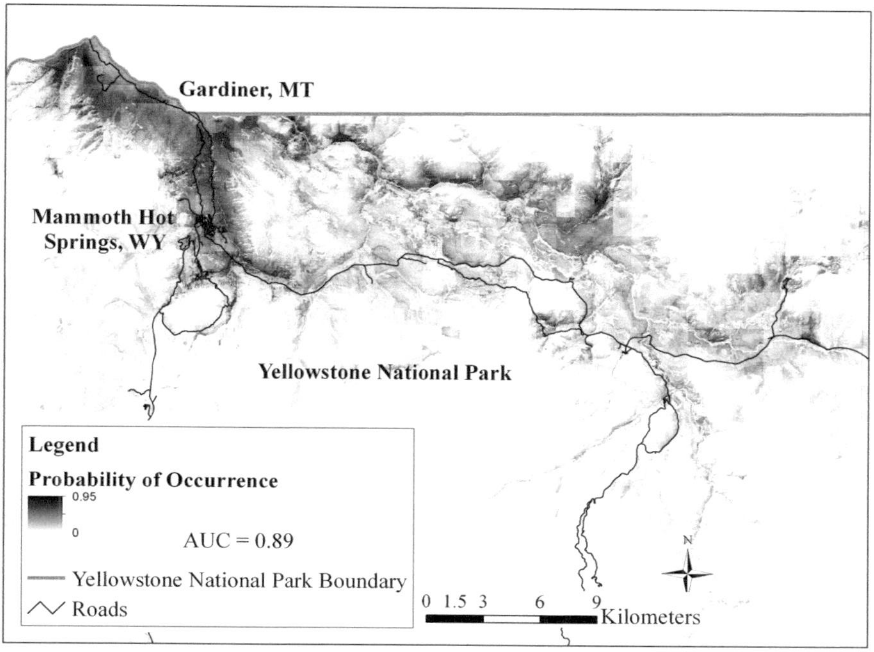

Fig. 5 Habitat suitable for cheatgrass invasion on the Northern Range of Yellowstone National Park

both refine models of vulnerability and differentiate presence from probability. Iterative cheatgrass assessments in the Jackson Hole region of the Greater Yellowstone Area contributed to the regional understanding of vulnerability (Fig. 6) and refined the envelope of potential occurrence. Distribution models were iteratively updated with new occurrence records found by sampling areas likely to be invaded, and data were used to refine the Whittaker (1967, 1970) species response curves that plot probability of occurrence as a function of environmental gradients. Elevation, likely a surrogate for a variety of highly correlated variables such as temperature and precipitation, dominated in Jackson Hole (see Table 1). Multi-scale vegetation plots and mapping survey efforts crossed elevation gradients and specifically targeted thresholds associated with occurrence of cheatgrass (Fig. 6). The resulting modeled surfaces changed only slightly, but the thresholds of the species response curves and associated understanding evolved and improved with each iterative effort.

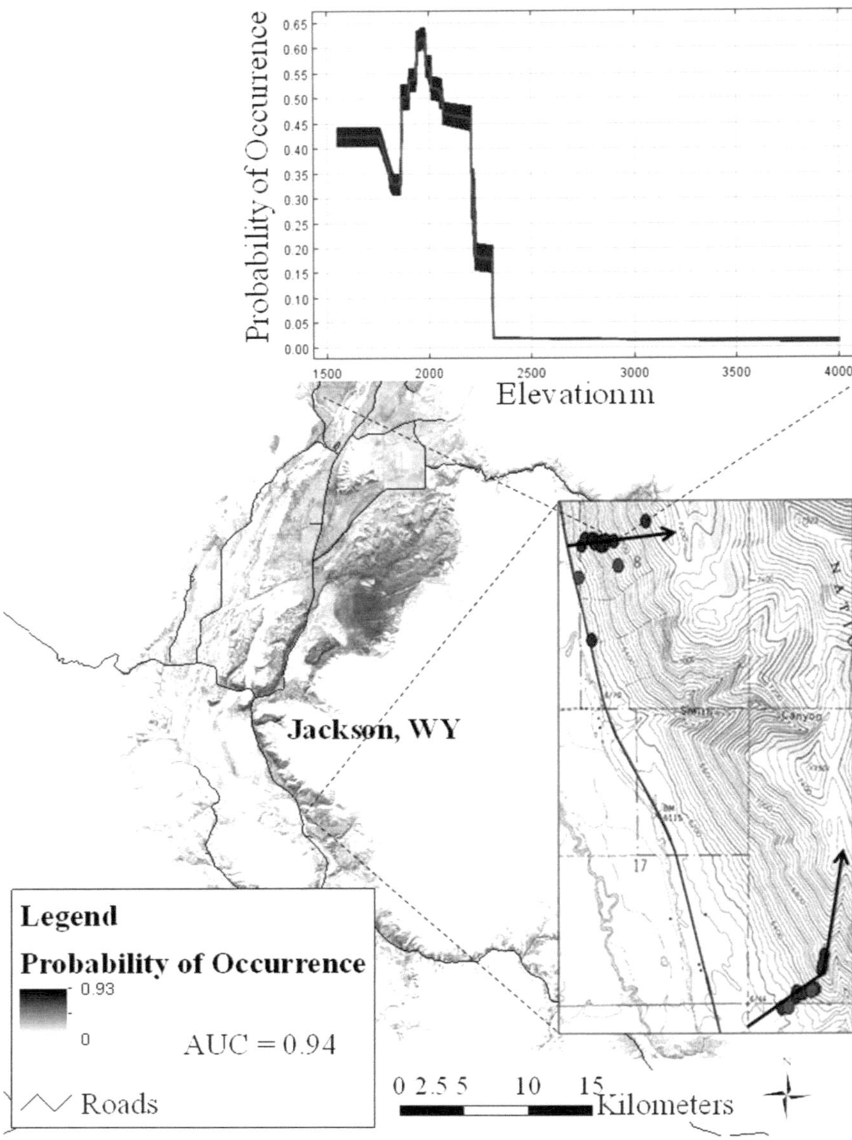

Fig. 6 The model of areas vulnerable to cheatgrass invasion in the Jackson Hole region was developed iteratively through repeated sampling efforts that targeted areas at the thresholds of variable values that contributed to the model (e.g., suitable elevation and brightness; Table 1)

5.7 Understanding Vulnerability at Multiple Spatial and Temporal Scales

The invasion of cheatgrass is progressed in specific and isolated locations of the GYA. History, readily available data, and the variability of extant distribution likely fueled a perception that higher elevation regions are immune to extensive invasion (Knapp 1996). For good reason: records of cheatgrass in the GYA date to 1927 (INVADERS database; http://invader.dbs.umt.edu/), yet until recently there has been little evidence of cheatgrass proliferation (Barnett et al. 2007) despite a history of extensive land management and scientific inquiry. The disparity between perceptions and demonstrated vulnerability may be an artifact of the bias associated with expectations, perhaps a comprehensive understanding of vulnerability required rigorous modeling and targeted sampling, or it may be that cheatgrass is beginning what could be substantial expansion in the GYA.

Habitat suitability, or the spaces that cheatgrass invades, is an important component of successful establishment (Stohlgren 2007; Chytry et al. 2008). At large scales, elevation and the truncated growing season associated with higher elevation, seem to limit establishment. All of our models suggested cheatgrass does better at lower elevations, and none of the data incorporated in this project found cheatgrass above 2300 m. Many other projects cite the importance of elevation (Pierson and Mack 1990; Bradford and Lauenroth 2006), and others (Chambers et al. 2007) found that cheatgrass did not do well above 2380 m even in the Great Basin.

Site-specific processes are also important for the inclusion of cheatgrass in community assembly. Disturbance from roads and bioturbation (for example, small mammal mounds) that are associated with successful invasion likely contribute to the invadable space described above and make soil resources like nitrogen and carbon available (Davis et al. 2000; Bradford and Lauenroth 2006; Stohlgren 2007). Similarly, the relationship between disturbance from fire and the subsequent increases in both cheatgrass and fire are well studied (D'Antonio and Vitousek 1992; Knapp 1996) across the west of the USA. The Jackson Hole investigation did find that fire contributed to the presence of cheatgrass, but the relationship deserves further study given the presence of natural and prescribed fire, patchy but locally abundant cheatgrass across the GYA, and animal movement that connects the landscape (Mack and Pyke 1984).

Temporal variability of disturbance and the timing of propagule arrival may be important to cheatgrass establishment but they are difficult to measure. Available resources and native plant species richness and cover fluctuate through time (Davis and Thompson 2000), and our results, as well as that of others, show that cheatgrass does better when these variables coincide (Bradford and Lauenroth 2006). Cheatgrass has also been shown to do well in very specific weather patterns that vary in time such as isolated showers in the late fall, warm early-spring air temperature, and snow free areas (Mack and Pyke 1984). We did find that cheatgrass tends to do better on steep, south-facing slopes that are more likely free of snow in the early spring and have longer growing seasons.

It is likely that cheatgrass is spreading in the Greater Yellowstone Area. The 'lag phase' (Hobbs and Humphries 1995) might be more pronounced at the margins of a species' optimal range (Pierson and Mack 1990). Much of the GYA is dominated by forested landscapes and high elevation areas with growing seasons which may currently be too short for sufficient growth and seed production (Bradford and Lauenroth 2006; Chambers et al. 2007). It is probably an extraordinary coincidence of favorable climate, habitat condition, disturbance, and propagule distribution that allows cheatgrass to overcome the spatial and temporal heterogeneity to establish new patches on the landscape (Sax and Brown 2000). Over time cheatgrass likely scraped out an existence in patchy and isolated populations until eventual synergy of favorable conditions resulted in numerous robust and viable populations that act as a seed source. Combine the increased number of source populations with increased disturbance (recreation, development, fire; (Pierson and Mack 1990) and a season of favorable climatic conditions, and cheatgrass may be poised for a legacy of persistence in the mountain west.

6 Conclusions

Modeling species in space and time is new and difficult for many reasons. Aside from the biases inherent in many datasets (i.e., few points, clumped distributions, only a portion of habitat sampled) it is difficult to differentiate intermittent presence from sustainable/persistent presence (Liang and Stohlgren 2011). Likewise, it is difficult to distinguish true absence from yet-to-be-invaded suitable habitat. Two model algorithms may provide different results. Data collected over multiple time periods may mask extirpation and persistence issues. Sample biases in the intensity, frequency, and pattern of sampling also affect results. Poorly surveyed areas and coarse-resolution predictor layers have routinely caused problems in modeling efforts. Projecting to future climates assumes environmental patterns found today will still exist in the future.

In addition to theoretical limitations, there are ecological limitations to the maps and models. Species reproduce, spread, hybridize, and back-cross. They adapt to new environments. Long-lived species may linger in poor environments, such as a deeply rooted tree tapping subsurface water and avoiding a drought at the surface. Species interact with many other species, including native and alien pathogens and diseases. Diseases and pathogens are adapting and evolving as a species' invasion continues. Competition for resources, herbivory, and predation vary in space and time (Tilman 1982). Dispersal, usually determined by wind or animals in plants, and by social systems or instincts in animals, can be facilitated by modern humans —trade and transportation in ships, planes, trucks, cars, and trains. We are in our infancy in understanding the ecological limitations of species distributions in a human-dominated world.

Acknowledgments We thank our many co-workers of the past several years, especially Greg Newman, Jim Graham, Sara Simonson, and Jeff Morisette. Logistical and data support were provided by the USGS Fort Collins and Northern Rocky Mountain Science Centers; the Natural Resource Ecology Laboratory, Colorado State University; and the Greater Yellowstone Coordinating Committee members. This research was supported by USDA CSREES/NRI 2008-35615-04666, the Yellowstone Park Foundation and the Upper Snake River Basin Sage-grouse Working Group. Any use of trade, product, or firm names is for descriptive purposes only and does not imply endorsement by the U.S. Government.

References

Barbour, M. G., Burk, J. H., & Pitts, W. D. (1987). *Terrestrial plant ecology* (2nd ed.). Menlo Park: Benjamin/Cumming.

Barnett, D. T., Stohlgren, T. J., Jarnevich, C. S., et al. (2007). The art and science of weed mapping. *Environmental Monitoring and Assessment, 132*, 235–252.

Beaumont, L. J., Gallagher, R. V., Thuiller, W., et al. (2009). Different climatic envelopes among invasive populations may lead to underestimations of current and future biological invasions. *Diversity and Distributions, 15*, 409–420.

Bradford, J. B., & Lauenroth, W. K. (2006). Controls over invasion of *Bromus tectorum*: The importance of climate, soil, disturbance, and seed availability. *Journal of Vegetation Science, 17*, 693–704.

Chambers, J. C., Roundy, B. A., Blank, R. R., et al. (2007). What makes Great Basin sagebrush ecosystems invasible by *Bromus tectorum* L?. *Ecological Monographs, 77*, 117–145.

Chytry, M., Jarosik, V., Pysek, P., et al. (2008). Separating habitat invasibility by alien plants from the actual level of invasion. *Ecology, 89*, 1541–1553.

D'Antonio, C. M., & Vitousek, P. M. (1992). Biological invasions by exotic grasses, the grass/fire cycle and global change. *Annual Review Ecological Systematics, 23*, 63–87.

Darwin, C. (1859). *The origin of species*. London: Reprinted by Penguin Books.

Davis, M. A., Grime, J. P., & Thompson, K. (2000). Fluctuating resources in plant communities: A general theory of invasibility. *Journal of Ecology, 88*, 528–534.

Davis, M. A., & Thompson, K. (2000). Eight ways to be a colonizer; two ways to be an invader. *Bulletin of the Ecological Society of America*, 81.

de Siqueira, M. F., Durigan, G., Junior, P. M., et al. (2009). Something from nothing: Using landscape similarity and ecological niche modeling to find rare plant species. *Journal for Nature Conservation, 17*, 25–32.

Elith, J., Graham, C. H., Anderson, R. P., et al. (2006). Novel methods improve prediction of species' distributions from occurrence data. *Ecography, 29*, 129–151.

Elton, C. (1927). *Animal ecology*. London: Sidgwick and Jackson.

Elton, C. (1958). *The ecology of invasions by plants and animals*. London: Meuthuen and Company LTD.

Franklin, J. (2009). *Mapping species distributions; spatial inference and prediction*. Cambridge, United Kingdom: Cambridge University Press.

Fuller, T., Morton, D. P., & Sarkar, S. (2008). Incorporating uncertainty about species' potential distributions under climate change into the selection of conservation areas with a case study from the Arctic Coastal Plain of Alaska. *Biological Conservation, 141*, 1547–1559.

Gallien, L., Munkemuller, T., Albert, C. H., et al. (2010). Predicting potential distributions of invasive species: Where to go from here? *Diversity and Distributions, 16*, 331–342.

Gleason, H. A. (1920). Some applications of the quadrat method. *Bulletin of the Torrey Botanical Club, 47*, 21–33.

Gleason, H. A. (1926). The individualistic concept of the plant association. *Bulletin of the Torrey Botanical Club, 53*, 1–20.

Grinnell, J. (1917). The niche-relationships of the California thrasher. *Auk, 34*, 427–433.
Guisan, A., Lehmann, A., Ferrier, S., et al. (2006). Making better biogeographical predictions of species' distributions. *Journal of Applied Ecology, 43*, 386–392.
Guisan, A., & Thuiller, W. (2005). Predicting species distribution: Offering more than simple habitat models. *Ecology Letters, 8*, 993–1009.
Hobbs, R. J., & Humphries, S. E. (1995). An integrated approach to the ecology and management of plant invasions. *Conservation Biology, 9*, 761–770.
Holcombe, T. R., Stohlgren, T. J., & Jarnevich, C. S. (2010). From points to forecasts: Predicting invasive species habitat suitability in the near term. *Diversity, 2*, 738–767.
Hutchinson, G. E. (1957). Concluding remarks. *Cold Springs Harbor Symposium on Quantitative Biology, 22*, 415–427.
Jarnevich, C. S., Holcombe, T. R., Barnett, D. T., et al. (2010). Forecasting weed distributions using climate data: A GIS early warning tool. *Invasive Plant Science and Management, 3*, 365–375.
Jarnevich, C. S., & Stohlgren, T. J. (2009). Near term climate projections for invasive species distributions. *Biological Invasions, 11*, 1373–1379.
Knapp, P. A. (1996). Cheatgrass (*Bromus tectorum* L.) dominance in the Great Basin Desert. *Global Environmental Change, 6*, 37–52.
Kumar, S., Spaulding, S. A., Stohlgren, T. J., et al. (2009). Potential habitat distribution for the freshwater diatom *Didymosphenia geminata* in the continental US. *Frontiers in Ecology and the Environment, 7*, 415–420.
Levine, J. M., & D'Antonio, C. M. (1999). Elton revisited: A review of evidence linking diversity and invasibility. *Oikos, 87*, 15–26.
Liang, C. T., & Stohlgren, T. T. (2011). Habitat suitability of patch types: a case study of the Yosemite toad. *Frontiers of Earth Science, 5*, 217–228. doi: 10.1007/s11707-011-0157-2
Mack, R. N. (1981). Invasion of *Bromus tectorum* L. into Western North America: An ecological chronicle. *Agro-Ecosystems, 7*, 145–165.
Mack, R. N., & Pyke, D. A. (1983). The demography of *Bromus tectorum*: Variation in time and space. *Journal of Ecology, 71*, 69–93.
Mack, R. N., & Pyke, D. A. (1984). The demography of *Bromus tectorum*: The role of microclimate, grazing, and disease. *Journal of Ecology, 72*, 731–748.
Miller, A. J., & Knouft, J. H. (2006). GIS-based characterization of the geographic distributions of wild and cultivated populations of the Mesoamerican fruit tree *Spondias purpurea* (Anacardiaceae). *American Journal of Botany, 93*, 1757–1767.
Moritz, C., Hoskin, C. J., MacKenzie, J. B., et al. (2009). Identification and dynamics of a cryptic suture zone in tropical rainforest. *Proceedings of the Royal Society B-Biological Sciences, 276*, 1235–1244.
Neerinckx, S. B., Peterson, A. T., & Gulinck, H., et al. (2008). Geographic distribution and ecological niche of plague in sub-Saharan Africa. *International Journal of Health Geographics, 7*.
Pawar, S., Koo, M. S., Kelley, C., et al. (2007). Conservation assessment and prioritization of areas in Northeast India: Priorities for amphibians and reptiles. *Biological Conservation, 136*, 346–361.
Pearson, R. G., Raxworthy, C. J., Nakamura, M., et al. (2007). Predicting species distributions from small numbers of occurrence records: A test case using cryptic geckos in Madagascar. *Journal of Biogeography, 34*, 102–117.
Phillips, S. J., Anderson, R. P., & Schapire, R. E. (2006). Maximum entropy modeling of species geographic distributions. *Ecological Modeling, 190*, 231–259.
Pierson, E. A., & Mack, R. N. (1990). The population biology of *Bromus tectorum* in forests: effect of disturbance, grazing, and litter on seedling establishment and reproduction. *Oecologia, 84*, 526–533.
Pulliam, H. R. (2000). On the relationship between niche and distribution. *Ecology Letters, 3*(4), 349–361.

Raxworthy, C. J., Martinez-Meyer, E., Horning, N., et al. (2003). Predicting distributions of known and unknown reptile species in Madagascar. *Nature, 426*, 837–841.
Richardson, D. M., & Pysek, P. (2008). Fifty years of invasion ecology—The legacy of Charles Elton. *Diversity and Distributions, 14*, 161–168.
Sax, D. F., & Brown, J. H. (2000). The paradox of invasion. *Global Ecology and Biogeography, 9*, 363–371.
Soberon, J. (2007). Grinnellian and Eltonian niches and geographic distributions of species. *Ecology Letters, 10*(2), 1115–1123.
Stohlgren, T. (2007). *Measuring plant diversity, lessons from the field*. New York: Oxford University Press.
Stohlgren, T. J., Barnett, D., Flather, C., et al. (2006). Species richness and patterns of invasion in plants, birds, and fishes in the United States. *Biological Invasions, 8*, 427–447.
Stohlgren, T. J., Barnett, D. T., & Kartesz, J. (2003). The rich get richer: Patterns of plant invasions in the United States. *Frontiers in Ecology and the Environment, 1*, 11–14.
Stohlgren, T. J., Chong, G. W., Kalkhan, M. A., et al. (1997). Multi-scale sampling of plant diversity: Effects of minimum mapping unit size. *Ecological Applications, 7*, 1064–1074.
Stohlgren, T. J., & Schnase, J. L. (2006). Risk analysis for biological hazards: What we need to know about invasive species. *Risk Analysis, 26*, 163–173.
Thill, D. C., Beck, K. G., & Callihan, R. H. (1984). The biology of downy brome (*Bromus tectorum*). *Weed Science, 32*, 7–12.
Thuiller, W., Albert, C., Araujo, M. B., et al. (2008). Predicting global change impacts on plant species' distributions: Future challenges. *Perspectives in Plant Ecology Evolution and Systematics, 9*, 137–152.
Thuiller, W., Richardson, D. M., Pysek, P., et al. (2005). Niche-based modelling as a tool for predicting the risk of alien plant invasions at a global scale. *Global Change Biology, 11*, 2234–2250.
Tilman, D. (1982). *Resource competition and community structure*. Princeton: Princeton University Press.
Whittaker, R. H. (1967). Gradient analysis of vegetation. *Biological Review, 42*, 207–264.
Whittaker, R. H. (1970). *Communities and ecosystems*. London: Macmillan Co., Collier-Macmillan Ltd.

Mapping the Universe

J. Richard Gott III

Abstract The varieties of astronomical mapping are presented, starting with maps of the celestial sphere. Important map projections are considered, including gnomonic projections which are particularly good for star charts. When telescopes arrived, astronomers began to make maps of other worlds, including influential maps of Mars by Schiaparelli and maps of the Moon made during the *Apollo* era. I discuss map projections specifically designed for Mars and the Moon. Map distortions are examined: area, local shape, flexion, lopsidedness, distances, and boundary cuts. Spacetime maps include Martin Kruskal's map of the interior of the black hole. Our map of the Sloan Great Wall of Galaxies, 1.37 billion light-years long, and the famous WMAP satellite map of the cosmic microwave background are described. Finally, my "Map of the Universe" displays everything in the universe on a logarithmic map: from the Earth to manmade satellites; to the Moon, Sun, and planets; to nearby stars; to our Milky Way, and other galaxies in the Local Group and Virgo supercluster; to the distant galaxies in the Sloan Digital Sky Survey; and finally, to the Cosmic Microwave Background, the most distant thing we can see.

Keywords Universe · Mars · Moon · Spacetime diagrams · Sloan Great Wall · Map projections

---

I am teaching a new course this year at Princeton University called "Mapping the Universe." Astronomical maps have had a long and interesting history, but are of particular interest today. When Mercator made his great map of the Earth in 1569 it was interesting not only because it portrayed local shapes and compass directions correctly, but also because for the first time it was able to show the coastlines of North America and South America more or less correctly. Today, astronomical maps are able to capture, for the first time, the entire visible universe on a single map. This has been due to the WMAP (Wilkinson Microwave Anisotropy Probe) satellite which measures the cosmic microwave background, the most distant thing we can observe, and the Sloan Digital Sky Survey (for a description see York et al. 2000) which has mapped the locations of a million galaxies in space. However let me start my story nearer the beginning.

J.R. Gott III (✉)
Department of Astrophysical Sciences, Princeton University, Princeton, NJ, USA
e-mail: jrg@astro.princeton.edu

S.D. Brunn and M. Dodge (eds.), *Mapping Across Academia*,
DOI 10.1007/978-94-024-1011-2_5

1 Mapping the Sky

The first astronomical maps were of the heavens. In some ways mapping the heavens was easier than mapping the Earth. To map the Earth you had to go on long voyages make astronomical observations and have good clocks, but the heavens you could see from wherever you lived. Go out at night and look at the sky. It is a dome overhead—a hemisphere. The Earth looks like a flat disk underfoot, and the horizon is a great circle bounding the hemispherical dome overhead. Now put on a spacesuit. Imagine the Earth vanishes. You are floating in space. Now you can see the other half of the sky as a hemispherical bowl beneath you. The sun is there: your view was just blocked by the Earth before. The whole sky is a sphere—the celestial sphere. Anaximander (610-547 B.C.) realized this (see Asimov 1976). Mapping the celestial sphere posed exactly the same mapmaking challenges as mapping the surface of the Earth—how to represent a curved surface on a flat piece of paper. Some distortion is inevitable because the curved surface of the sphere does not obey the laws of planar Euclidean geometry. However, you can preserve some properties. The Mercator projection preserves local shapes—but areas are shown badly. The Mollweide projection, named after Karl Brandan Mollweide (1774–1825) who invented it in 1805, preserves areas but local shapes are distorted. What is the best projection for star charts? For star charts one wants the constellations on the chart to look to your eye like they look in the sky.

The Gnomonic projection, named after the gnomon on a sundial, does this perfectly. The Gnomonic projection uses a projection light at the center of the sphere to project (as shadows) what is on the sphere onto a flat plane. If you put your eye at the center of the sphere, your view of the flat map will be exactly the same as the view you have of the celestial sphere. You just have to put your eye at the right distance in front of the map. Imagine that you look at the sky and hold a transparent pane of glass in front of your eye. With magic marker put a dot on the plane of glass in front of every star you see. On that pane of glass, you will now have a Gnomonic map of that part of the heavens. The Gnomonic projection plots great circles on the sky as straight lines on the flat map. Great circles on the sky look straight to your eye, and they look straight to you on the Gnomonic map. The iPhone makes available Gnomonic star charts. One of my students showed me her iPhone app that enables the viewer to hold the phone up and display the constellations that are up in that direction at that very moment. The phone knew the time, and its GPS told it where it was and which way north was, so it could put on the screen just what your eye would see. Hold it up in a certain direction on a winter's night, and you would see Orion on your screen and behind it in the sky in that very direction would be the real Orion. It is pretty impressive. My student asked what map projection the app used for its star charts. I could move it around and notice that the lines of celestial longitude (called Right Ascension) were straight as was the celestial equator and the ecliptic (the path the sun takes in the sky during the course of the year)—great circles all—so the projection had to be the Gnomonic, just as it should be.

A disadvantage of the Gnomonic projection is that it can cover in a single plane only half the celestial sphere at most. However, if you want to show the whole sky you can encase the celestial sphere inside a cube, and using a projection light in the center, project the sphere onto the six faces of the cube. Then you can open the cube and spread it out on the plane in a cross-shaped pattern to produce six Gnomonic star charts. The top of the cube includes all the North circumpolar stars (always visible in the northern sky to an observer at 45° North Latitude). The four sides of the cube show the four seasonal charts for Winter, Spring, Summer, and Autumn. These are stars that appear over the southern horizon for an observer at 45° North Latitude at midnight on December 21, March 21, June 21, and September 21, respectively. As the Earth circles the sun once a year, the sun circles the sky leaving the Winter, Spring, Summer, and Autumn stars opposite it to appear in sequence through the year at midnight. The bottom of the cube includes the South Circumpolar stars that are not visible from 45° North latitude. I thought this would be a particularly elegant way to show the heavens. I made up such a set of star charts and entered them as my 8th grade science project, winning first prize in the discussion division of the Kentucky State Science Fair (Fig. 1). Years later when I was in graduate school, I found a book in the library: *Six Maps of the Stars*, with charts by W. Newton (Society for the Diffusion of Useful Knowledge Great Britain 1831). It was exactly the same idea. It said similar charts had been made dating back as far as 1674! Such Gnomonic star charts had pretty much gone out of use in books before my time, but Bob Vanderbei and I hope to bring them back by

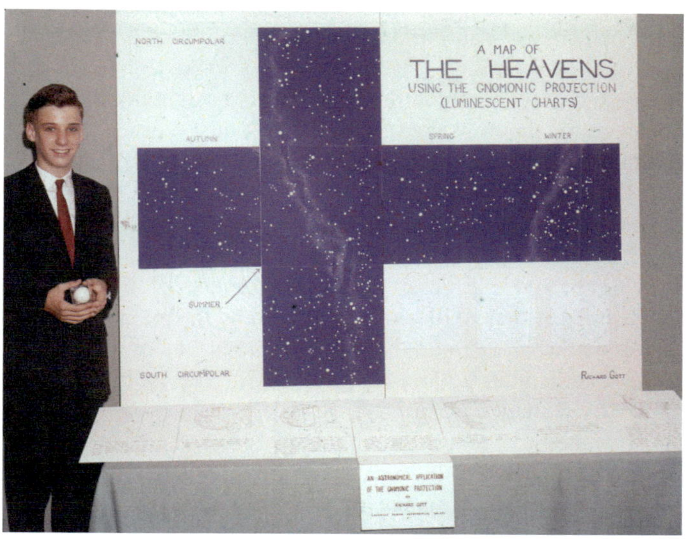

Fig. 1 Gnomonic Cube star charts (my 8th grade science project). Cube is unfolded into a cross. The *top square* shows the North Circumpolar stars; the four sides show the Autumn, Summer, Spring, and Winter Stars; and the *bottom* shows the South Circumpolar stars

including six modern Gnomonic star charts covering the celestial sphere in our 2010 National Geographic book, *Sizing up the Universe*.

2 Mapping the Solar System

When telescopes arrived, astronomers were able to map other worlds. Maps of the Moon were made, mostly using the orthographic projection, which just gave a view of the Moon as seen from the Earth. But eventually, starting with the Soviet space probe *Luna III*, the far side was seen. In 1969, National Geographic presented a large map of the far side as well as the visible side of the moon using the Lambert Azimuthal Equal Area projection in two hemispheres (National Geographic Staff 1969). Venus presented a cloud covered blank face and so was not a good subject for early maps, but now its surface has been mapped in great detail by radar from orbiting spacecraft (Saunders et al. 1992). On Jupiter, one could see equatorial cloud belts and a great Red Spot (a storm that has been raging for over 300 years now). On Jupiter one is seeing variable cloud features rather than features on a solid surface. Saturn had dramatic rings, and the Cassini division in the rings was noticed, but the globe of Saturn itself had less prominent features than Jupiter. Historically, after the Moon, Mars was the solar system object that received the most attention from mapmakers. Mars rotates on its axis once every 24 h and 39.6 min, showing us all its faces—and surface markings can clearly be seen, so mapping was of great interest. Mars has both northern and southern polar caps. Most of Mars is covered by red dust but it has dark markings as well. Most prominent is *Syrtis Major*—an Africa-shaped region where darker basalt rock not covered by dust shows through. These dark areas looked green to the eye, contrasting with the red deserts around them. They were originally thought to be areas of vegetation because they changed seasonally as dust storms changed their outlines. The Italian astronomer Schiaparelli announced in 1877 his discovery of "canals" on Mars. His map of Mars was published in the German encyclopedia *Meyers Konversations-Lexikon* in 1888. American astronomer Percival Lowell (1855–1916) famously argued that these were canals built by a dying Martian civilization trying to irrigate its desert planet (see Asimov 1976). We now know that these "canals" were just optical illusions, produced by the brain's natural propensity to see linear features when visibility is poor. A famous map of Mars showing canals was made by Antoniadi in 1896 (cf. Kim 1993). The map was reproduced in *Exploring Mars* by Gallant (1956).

For Mars, I wanted to produce an equal-area projection that showed the polar areas well, since the planet's polar caps are so important. They are made primarily of water-ice and are a major source of water for any potential future Martian colonists. The Winkel-Tripel projection, popular for the Earth, stretches out the polar caps, as does the equal-area Mollweide projection. I started by halving each longitude, so that the entire surface of Mars is plotted in its western hemisphere. I rotated the globe by 90° and halved the "new longitudes" creating a quadrant of the sphere, and then I mapped this quadrant onto a flat piece of paper with a Bromley-Mollweide equal-area projection. The Bromley-Mollweide projection is a stretched Mollweide projection that produces perfect local shapes on the equator—but I have rotated the globe so that "equator" now falls on Mars' prime meridian. Since regions on the prime meridian of Mars have been squashed horizontally by a factor of two and then squashed vertically by a factor of two, they have perfect shapes on the quadrant of the globe and they retain these shapes as they are projected by the Bromley-Mollweide projection back onto the plane. This produces a nice looking elliptical equal-area map where the shapes on the prime meridian are perfect and the scale on the prime meridian is linear. The Bromley-Mollweide has root-mean-square logarithmic distance errors (between pairs of random points) of $\sigma = 0.420$, the Mollweide $\sigma = 0.390$, the Hammer $\sigma = 0.388$, and the Gott elliptical equal area $\sigma = 0.365$. So the Gott elliptical has smaller distance errors than these other equal-area projections; its map coordinates are given in Gott et al. (2007). We published a map of Mars using this projection in the journal *Cartographica* (Gott et al. 2007). The dark areas *Syrtis Major* and *Solis Lacus* are shown nicely as are the polar caps (Fig. 2). The significance of this map for how astronomy understands and represents space through maps is that we should tailor the map projection to the features of the planet being depicted.

For a map of the Moon I developed a circular azimuthal map projection (the Gott-Mugnolo, with Charles Mugnolo) that minimizes root-mean-square (rms) logarithmic distance errors between pairs of points on the globe. This measure has the property that distances between pairs of points that are plotted a factor of 2 too far apart on the map relative to the scale bar are assigned an error equal to that if

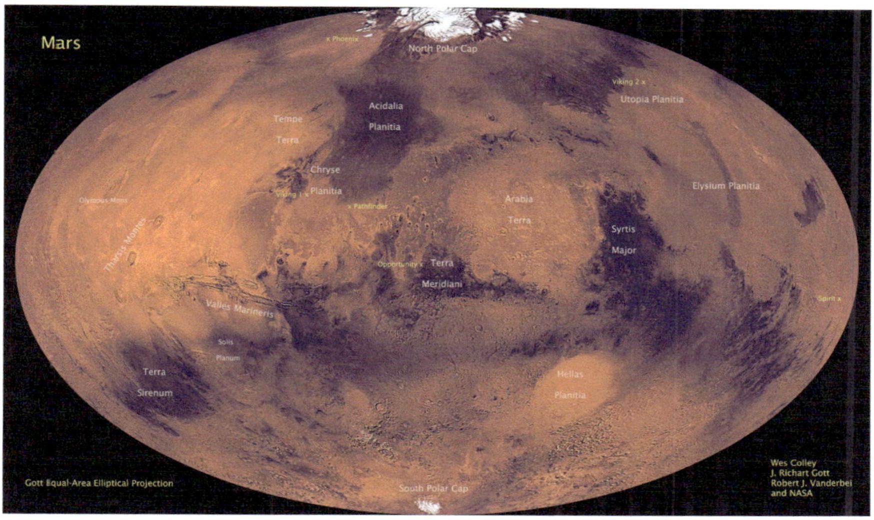

Fig. 2 The Gott-Colley Map of Mars using the Gott-Equal Area Elliptical Projection which shows perfect shapes on the central meridian and illustrates the polar caps well (Gott et al. 2007)

they were plotted at ½ the correct distance. Other distance errors measures by Gilbert (1974) and Peters (1975, 1978, 1984) share this feature, but ours is simpler (see discussion in Gott et al. 2007). This map has RMS logarithmic distance errors of $\sigma = 0.341$, the smallest for any map we have studied. It has no boundary cuts. Boundary cuts tend to increase distance errors by taking pairs of points that are close on the globe and widely separating them on the map. Our lunar map is centered on 0° longitude, 0° latitude. In addition to showing distances well, it has the advantage that the visible face of the moon is shown as a circular disk (Fig. 3). The projection thus gives a good rendition of the moon as we see it, while allowing us to peek around the visible face in any direction to see the back side of the moon as well. Again, we seek to tailor the map projection to the object being depicted.

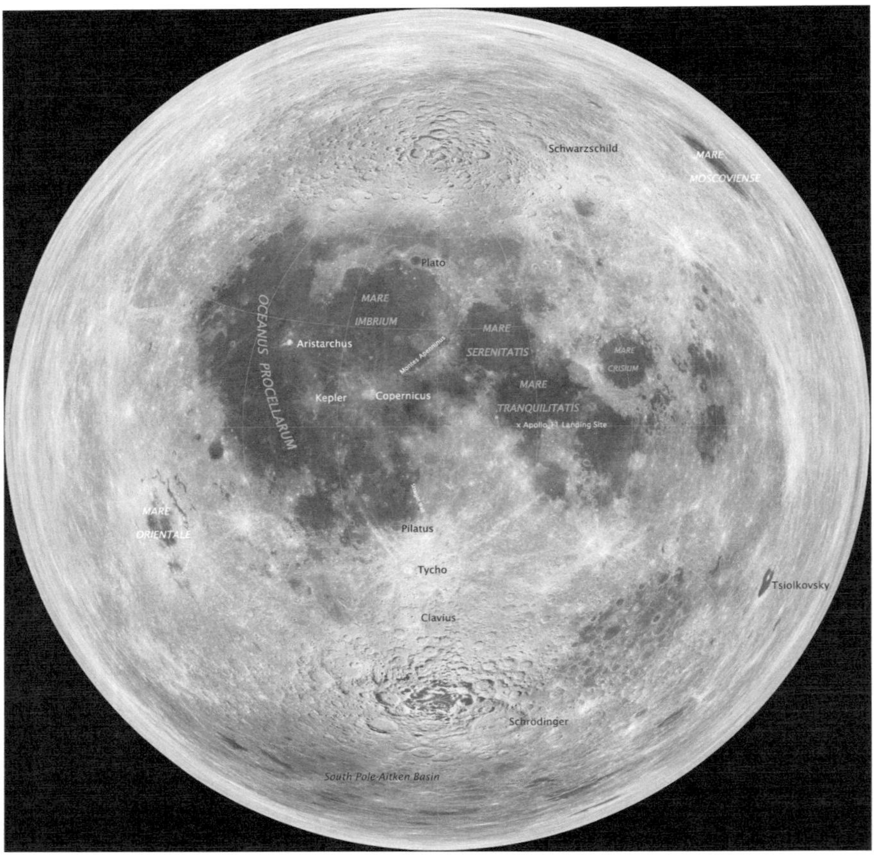

Fig. 3 The Moon on the Gott-Mugnolo projection, minimizing distance errors and showing the visible face of the moon as a circle in the center, with the far side around the perimeter (Gott et al. 2007). The North and South Pole of the moon appear at low solar illumination and show craters with shadows

Since most astronomical objects are spherical it is interesting to find map projections of the sphere that minimize total distortions. Distortions of map projections have typically been evaluated only on isotropy and area errors as measured by Tissot Ellipses (Tissot 1881) which show the shape and area of an infinitessimal circle on the globe as projected onto the map. Laskowski (1997a, b) for example has rated map projections on the sum of the squares of their isotropy and area errors integrated over the globe (normalized to the simple equirectangular projection). Distance errors should be included also, since measuring distances between pairs of points is one of the primary functions of a map. David Goldberg and I have also added errors due to flexion (bending) and skewness (lopsidedness) (Goldberg and Gott 2007). Area and isotropy errors depend on the metric. Flexion and skewness depend on the first derivatives of the metric. Second derivatives of the metric tell you about the curvature of the surface—you are not going to get rid of them. The Mercator map has flexion and skewness errors off the equator. The United States is bent downward in the Mercator map because the border with Canada is shown as a straight line on the map whereas on the globe it is a line of latitude that is concave toward the North Pole (it is not a great circle). North America on the Mercator Map is lopsided because Canada is shown too large and Mexico too small relative to the U.S. Goldberg and I developed error measures for (1) local isotropy, (2) area, (3) distances, (4) flexion, (5) skewness, and (6) boundary cuts. Following Laskowski (1997a, b) we normalize these errors relative to those found on the simple equirectangular projection ($x = \lambda$, $y = \varphi$). Thus, our normalized boundary cut error is the length of the boundary cut relative to that in the equirectangular projection, which is 180°. The equirectangular projection has a sum of squares of errors $\Sigma = 6.0$ by definition. The known map projection with the lowest value of Σ was the Winkel-Tripel with $\Sigma = 4.5629$. This is interesting since this had *already* been chosen by *National Geographic* as the best map projection for their overall world maps! It is neither equal area, nor conformal, but a good compromise. It has north and south poles that are line segments, shorter than the equator. The map with the least overall distortion that has straight latitude lines is the Kavrayskiy VII projection ($\Sigma = 4.8390$). This therefore makes a good map projection for Jupiter, whose cloud belts that stretch along latitude lines look best when shown as straight (Fig. 4). Can one find a map with less overall distortion than the Winkel-Tripel? It is hard—the Winkel-Tripel is very good. Map projections with linear scales along both the prime meridian and the equator usually did well. The Hammer-Wagner projection with a curved pole line was the best equal area projection with $\Sigma = 5.7847$. So I thought the Aitoff-Wagner projection might do well. In this projection, Wagner multiplies each latitude by 7/9ths on the globe. The North Pole is then plotted as the 70° latitude line. Then halve all longitudes and project the hemisphere of the globe on the plane with the azimuthal equidistant projection. Finally multiply the x- and y-coordinates by 18/5 and 9/7, respectively. It did not beat the Winkel-Tripel, but I found I could improve its results if I started with the globe and multiplied latitudes by 6/9ths, longitudes by $(3)^{1/2}/4$, then mapped to the plane with the azimuthal equidistant and multiplied the x-coordinate by 2, as Aitoff had done. I chose these parameters to produce a pole line about as long as in the

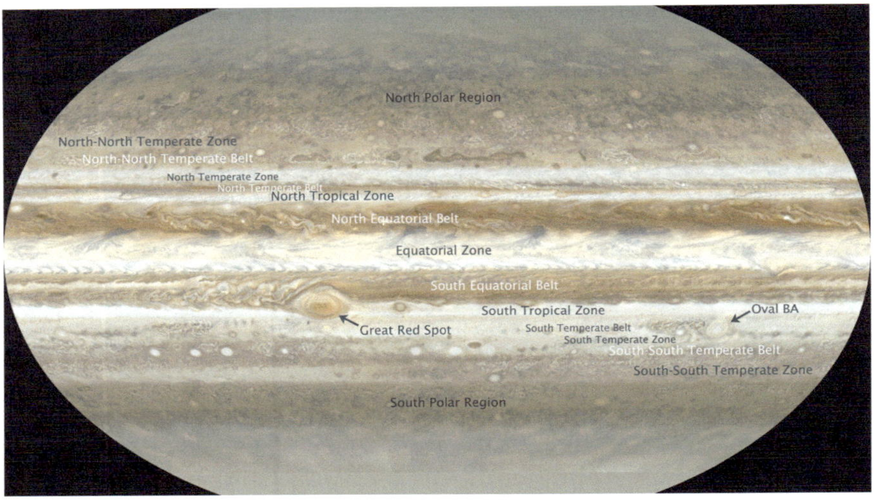

Fig. 4 Jupiter on the Kavrayskiy-VII projection. This is the projection with the lowest overall area, isotropy, flexion, skewness, and boundary cut errors (Goldberg and Gott 2007) that has straight latitude lines. This property is useful for depicting Jupiter's cloud belts

Winkel-Tripel and have Africa no more squashed than on the Kavrayskiy-VII, which was acceptable. I was anxious to get a low $\sum$ score but not 'teach to the test,' so I only tried this one projection: it did beat the Winkel-Tripel with a score of $\sum = 4.5338$. It has better distances than the Winkel-Tripel, and better shapes, but somewhat larger area errors. Figure 5 shows Venus on this Aitoff-Wagner-Gott projection. Its (x, y) map coordinates are given by:

$$
\begin{aligned}
x &= 2z\ \cos(2\varphi/3)\sin(\lambda/2\sqrt{3})/\sin(z)\\
y &= z\sin(2\varphi/3)/\sin(z)
\end{aligned}
$$

where $z = \arccos[\cos(2\varphi/3)\cos(\lambda/2\sqrt{3})]$, φ is latitude, and λ is longitude.

This map projection is good for Mercury, Venus, Titan, and Io where the important features are scattered over the sphere. It also makes an excellent map of the Earth. Australia in particular has a better shape than on the Winkel-Tripel (see Gott and Vanderbei 2010).

3 From the Solar System to the Galaxy

Copernicus (1543) was the first to make good maps of the solar system—because he had the right (heliocentric) idea of how it worked. By measuring line of sight angles to the planets as a function of time as they moved, and by assuming they all

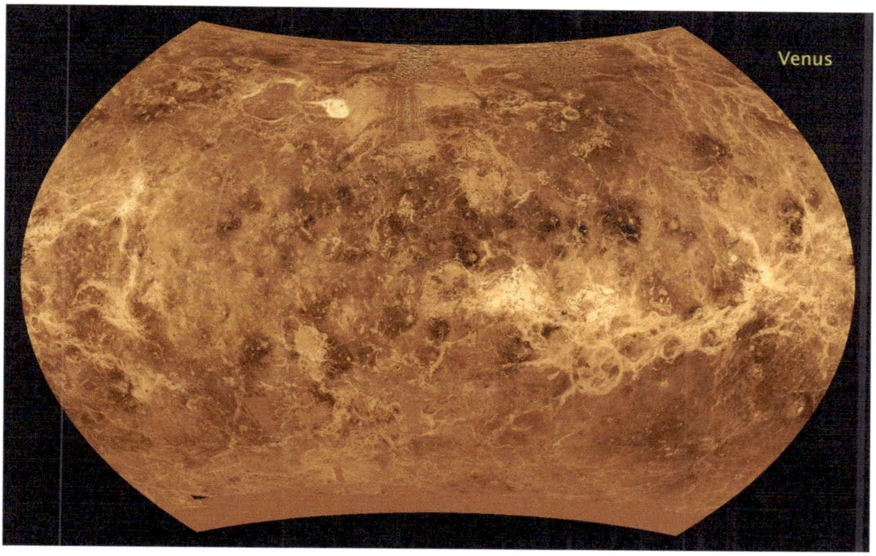

Fig. 5 Venus using the Aitoff-Waggener-Gott projection which minimizes errors in isotropy, area, flexion, skewness, distances and boundary cuts. Highlands are shown lighter, and lowlands darker. This projection is useful for astronomical objects like Earth, Mercury, Venus, Io and Titan where important features are scattered around the globe. The North Pole and South Poles are the *curved lines* at the *top* and *bottom*. The equator is a straight horizontal line bisecting the map. The central meridian is a straight vertical line bisecting the map. Longitude is plotted linearly on the equator; latitude is plotted linearly on the prime meridian

moved uniformly on circular orbits Copernicus was able to measure the relative radii of the orbits of Mercury, Venus, Earth, Mars, Jupiter, and Saturn to pretty good accuracy (see "Copernican Derivations," http://astro.unl.edu/naap/ssm/ssm_advanced.html). As the Earth orbits the sun, stars alter their positions in the sky slightly over the course of each year as the viewing angle from Earth to the star slowly changes. This causes a parallax motion of each star in the sky during the year (of less than one second of arc). By measuring these stellar motions during the year we can measure their distances. When such parallax distances to the stars could be plotted, we could begin to map the solar neighborhood. The stars are other suns like our Sun—they are dim because they are so far away, the nearest being over 200,000 times as far away from us as the Sun. Sir William Herschel (1738–1822) made a map of the Milky Way galaxy by counting stars in different directions (see Lemonick 2009, for a discussion of Herschel's method and archival resources for his work). It was the first attempt to show our place in the Galaxy. Eventually, when variable stars provided new distance clues, Harlow Shapley (1918) was able to show that our Sun was about halfway from the center of our disk-like Milky Way galaxy, which measures about 100,000 light-years across (see Gott and Vanderbei 2010). That means it takes light traveling at 186,000 miles per second 100,000 years to cross it!

4 Mapping Spacetime

In his theory of Special Relativity in 1905, Einstein postulated that motion was relative and the speed of light was a constant, c = 186,000 miles per second, as seen by all observers. With this he was able to prove a number of remarkable theorems, including $E = mc^2$. A little bit of mass could be converted into an enormous amount of energy (because c is very large). Einstein's work showed that time could be plotted as a dimension, along with the three dimensions of space. This opened the way for space-time diagrams (Minkowski 1908) with space as the horizontal axis and time as the vertical axis. Light waves were plotted going diagonally at 45°, moving at one light-year per year. Since the speed of light was a constant, one always wanted light beams plotted at 45° in one's space-time diagrams. Thus conformal map projections which preserve angles (just like the Mercator projection preserves compass directions) are used for space-time diagrams, such as the Kruskal and Penrose diagrams discussed below.

Perhaps the most influential space-time map was developed by Martin Kruskal in 1960, a map of the interior of a black hole. Black holes were a byproduct of Einstein's greatest idea—General Relativity—his theory of gravity as due to curved spacetime. In the solar eclipse of 1919, Einstein's prediction of the amount of light bending around the Sun was confirmed. One of the first solutions of Einstein's equations was the Black Hole solution found by Karl Schwarzschild in 1916, but its coordinate system broke down at a particular radius, now called the Schwarzschild radius (Schwarzschild 1916). Kruskal (1960) found a new coordinate system that allowed one to map in a complete way what happened inside that radius. There was a singularity at the center of the black hole of radius zero. If you fell inside the Schwarzschild radius, you were trapped inside with no way out, drawn inexorably to the center where tidal forces ripped you apart. Kruskal's map showed that the geometry inside the black hole was so twisted that the singularity at the center was bent around into the future. Once you were inside the black hole the singularity was a giant hyperbola that loomed ahead in the future, blocking your way. You could no more miss it than you could miss next Tuesday. Other spacetime maps pioneered by Penrose (1964), Carter (1966, 1968), and Hawking and Ellis (1975) have helped us understand what might happen inside rotating black holes. The complete solution to Einstein's equations for a rotating black hole shows a region of time travel hidden inside where you may visit an event in your own past. However, you cannot come back out of the black hole to brag to your friends about your adventures. Instead, you may be able to exit the black hole into another universe. But photons falling into the hole may form a singularity blocking your way into the time travel region. Some recent studies suggest that this singularity may be weak, so that you may be able to survive passing through the singularity (like passing over a speed bump in the road) to reach the time travel region after all. To know for sure what happens we may need to understand the laws of quantum gravity—how gravity behaves at small scales. For more discussion see Gott (2001).

Einstein's General Relativity has also been important in understanding cosmology. Edwin Hubble (in 1922–1923) discovered that the nebula M31 was actually an entire galaxy like our own, 2.5 million light-years away. Further away, in every direction he looked, Hubble (in 1929) found galaxies moving away from us, like debris from a giant explosion—galaxies twice as far away were moving twice as fast. Friedmann's (1922, 1924) big bang solutions to Einstein's equations of General Relativity had actually predicted such an expanding universe ahead of time. Gamow (1948a, b) and his students Alpher and Herman (1948) predicted that if such a big bang explosion occurred, hot radiation left over from the big bang should still be travelling around in the universe today and should be visible in the radio wavelengths. This cosmic microwave background radiation was found by Penzias and Wilson (1965), for which discovery they received the Nobel Prize in 1978. The WMAP satellite has made a detailed map of this cosmic microwave background radiation which comes to us from all over the sky. The WMAP data (Fig. 6) shows temperature fluctuations of one part in 100,000 depicted here on a grey scale. White is hottest and black is coldest. The WMAP team used a Mollweide equal-area projection to plot it. From these small fluctuations (present just 380,000 years after the big bang) the giant clusters of galaxies we see today have grown by the action of gravity over the course of 13.7 billion years. By careful analysis of this map (i.e., the amplitude of fluctuations at different angular scales) the WMAP team has been able to set the cosmological model with high accuracy, determining that about 4 % of the matter in the universe is ordinary matter, 23 % is dark matter made of yet unknown elementary particles, and 73 % is a gravitationally repulsive quantum vacuum state known as dark energy (Spergel et al. 2003). They also made an accurate determination of the age of the universe—13.7 billion years (Spergel et al. 2003).

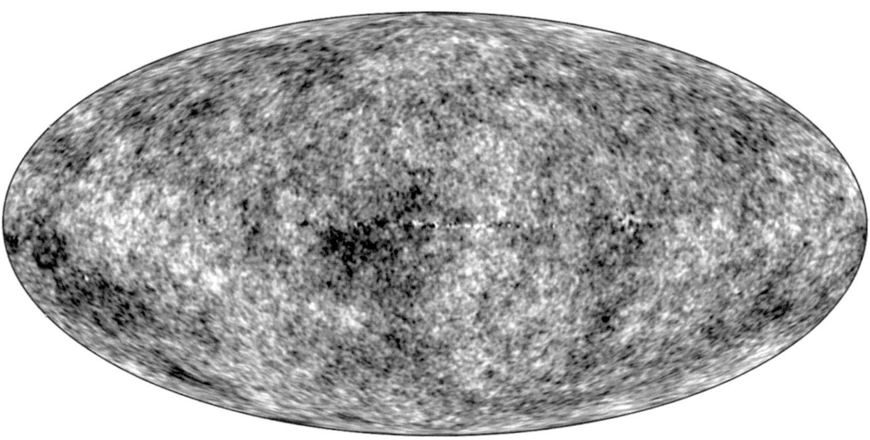

Fig. 6 The WMAP map of the cosmic microwave background sky. Temperature fluctuations of one part in 100,000 are shown on a grey scale. A Molleweide Equal Area Projection is used (Courtesy of Wes Colley)

The cosmic microwave background is the most distant thing we can detect, coming to us from a distance of 13.7 billion light years. We can't see farther than that because light from those regions has simply not had time to reach us yet in the 13.7 billion years the universe has been in existence. The region we can see includes about 140 billion galaxies and is centered on us. That does not mean we are at the center of the universe. It's like when you go to the top of the Empire State Building in New York City; you can see a circular region extending out to the horizon (including parts of New Jersey and Connecticut) that is centered on the Empire State Building. If you went to the top of the Eiffel Tower you would see a different region centered on it. From the theory of inflation (whose predictions fit the WMAP observations), we have reason to believe that the universe is actually much, much larger than the part we can see.

Within the visible universe we can map galaxy locations. The distribution of galaxies in three dimensions can be used to test the theory of *inflation*, Alan Guth's (1981) theory that the big bang explosion was caused by an early epoch of accelerated expansion. Inflation predicts that the distribution of galaxies should be sponge-like (Gott et al. 1986), with chains of galaxies connecting great clusters in a sponge-like pattern we have now observed in many surveys (e.g. Gott et al. 2009) and now call the *cosmic web*. Tulley and Fisher (1987) made a beautiful *Nearby Galaxies Atlas* showing the positions of 2367 nearby galaxies both in the sky and in three dimensions using a series of slices. Because of the expansion of the universe, galaxies that are receding faster from us are further away. By measuring the recession velocity of individual galaxies one can learn their distances and plot them on a map. This elucidates the three dimensional arrangement of the galaxies in space. Geller and Huchra (1989) made such a slice map. It appears as the bottom fan in Fig. 7. The Earth is at the bottom of the fan. They also discovered a great chain of galaxies (called the [CfA2] Great Wall) stretching from one side of their survey to the other. It was 760 million light-years long and made the *Guinness Book of Records* as the largest structure in the Universe. Later, Mario Juric and I measured an even larger chain of galaxies in the Sloan Digital Sky survey, one structure containing over 10,000 galaxies (which we called the Sloan Great Wall; Gott et al. 2005). It was 1.37 billion light-years long (Fig. 7 top) and appeared in both the 2006 and 2012 *Guinness Book of Records*, so it has held the record for 6 years. There are many "largest" things in the *Guinness Book*; this is the largest of the largest.

Figure 8 shows a cross-sectional map of the entire visible universe in the plane extending outward from the Earth's equator. The Earth is at the center. Around the perimeter is the cosmic microwave background as measured by WMAP. The map shows 126,594 Sloan Survey galaxies—and quasars (very luminous accretion disks around black holes in galactic nuclei) that lie within 2 degrees of the plane of Earth's equator. The black wedges are regions obscured by our galaxy and not covered by the Sloan Survey. This is the first time that we have had the data to plot such a map of the entire visible universe. One of man's intellectual goals has been to understand his place in the universe. This map illustrates it for the first time.

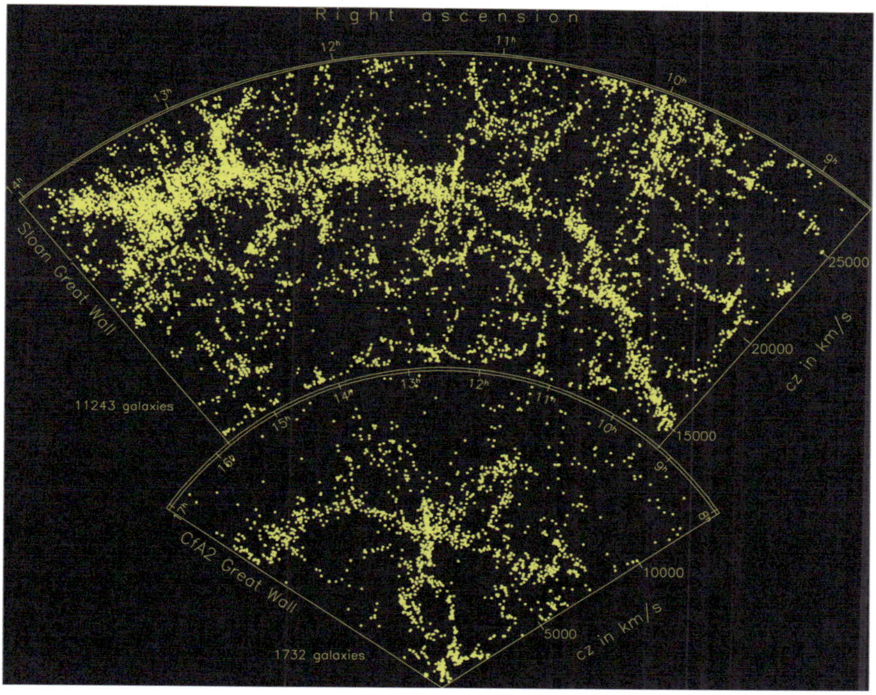

Fig. 7 Slice diagrams of the Sloan Great Wall of galaxies (*upper*) and the CfA2 Great Wall of galaxies shown to scale. The Earth is at the *bottom* of the fan

However, we would like to have a map of the universe that shows *all* the things interesting to us in space, everything from the Moon and planets, to other stars and galaxies, all the way out to the cosmic microwave background radiation. The challenge for such a map is to cover the enormous range of scales involved. The visible universe is huge, the solar system is tiny and yet we are very interested in things in our solar system, so we would like our map to show them as well. What we really want is a map like that famous Saul Steinberg *New Yorker* cover of March 29, 1976, "View of the World from 9th Avenue," which humorously depicts a New Yorker's view of the world. Streets in New York City are large in the foreground, across the Hudson River is a narrow strip of land marked "Jersey," and the rest of the United States is not much wider than the Hudson River, with the Pacific Ocean about as wide as the Hudson and bounded by tiny Asia beyond. This is a parochial view, but it is just the kind of view that we are looking for in our map of the universe. I designed such a map of the universe in 1972 when I was a graduate student. Confronted with the Sloan data, Mario Juric and I decided to use my map projection to show not only the Sloan data but all the other interesting things astronomers have been discovering. The Map of the Universe is a conformal map (like the Mercator map) which preserves shapes locally. Objects that are 10 times

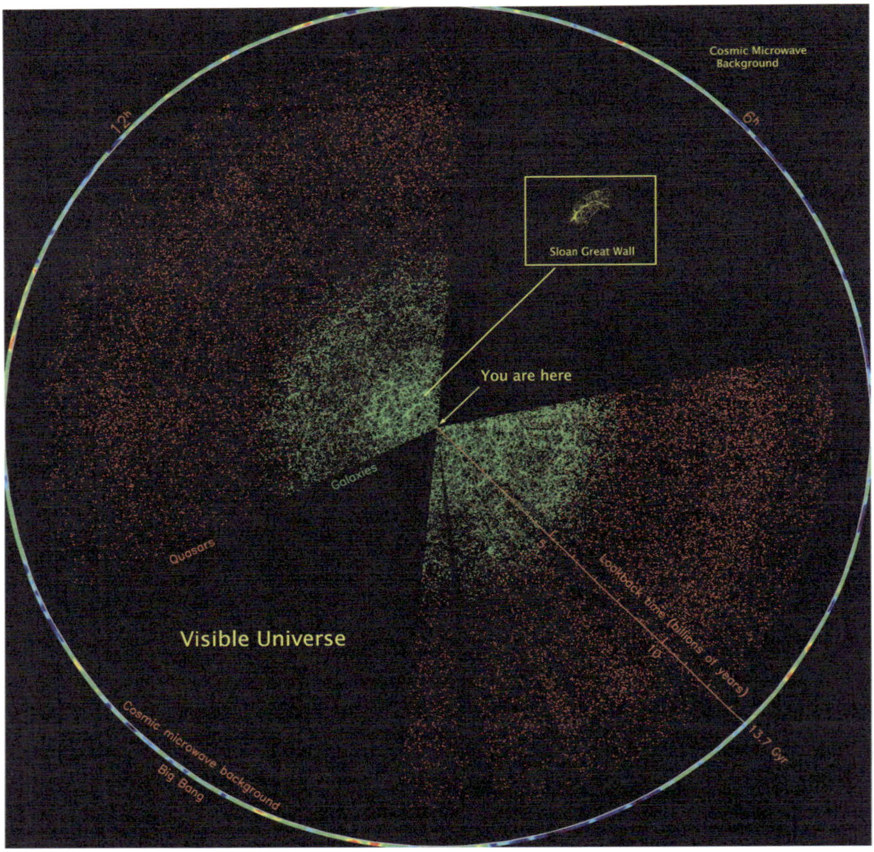

Fig. 8 The visible universe shown in a cross section through a plane extending from Earth's Equator. Galaxies and quasars from the Sloan Digital Sky Survey are shown. The two black wedges lie near the plane of our own Milky Way galaxy which the survey did not cover. Around the perimeter of the map is the cosmic microwave background at a distance of 13.7 billion light years. The Earth is in the center of the map—we are at the center of the region we can see. The Sloan Great Wall of galaxies, 1.37 billion light years long, is indicated

further away are shown at 10 times smaller scale. This allows us to plot everything from satellites orbiting the Earth to the cosmic microwave background all on one map.

Mathematically, it is a conformal logarithmic map of the complex plane. An earlier pocket version of the Map of the Universe I made in the 1990s is shown in Fig. 9. (The Gott-Juric Map, published in the *Astrophysical Journal* in 2005, was 36 in. tall and much more detailed—too large to reproduce here. In it we showed all known objects in a narrow equatorial slice, plus famous objects that lie above or below that slice.) From left to right in Fig. 9 my 1990s pocket Map of the Universe is a 360° panorama looking out from Earth's equator. Vertically it shows distance

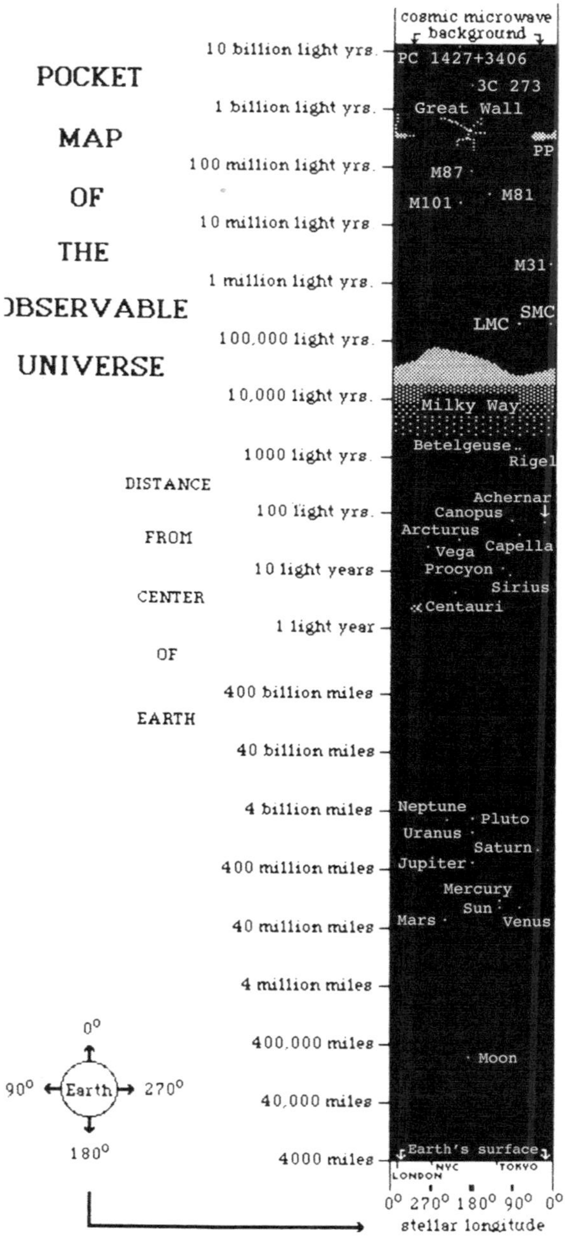

Fig. 9 Gott's Map of the Universe from the 1990s. From left to right, this represents a 360° panorama looking out from Earth's equator. The vertical logarithmic coordinate shows distance from the center of the Earth. Each equally spaced tick mark shows a distance 10 times further away from the center of the Earth. This is a conformal map that preserves shapes locally, like the Mercator map. It allows us to plot interesting objects all the way from the Moon, to distant stars and galaxies, and to the cosmic microwave background

from Earth on a logarithmic scale: each equally spaced large tick mark shows a distance 10 times further from Earth. Objects are shown at the time of Neil Armstrong's and Buzz Aldrin's Moon landing in 1969. At the bottom is a straight

Fig. 10 Carpet Map of the Universe in Peyton Hall, Princeton University approximately 50 feet long

line representing the Earth's surface. The Moon, Mars, Sun, Mercury, Venus, Jupiter, Saturn, Uranus, Neptune and Pluto are shown (Pluto was still classed as a planet then!). Nearby stars, like Alpha Centauri and Sirius, are shown. Beyond the Milky Way, M31 (the Andromeda Galaxy) appears—the other large galaxy in our Local Group. The galaxy M87 has now been found to harbor a 3 billion-solar-mass black hole, the largest discovered so far. The [CfA2] Great Wall of galaxies is shown since it was known in the 1990s whereas the Sloan Great Wall does not yet appear, as it had not been discovered yet. At the top is the line marking the Cosmic Microwave Background radiation—the most distant thing we can detect. A new beautiful, up-to-date, 40-in.-long, fold-out color version of the Map of the Universe appears in *Sizing up the Universe* that I co-authored with Bob Vanderbei. We have also made a 50-foot-long carpet version of the Map of the Universe in the Astrophysics Department at Princeton (Fig. 10). Every step you take down that hallway takes you another factor of 10 further away from the Earth.

Finally, for fun, a new version of the Map of the Universe, never seen before, based on the 12th root map of the complex plane is shown in Fig. 11. This is also a conformal map that preserves shapes locally. It is a 30°-wide fan (1/12th of 360°). Angles as seen from Earth are thus compressed by a factor of 12 and the relative radial distances from Earth are replaced by the 12th root of those distances. At the bottom, a circular arc shows the surface of the Earth. At the top is the cosmic microwave background radiation. The Subaru galaxy (the most distant galaxy

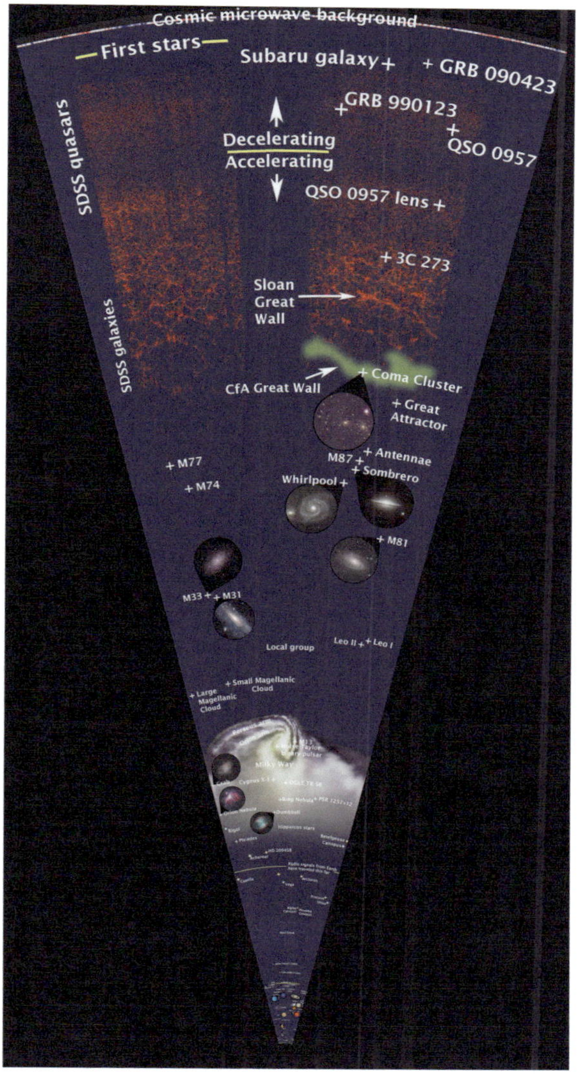

Fig. 11 Fan map of the universe. This 30° wide fan with vertex at the center of the Earth, shows a 360° panorama looking out from Earth's equator with angles compressed by a factor of 12. The radial coordinate measures the 12th root of the actual distance. This is also a conformal map that preserves local shapes. At the bottom of the map is the surface of the Earth, and at the top is the cosmic microwave background

known) is shown along with the gamma ray burster GRB 090423 the most distant individual object known (probably a supernova). 3C273 is the first quasar to have its distance measured. The Sloan Great Wall, and the CfA2 Great Wall can be seen. The galaxies M81 and M31 are shown. Spanning the slice lower down one can see the spiral arms of our own Milky Way galaxy. Just as the rectangular logarithmic map of the universe in Fig. 9 can be seen as a cousin of the rectangular Mercator map of the Earth, this map can be seen as a cousin of the fan-shaped Lambert conformal conic projection of the Earth.

We hope that the Map of the Universe can serve, as Mercator's map did, to chronicle the great discoveries of the era. When I started to study astronomy at the age of 8 in 1955, there were no artificial satellites orbiting the Earth, no Kuiper Belt objects (other than Pluto), no black holes, no great walls of galaxies and no cosmic microwave background radiation. Many of the salient features of the present map have been discovered in one astronomical generation. (Many of the maps in this chapter can be found in our book *Sizing Up the Universe* along with many other maps and scaled pictures showing sizes of things in the universe.)

5 The Next 25 Years: Turning Maps into Apps

What course is astronomical visualization likely to take in the next 25 years? Maps of the Earth are already tending toward a photographic mode of presentation. Atlases used to have finely drawn large scale maps, with delicate coloring to illustrate altitude. Today we tend to see maps with the appearance of photography from space. This trend will be even truer in astronomy. Photographic maps are simply easier to produce and show the features most apparent to spacecraft. We can make altitude maps of Mars, but we more commonly see ones that reproduce the topographic features and landscape colors as seen from space. For Jupiter, where we are observing the top of the atmosphere, only a photographic map makes sense. We see this trend in Google maps, with a photographic Mercator map projection of the entire Earth where you can zoom in until you can see your own house.

Browsers will, I think, ultimately supplant atlases. I love atlases. The feel of a finely made atlas, showing beautiful large scale maps as you turn the pages, has a special magic. But astronomical atlases are much less frequently seen than atlases of the Earth which have larger audiences. Recently a browser has been produced for inspecting the entire moon. It uses images taken by NASA's Lunar Reconnaissance Orbiter Camera, and the images are shown with low solar illumination to make the shadows of the craters and mountains stand out. It starts with the visible face of the moon as seen from Earth (http://wms.lroc.asu.edu/lroc_browse/view/wac_nearside).

This is extremely useful for amateur astronomers. You see the whole face of the moon on the computer screen. Then you can pick any particular point of interest and zoom in for a close-up. This effectively produces a map of the visible face of the moon about 16 feet across with a resolution of 145 m/pixel. It would be costly to make an atlas showing the entire visible face of the moon at this scale, and it would be hard to convey the relation of each small feature to the others around it—you would always find craters split between pages. The browser is more efficient and seamless. You don't waste time looking at blown up regions you are not interested in; you focus in on the object of your interest right away. For astronomy in particular, it would seem that atlases will have a hard time competing with browsers.

With modern technology we might ask again the cartographer's question of how best to view the curved surface of the Earth (or another planet). Any map of the Earth on a flat piece of paper involves some distortion. One approach is to make a spherical television that is capable of showing the surface of any spherical astronomical body or the Earth or the sky. The Magic Planet is one such television today. Our department has one with a diameter of 18 inches. You can show any planet or moon, and walk around it to view it as a globe. Such digital spheres will replace globes in libraries, museums, and universities. They also allow you to show movies of planets. You can show a movie of the weather on Jupiter, for example, or follow the course climate change on the Earth, or track continental drift. On the iPad or iPhone you can get pictures of spheres of planets and moons that you can freely rotate with your cursor. Google Earth adopts such a viewpoint. You see a globe of the Earth, which you can rotate and then zoom in for a closeup of any point you want.

How can one improve on a globe of a planet? When you look at such a globe you do not get an undistorted view; things near the edge of the visible hemisphere are foreshortened. Areas near the edge are diminished, and great circles do not appear as straight lines. Is there a better way to observe the Earth or other spherical bodies?

Yes! You can get an undistorted view if you observe the spherical surface from inside—from the center of the sphere. There is a map like this of the Earth in the Mapparium at the Christian Science Church in Boston. It is a spherical map of the Earth, 30 feet in diameter, with its map back lit in stained glass on the interior of the sphere. A glass bridge crosses the diameter of the sphere allowing you to stand in the middle. Look straight up and you will see the North Pole; look straight down and you will see Antarctica. The equator appears around the perimeter at eye-level. The map is mirror reversed, so that it looks right as seen from the inside. Look up at the continental United States, and you will see Florida and Maine on the right and California on the left. Your view preserves both local shapes and areas. Countries that take up twice the area on the globe take up twice the area as seen by your eye. Great circles appear as straight lines, and local shapes are perfect because you are looking at each small region straight on from the center. Such a view of Earth can be produced as a virtual view. With the right set of digital goggles and motion sensors you can create a virtual map that looks just like the view inside the Mapparium. Look up and you will see the North Pole, look straight down and you will see the South Pole, scan around the horizon and you will see the equator. The human eye has a peak resolution of about 1 min of arc, corresponding to 1 nautical mile on a globe of the Earth. You can get an undistorted view of the entire Earth with a resolution of 1 nautical mile. Goggles available today have a 35° field of view with 852×480 resolution, for a resolution of about 2.5 min of arc. Resolutions of 1 min of arc should be available soon. Google is said to be planning to equip glasses with motion sensors; we may soon be able to look up in the night sky and see the constellations labeled.

Apps are available today for viewing the celestial sphere using the iPad or iPhone. The map projection required on the iPad, and iPhone is the gnomonic projection, and you must view the map from the right distance (reading distance) to get the undistorted view. (Recall my discussion of the Gnomonic projection at the beginning of this article.) Hold your iPad over your head and you will see the zenith. At a reading distance of 14 in., the iPad gives a resolution of 1.9 min of arc and a 32° field of vision. In the future, resolution should increase to take advantage of the 1 min of arc resolution of the human eye. Such apps made to view spherical planets could also include a feature that allows you to zoom into view a close-up of any region of interest.

To view the three-dimensional (3D) arrangement of orbits in the Solar System, or the positions of stars in the solar neighborhood, or the 3D arrangement of galaxies in space, 3D viewers are called for. Today the Nintendo 3D game hand-held unit has a screen showing 3D images you can see without glasses. Bob Vanderbei has uploaded astronomical images, such as those appearing as stereo pairs in our book *Sizing up the Universe* (Gott and Vanderbei 2010) on this viewer and they are spectacular. Toshiba now has a laptop with a similar 3D screen. In the next 10–25 years, I think 3D viewers will be commonplace on computers of all types and on televisions at home. Ultimately holographic screens will produce perfect 3D images. Then one will find apps that will allow you to follow your imagination as you fly through the universe, stopping to admire a planet from a distance or descending for a closeup of its surface.

References

Alpher, R. A., & Herman, R. (1948). Evolution of the universe. *Nature, 162*, 774.

Asimov, I. (1976). *Isaac Asimov's biographical encyclopedia of science and technology*. New York: Avon Books.

Carter, B. (1966). Complete analytic extension of the symmetry axis of Kerr's solution of Einstein's equations. *Physical Review, 141*, 1242.

Carter, B. (1968). Global structure of the Kerr family of gravitational fields. *Physical Review, 174*, 1559.

Copernican Derivations. University of Nebraska-Lincoln Astronomy Education Group. http://astro.unl.edu/naap/ssm/ssm_advanced.html

Friedmann, A. Z. (1922). Uber die Krümmung des Raumes (On space curvature). *Zeitschrift für Phys, 10*, 377.

Friedmann, A. Z. (1924). Über die Möglichkeit einer Welt mit konstanter negativer Krümmung des Raumes (On the possibility of a world with constant negative curvature of space). *Zeitschrift für Phys., 21*, 326.

Gallant, Roy. (1956). *Exploring Mars*. Garden City, NY: Garden City Books.

Gamow, G. (1948a). The origin of elements and the separation of galaxies. *Physical Review, 74*, 505.

Gamow, G. (1948b). The evolution of the universe. *Nature, 162*, 680.

Geller, M. J., & Huchra, J. P. (1989). Mapping the universe. *Science, 246*, 897–903.

Gilbert, E. N. (1974). Distortion in maps. *Society for Industrial and Applied Mathematics Review, 16*(1), 47–62.

Goldberg, D., & Gott, J. R. (2007). Flexion and skewness in map projections of the Earth. *Cartographica, 42*, 297–318.

Gott, J. R. (2001). *Time Travel in Einstein's Universe*. Boston: Houghton Mifflin.

Gott, J. R., Choi, Y.-Y., Park, C., & Kim, J. (2009). Three-dimensional genus topology of luminous red galaxies. *Astrophysical Journal Letters, 695*, L45–L48.
Gott, J. R., Juric, M., Schlegel, D., Hoyle, F., Vogeley, M., Tegmark, M., et al. (2005). A map of the universe. *Astrophysical Journal, 624*, 463–484.
Gott, J. R., Melott, A., & Dickinson, M. (1986). The spongelike topology of large scale structure. *Astrophysical Journal, 306*, 341.
Gott, J. R., Mugnolo, C., & Colley, W. N. (2007). Map projections minimizing distance errors. *Cartographica, 42*, 219–234.
Gott, J. R., & Vanderbei, R. J. (2010). *Sizing up the universe*. Washington, DC: National Geographic Society.
Hawking, S. W., & Ellis, G. F. R. (1975). *The large-scale structure of space-time, Cambridge monographs on mathematical physics*. Cambridge: Cambridge University Press.
Kim, R. (1993). The life & times of E. M. Antoniadi (1870–1944), Part I: An astronomer in the making. *Journal of the British Astronomical Association, 103*, 4.
Kruskal, M. (1960). Maximal extension of Schwarzschild metric. *Physical Review, 119*, 1743–1745.
Laskowski, P. (1997a). Part 1: Distortion-spectrum fundamentals: A new tool for analyzing and visualizing map distortions. *Cartographica, 34*(3), 3–18.
Laskowski, P. (1997b). Part 2: Distortion-spectrum applications: Projections cross-breeding. *Cartographica, 34*(3), 19–39.
Lemonick, M. (2009). *The Georgian Star: How William and Caroline Herschel revolutionized our understanding of the cosmos*. New York: Norton.
Minkowski, H. (1908). Raum und Zeit, 80. Versammlung Deutscher Naturforscher (Koln, 1908). *Physikalische Zeitschrift, 10*, 75–88.
National Geographic Staff. (1969). "The earth's moon," printed map included with *National Geographic* magazine (February).
Penrose, R. (1964). Conformal treatment of infinity. In C. M. de Witt (Ed.), *Relativity, Groups and Topology, 1963 (Les Houches Lectures) Relativite, Groupes et Topologie*. New York: Gordon and Breach.
Penzias, A., & Wilson, R. W. (1965). A measurement of excess antenna temperature at 4080 Mc/s. *Astrophysical Journal, 142*, 419.
Peters, A. B. (1975). Wie man unsere Weltcarten der Erde ahnlicher machen kann. *Kartographische Nachrichten, 25*(5), 173–183.
Peters, A. B. (1978). Uber Weltkartenverzerrungen und Weltkartenmittelpunkte. *Kartographische Nachrichten, 28*(3), 106–113.
Peters, A. B. (1984). *Distance related maps. The American Cartographer, 11*(2), 119–131.
Saunders, R. S., et al. (1992). Magellan mission summary. *Journal of Geophysical Research, 97* (E8), 13067–13090.
Schiaparelli, A. (1888). Map of Mars. In: *Meyers Konversations-Lexikon,* 4th ed.
Schwarzschild, K. (1916). *Sitzungsberichte der Deutschen Akademie der Wissenschaften zu Berlin* (p. 189). Physik und Technik: Klasse für Mathematik.
Shapley, H. (1918). Globular clusters and the structure of the galactic system. *Publications of the Astronomical Society of the Pacific, 30*(173), 42.
Society for the Diffusion of Useful Knowledge (Great Britain). (1831). *Six maps of the stars*. London: Baldwin and Cradock.
Spergel, D. N., Verde, L., Peiris, H. V., Komatsu, E., Nolta, M. R., Bennett, C. L., et al. (2003). First-tear Wilkinson microwave anisotropy probe (WMAP) observations: Determination of cosmological parameters. *The Astrophysical Journal Supplement Series, 148*, 175–194.
Tissot, N. A. (1881). *Mémoire sur la répresentation des surfaces et les projections des cartes géographiques*. Paris: Gauthier Villars.
Tulley, R. B., & Fisher, J. R. (1987). *Nearby galaxies atlas*. Cambridge: Cambridge University Press.
York, D. G., et al. (2000). The sloan digital sky survey: Technical summary. *Astronomical Journal, 120*(3), 1579.

Mapping the Human Body: A GIS Perspective

Jane L. Garb

Abstract This chapter describes a novel application and interpretation of geography and mapping, applying the concepts and technology of Geographic Information Systems (GIS) at the scale of the human body. We also explore the use of spatial statistics as an adjunct to GIS to gain further understanding of the patterns found in maps. Despite the technical computational challenges, there are several reasons why a spatial or "geographic" perspective in this context is particularly useful. The concept of spatial contiguity is central to our understanding of disease. Different body systems, like the gastrointestinal, nervous or cardiovascular system, have a spatial organization. Disease in one region often spreads outward to neighboring regions, or along networks like the lymphatic system or vascular system. The spatial location of disease affects the ability to detect and treat it. Exploring and analyzing the particular effect of location on clinical outcomes thus becomes a crucial tool in understanding the disease process and improving diagnosis and treatment. The examples given and issues raised will hopefully stimulate further inquiry, and generate additional solutions to the challenge of representing and analyzing the human anatomy from a spatial perspective.

Keywords Anatomy · Human body · Spatial statistics · Mapping · GIS · Geographic information systems

1 Introduction

This chapter will explore a novel application and interpretation of geography and mapping, applying the concepts and technology of Geographic Information Systems (GIS) at the scale of the human body. As part of this, the use of spatial statistics as an adjunct to GIS will be explored to gain further understanding of the patterns found in maps.

J.L. Garb (✉)
Baystate Health, Springfield, MA, USA
e-mail: jane.garb@baystatehealth.org

S.D. Brunn and M. Dodge (eds.), *Mapping Across Academia*,
DOI 10.1007/978-94-024-1011-2_6

Despite the technical challenges, there are several reasons why looking at the human body and the anatomy from a spatial or "geographic" perspective is particularly useful. Spatial contiguity is an important concept in our understanding of disease. Different body systems, such as the gastrointestinal, nervous or cardiovascular system, have a spatial organization. Disease in one region often spreads outward to neighboring regions, or along networks like the lymphatic system or vascular system. The spatial location of disease such as a cancerous lesion or an aortic aneurysm affects the ability to detect and treat it. For example, in treating a diseased organ, care must be taken not to affect nearby organs. Disease in certain locations may make it inaccessible to surgical or radiation treatment.

Exploring and analyzing the particular effect of location on clinical outcomes thus becomes a useful tool in understanding the disease process and improving diagnosis and treatment. The perspective of this chapter is epidemiologic rather than biological, anatomical or clinical—that is, population-level rather than individual-level. Epidemiologists are used to looking at patterns of disease. Using GIS to summarize information across individuals and examine disease patterns at any scale—across the globe or within the human body—therefore makes perfect sense to an epidemiologist. It is a tool ideally suited to address questions about disease origin, progression and treatment.

With this in mind, the following chapter will describe the use of GIS and spatial analytic techniques to explore a "geographic" perspective of the human body. The examples given and issues raised will hopefully stimulate further exploration and solutions in this field.

2 Background

2.1 A Brief Description of GIS

A Geographical Information System (GIS) comprises the hardware and software tools for managing, summarizing and analyzing spatial data, i.e. information about a geographic location. The GIS stores data electronically in "layers" associated with one another geographically. For example, states, cities, and mountain ranges in the United States might be three layers stored in a GIS. These are associated with each other geographically because they all encompass the same geographic extent.

One prime contribution of a GIS is that it allows the overlay of spatial data. This overlay is not merely visual. It allows linkage of the layers and examination of the interrelationship between them, including spatial analysis. Overlaying the different layers of information and viewing them together can reveal patterns and relationships that might not otherwise be seen.

Each spatial layer in the GIS is associated with a series of attributes. This information is also stored in the GIS. Examples of attributes for US states or cities would be name and population size. Attributes for mountain ranges might be name, maximum elevation and area.

There are two types of geographic data used in the GIS: raster and vector. Raster data represent locations with continuous attributes and no discrete boundaries. Rather than being arranged as a table, data are organized in a grid. The cells of the grid are akin to pixels in a photograph (or voxels in 3D imagery). Raster datasets can only store one attribute at a time in each cell. For example, elevation levels across the United States would be represented most effectively as raster data, and elevation value stored in each raster cell.

Figure 1, a CT scan image of the liver, is an example of raster representation of the human body. This is a 2D representation in pixels. The value of each pixel represented by the intensity of the grayscale color. In a CT scan, this value represents the amount of contrast medium present in the tissue. The darker pixels have taken up more of the contrast medium. In 3D imagery, voxels represent a point (actually a tiny area) in three-dimensional space. In a diffusion MRI, for example, the value of each voxel would represent the diffusion density of water through the tissue.

Vector data use points, lines, and polygons (areas) to represent real-world (physical) features with discrete boundaries. Feature attributes of vector data are stored as tables in the GIS. US cities and states, for example, are most commonly represented as vector data. In contrast to raster data storage, multiple attributes can be stored in a vector data table. Vector data can also be used to represent the human body. An example is given in Fig. 2, depicting the human liver.

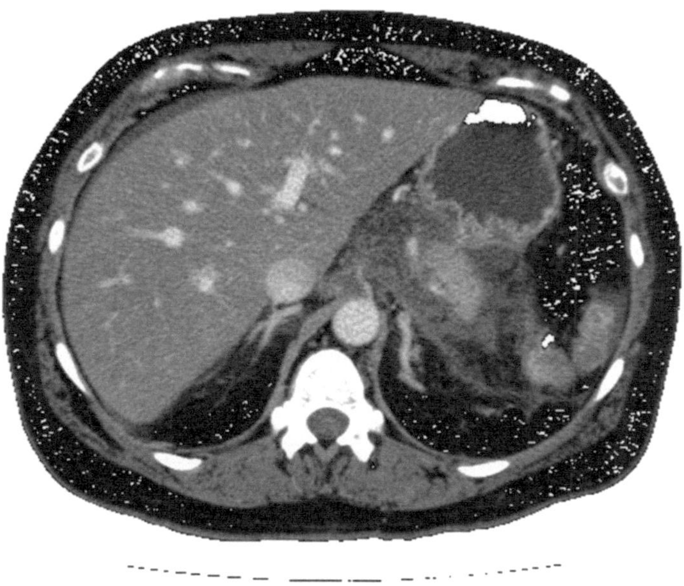

Fig. 1 CT scan image of the liver

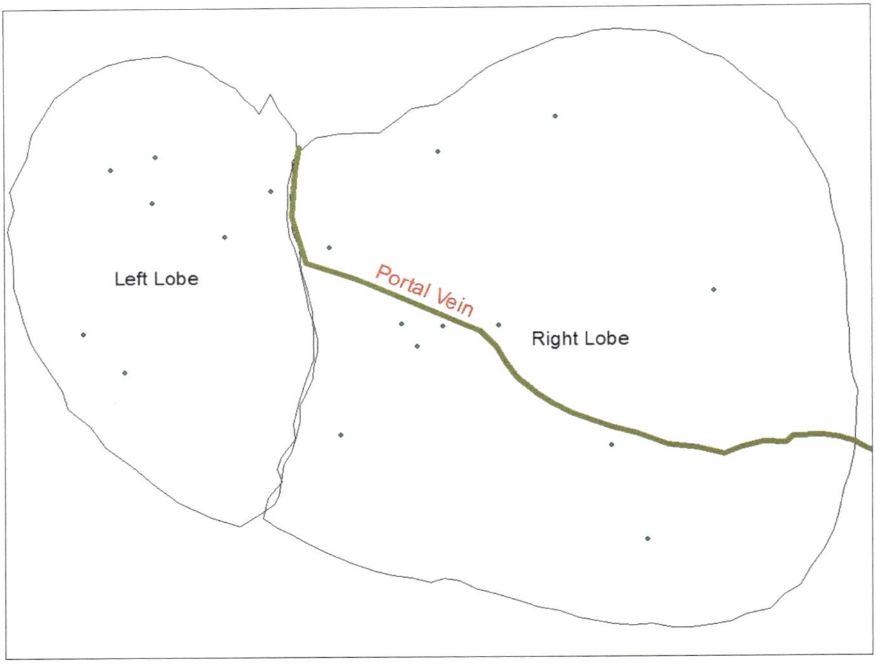

Fig. 2 Vector representation of the liver

A GIS has three main functions: visualization, exploratory analysis, and mathematical modeling (Bailey and Gatrell 1995). Visualization of spatial data is accomplished through mapping. Exploratory analysis investigates visual patterns seen in maps, using spatial statistics to detect clustering or other spatial patterns. This analysis is used to generate hypotheses about the spatial relationship of various attributes, which can be tested with mathematical modeling techniques such as spatial regression. Examples of each of these three GIS functions will be given later in the chapter.

It is important to make a distinction between *spatial analysis* and *spatial statistics*, both of which are applied to geographic data. *Spatial analysis* refers to GIS operations. These can be performed on separate spatial data layers (e.g. calculating distance, drawing a buffer around a geographic entity for vector data; or calculating surface characteristics of raster data such as slope, aspect or hill shade). Operations can also combine separate spatial data layers with one another. For example *overlay* operations such as the union of two layers, subtraction of one layer from another, or the transfer of the attributes of one layer to another one with a different geographic configuration (called areal interpolation) are all GIS operations.

Spatial statistics (or spatial statistical analysis) summarizes spatial data, tests hypotheses about the geographic distribution of feature attributes, and develops mathematical models to predict some geographically distributed outcome such as disease density. The use of spatial analytic techniques differs from traditional statistics in two related ways. First, it avoids the assumption of independence of observations. Second, it accounts for the spatial proximity of the observations to one another.

Two examples of spatial statistics are spatial regression and cluster analysis. Spatial regression is a form of mathematical predictive modeling which assesses the contribution of various factors to a continuous polygon attribute. It differs from ordinary least squares regression, the more commonly used regression method, in that it accounts for and models interrelationships of neighboring polygons known as spatial autocorrelation. This is the tendency of things near each other in space to be more alike than things far apart, known as Tobler's (1970) First Law of Geography. In spatial regression, autocorrelation is treated as a nuisance factor which must be accounted for before examining an attribute of interest. For example, in examining social or economic factors in the geographic distribution of neighborhood breast cancer rates in a city, the spatial contiguity of the neighborhoods must be corrected for with the spatial regression. On the other hand, if the autocorrelation itself is the factor of interest, i.e., the aim of the analysis is to locate areas of spatial autocorrelation—points or polygons that are more similar than would be expected by chance. These collections of spatially contiguous areas which are similar in the outcome of interest are called "clusters." Identifying these clusters and examining the reasons for the clustering is the focus of a variety of cluster analytic techniques such as the *Spatial Scan Statistic* (Kulldorff 2010) or *Moran's I* (Moran 1948).

Until recently, neither of these areas of analysis—spatial analysis nor spatial statistics—has been applied to investigating the human body. In fact, use of GIS in any manner at the anatomic scale is rare. The following discuss a few examples. Some of these are developed on animals, but could apply to humans as well.

2.2 *Previous Examples of Mapping the Anatomy*

2.2.1 Raster Mapping

There are several ambitious initiatives, primarily in neuroinformatics, which have established digital repositories for image (raster) data. As part of this initiative, publicly accessible digital atlases have been compiled. Researchers, clinicians and educators can contribute to these and utilize them. These atlases provide sophisticated 3D visualization and querying capabilities. Examples are the *Human Brain Project* (Brinkley and Rosse 2002) and the *SMART Atlas* (Zaslavsky et al. 2004; Zaslavsky 2006; Bug et al. 2007) which pertain to the brain. *EMAP* (Baldock et al. 2003) provides a spatial conceptualization of the mouse embryo using 3D raster data in voxels. It develops ontology for mapping data in discrete regions or tissues and provides an atlas of different stages of development for visualization and query.

Image registration (i.e., alignment of all points) is a major issue when overlaying different images on one another for comparison or summarization. It is a particular challenge when images are from different individuals, rather than different "slices" of the same individual as in CT or MRI scans. One is not dealing with a common "geography" in human anatomy. Each individual's "geography" or base map is different. Each has a different spatial configuration and orientation of irregular anatomic features. In other words, the anatomy of each individual, while generally similar, is unique. Not only is the shape and size of each organ different in different people, but the spatial relationship of organs and other features to one another can vary between individuals, sometimes a great deal. Thus, no two images of the same organ or body region from different individuals are alike in shape or size (and therefore in pixel or voxel arrangement or number).

Registration has been addressed by the architects of these atlases and others in the field through a variety of transformation, interpolation and smoothing methods. In the *Human Brain Project*, one method used was "warping" or reshaping individual images to a single brain chosen as a template using a nonlinear transformation based on matching the value of voxels at different locations or proximity to recognizable anatomical landmarks (Brinkley and Rosse 2002). In another example, Smith et al. (2006) used non-linear smoothing to project images onto a common registration template in order to combine images of spatial diffusion patterns in the mouse brain. They derived centerlines of white matter tracts which consisted of the average location of the centers of tracts in all subjects visualized to derive a "group mean … skeleton."

All such methods provide only approximate solutions to the problem of registration. Some degree of misalignment and image distortion is always present to a greater or lesser degree, resulting in a given voxel or pixel not representing the same anatomic location in every subject. And the choice of registration or smoothing algorithm can drastically alter the results. This leads to errors and inconsistent results when voxel or pixel data from different individuals are summarized, for example in calculating the mean diffusion rate for individual voxels to create a summary map of diffusion patterns. These errors, while significant, are of less consequence in visualization than in statistical analysis.

Visualization and querying are the primary aims of most raster applications used in the context of mapping the human body. Statistical analysis of outcomes, when used, primarily consists of descriptively summarizing or comparing groups, rather than formal hypothesis testing or modeling. Furthermore, the contribution of distance or topology (spatial relationships) to clinical outcomes was not explored in any of the applications reviewed.

2.2.2 Vector Mapping

Other initiatives in mapping anatomy have relied on two-dimensional vector data. Uren and colleagues mapped patterns of lymphatic drainage for melanomas (Uren et al. 2003, 2006). The location of lesions was only approximate, based on a hand-drawn sketch and then randomly allocated to a point on a grid. No reference system was used to register the points.

In an ambitious mapping project called ANET, Roth and Kiani (1999) used a GIS to digitize microvascular networks as lines and link this to attribute data on vessel diameter, red blood cell (RBC) velocity and other vessel characteristics using GIS software. Digitizing allowed the establishment of topolology (spatial relationships between vessels), enabling investigators to perform spatial analysis functions such as distance measurement, calculation of vessel structural characteristics and flow rates. They then used traditional statistical analysis, rather than spatial statistical analysis, to examine the effect of radiation on RBC transit time and vessel structure. None of the other studies mentioned relied on spatial statistical analysis to examine outcomes based on spatial proximity.

Use of vector data for mapping the anatomy has the advantage of being simpler, avoiding the use of customized computer programming. It has the added benefit of access to a variety of spatial statistical methods (some of which were described above) to analyze the impact of distance and spatial proximity on various outcomes which have been developed for the more conventional uses of GIS applied to the surface of the earth and are not available for raster data.

Vector mapping of the anatomy requires a different approach to the problem of image registration than for raster data. No suggestions for the construction of a common vector geography for the human (or mammal) have as yet been proposed. A possible solution is the identification of common anatomical *reference points*. The best way to explain how this addresses the problem is to provide an example.

3 Mapping the Colon

The colorectal anatomy presents a unique opportunity to apply GIS to the human anatomy for analysis. Unlike other organs, the colon can be represented as a simple cylinder (think of the colon as stretched out end-to-end). As described in Garb and colleagues (Ganai et al. 2006; Garb et al. 2007), clock-face (radial degrees) coordinates were used as x and y-coordinates and distance from the anal verge (anus) was used as the z-coordinate (height) to form a cylinder that accurately represented the colon. Reference points based on *side* (left/right), *spatial orientation* (anterior/posterior, medial/lateral) and the location of recognized anatomic *features* such as the splenic flexure or peritoneal reflection were plotted on the cylinder. These reference points were used to divide the colon into anatomic regions that would have meaning to the clinician (Fig. 3).

The investigators then took advantage of an existing technology, transanal endoscopy (TEM), which enables the precise location of rectal lesions. In this procedure, lesion locations are recorded according to the same cylindrical coordinates used for the generic colon base map described above. Data on lesion locations in patients undergoing TEM was overlaid onto the map of anatomic regions in the GIS, as shown in Fig. 4.

Employing a simple geometric conversion, the cylindrical locations of both anatomic regions and lesions of the colon were transformed into a planar

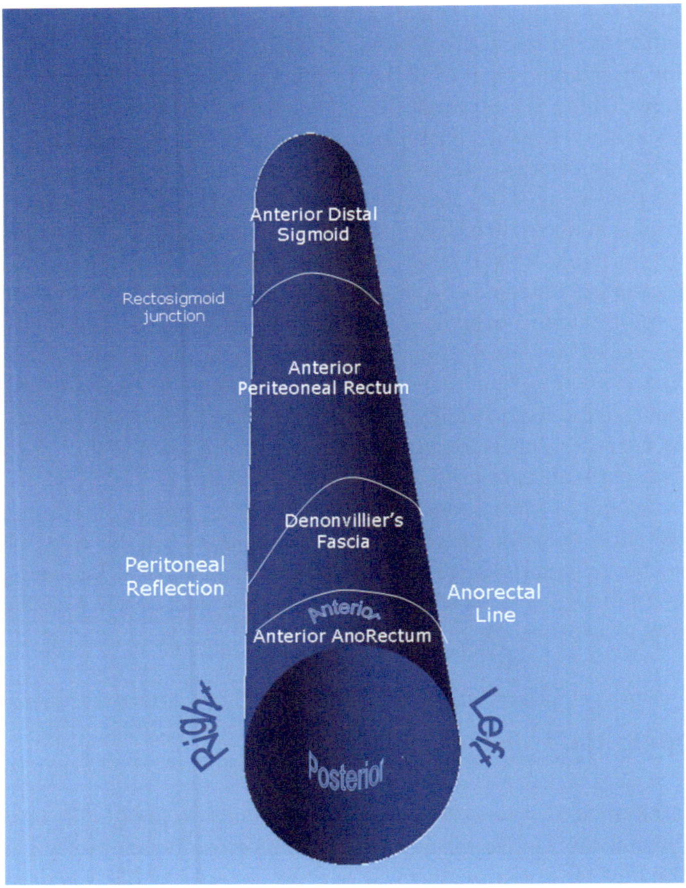

Fig. 3 Anatomic regions of the colon

(two-dimensional) map. Similar map projection techniques are used to transform spherical coordinates of longitude and latitude into a two-dimensional plane. In this process, the three-dimensional anatomic regions became two-dimensional polygons and the lesions became points on a schematic map of the colon, shown in Fig. 5.

Once the 2D map of the colon regions and lesions was derived, spatial analytic techniques could be applied to test a variety of clinical hypotheses. Cluster analysis was conducted to determine if lesions (points) tended to occur more frequently together in certain regions of the colon. Next, lesion attributes were summarized or aggregated across regions and mapped to display the spatial distribution of various outcomes (e.g. mean operative time, operative complication rates or mortality rates) by anatomic region. Figure 6 is a choropleth map (i.e., colors represent values) generated from this process. This figure depicts the rate of conversion from one operative procedure to another during surgery due to an inability to perform the

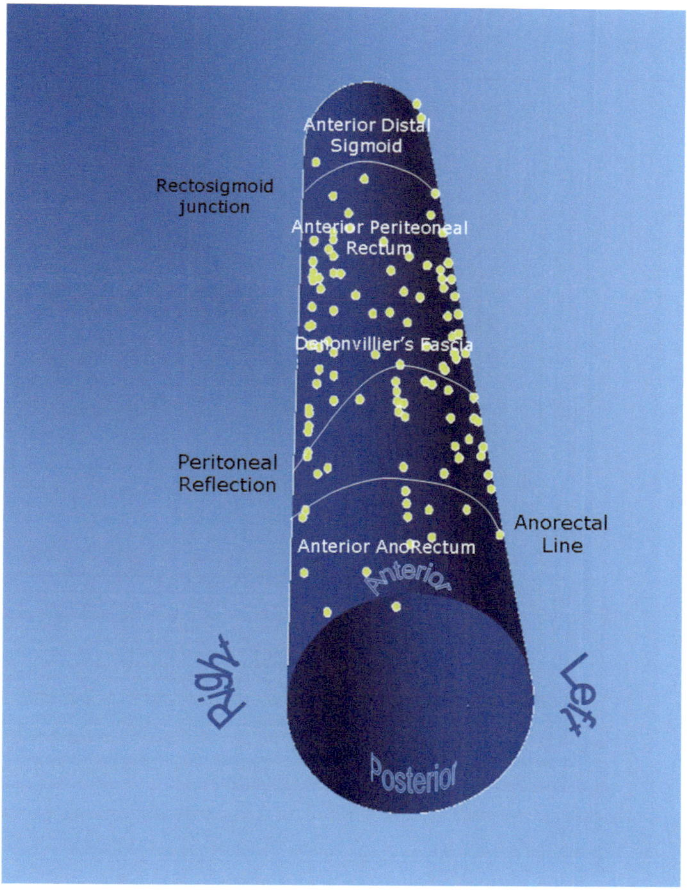

Fig. 4 Colon regions overlaid with lesions as points

originally-planned procedure. Conversions were more likely in the uppermost regions of the colon, furthest from the anal verge and therefore less surgically accessible.

Rendering the anatomy in this way circumvents many of the issues involved in mapping other organs or anatomic features. Four primary issues are (1) choice of a coordinate system (2) choice of a reference point (3) measuring distance and (4) regionalization.

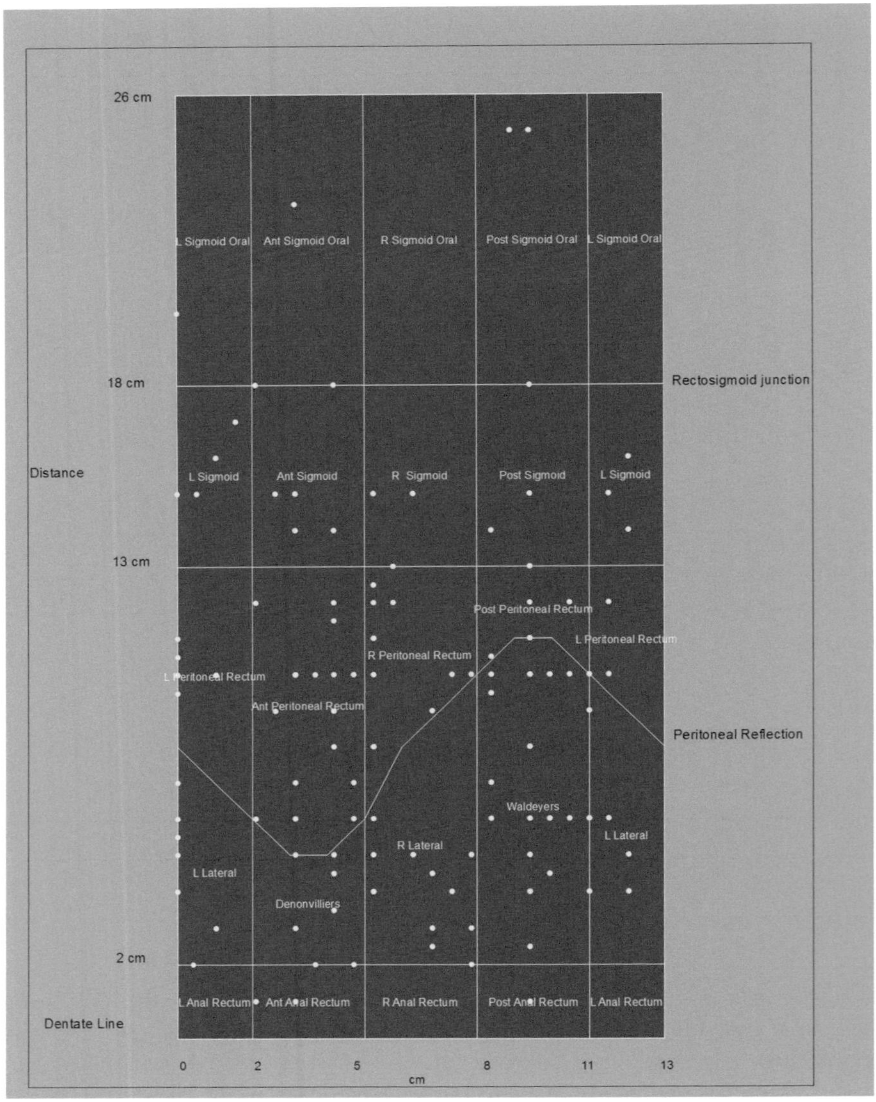

Fig. 5 2D colon regions and lesions

3.1 *Choice of a Coordinate System*

Representing the location of vector data on a two-dimensional plane requires the choice of a coordinate system—a reference framework used to define the position of points in space (Wade and Sommer 2006). One coordinate system commonly

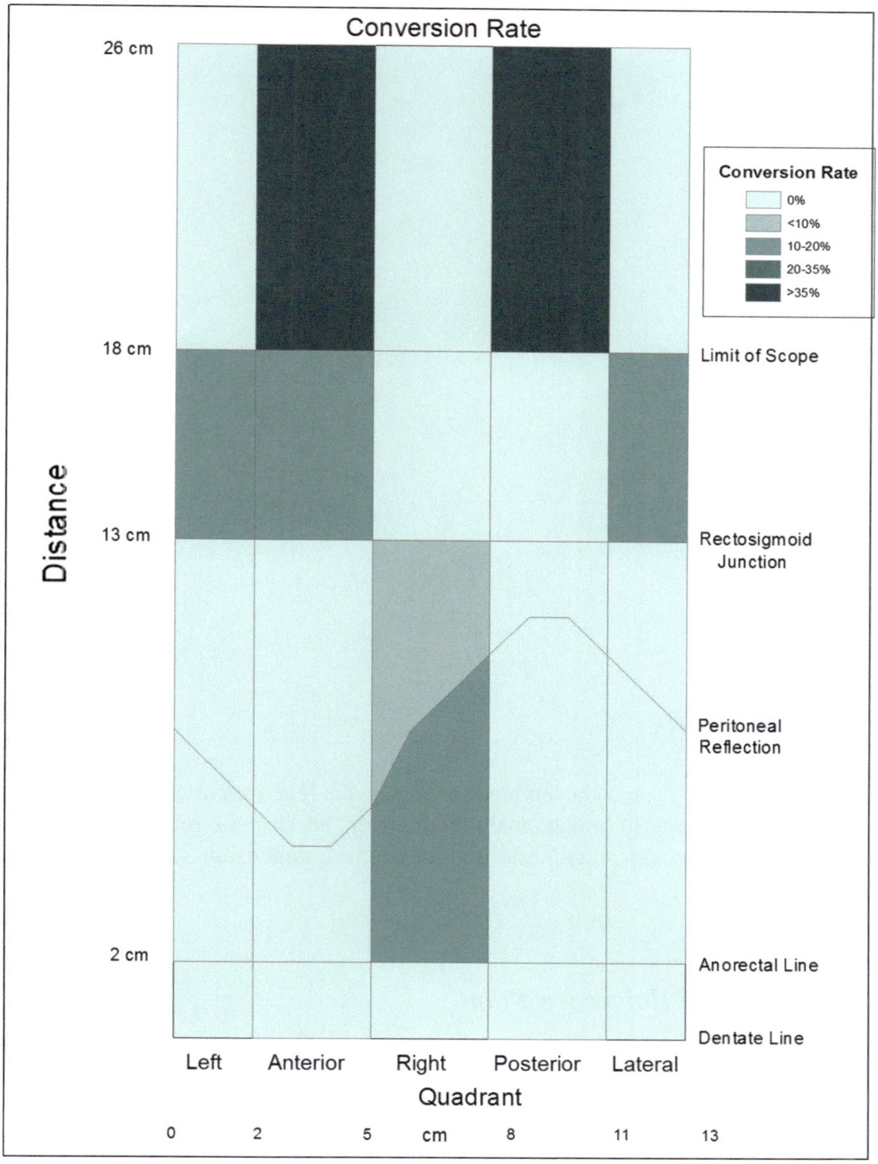

Fig. 6 Choropleth map showing operative conversion rates. *Darker shades* represent higher conversion rates

used to represent the surface of the earth is the geographic coordinate system (latitude/longitude). Another common system, the Cartesian coordinate system (Fig. 7), is based on distance from some reference point (0-point) in two

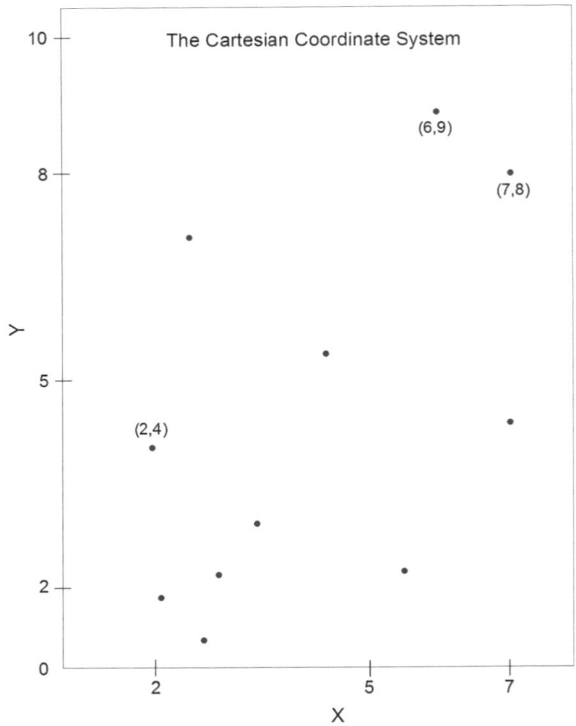

Fig. 7 Example of a Cartesian coordinate system

dimensions (x and y) on a chosen measurement scale (e.g. mm or cm) and would be better suited to represent human anatomic features, because the reference point and measurement scale can vary. Addition of a third dimension (z) also becomes possible.

3.2 Choice of Reference Point

Using a Cartesian coordinate system allows flexibility in the choice of reference point depending on the anatomical feature being mapped. In our example of the colon, choice of the anal verge as a reference point enabled calculation of the x, y *and* z coordinates. This also corresponded to the lower end of the colonoscope, along which distance was measured. However, other anatomic organs seldom lend themselves to such an easy choice of a reference point, because of their asymmetric shape. It would seem most logical to choose an established reference feature (or features), if available, when mapping a body organ, system or region. For example, the nipple could be chosen for the breast, as the center of the organ.

Besides allowing the placement of points or polygons in the body on a grid (x and y axes) with constant units, reference points can also serve as registration points, enabling the meaningful overlay of different images of some anatomic feature from different individuals, as previously discussed in the context of raster data. If we can derive a common and clinically meaningful spatial configuration of polygons for an anatomic feature or group of features, we can treat these polygons like any other geographic polygons in the "real world." We can apply a whole variety of established spatial analytic and spatial statistical techniques to explore and analyze these features in a GIS. This is based on the ability to *summarize* or combine point or polygon attribute information across subjects.

3.3 Measuring Distance

Distance plays a key role in many GIS and spatial statistics functions. Distance from a reference point is not only the basis for constructing a coordinate system for mapping but is the basis of many spatial analytic operations. For example, the GIS operation of buffering creates a new zone of a given diameter (measured in kilometers, miles, etc.) around a central point. Distance is also important in spatial statistics. In spatial interaction analysis, for example, the gravity model is based on the impact of distance on the flow (of people, goods, body fluids, etc.) from one place to another. In the colon example, distance from the anal verge was found to be a significant factor in operative time, as found by spatial regression, another spatial statistical technique.

If the unit of measurement of the coordinate system (e.g. centimeters, kilometers, etc.) used in a map is known, the GIS can calculate distance between any two entities or points. Alternatively, if distance cannot be calculated within the GIS, distance measurements can be obtained independently (outside the GIS) and read into the GIS as attribute information.

3.4 Regionalization

The concept of regionalization allows polygon mapping, either bypassing the need for registration points, or using them to define regions. If we can derive a series of regions, preferably bound by visible and clinically meaningful and recognizable anatomic features as in the colon example, we can analyze these regions in the GIS. Since these regions will be bounded by the same anatomic features in every individual, they will allow summarization of attributes across individuals. Attributes of points within the regions (e.g. presence of cancer, operative time to repair or size of a lesion) or areas within these regions (e.g. dimensions of metastatic tumors or areas of defect) can be summarized across individuals within regions and related to clinical outcomes in analysis. In our colon example, we computed the mean operative time to repair all lesions within each region. We also computed the mean conversion rate of each region.

Summarization within regions allows for application of all of the GIS functions available for vector data: visualization (mapping), exploration and modeling with spatial statistics and spatial analysis. In addition to solving the problem of multiple geographies among individuals and circumventing the problem of registration, regionalization enables the determination of spatial proximity (spatial neighbors) required for such spatial analytic techniques such as cluster analysis and spatial regression analysis.

4 Considerations in Mapping Other Organs

The appropriate choice of reference points for regionalization and distance measurements depends on both the structure and location of the particular organ in question and the clinical question at hand. For example, the Couinaud classification of liver anatomy (Standing et al. 2008; Fig. 8) divides the liver into eight segments, each with its own independent vasculature and biliary drainage. Each segment intersects the following anatomic features: portal vein, hepatic artery and bile duct, with peripheral outflow through the hepatic vein. The independent vasculature of these segments makes this classification system a good choice for regionalization of the liver. As a clinical question, suppose we are interested in efficacy and ease of use of Radiofrequency Ablation (RFA), a procedure which can be used to destroy

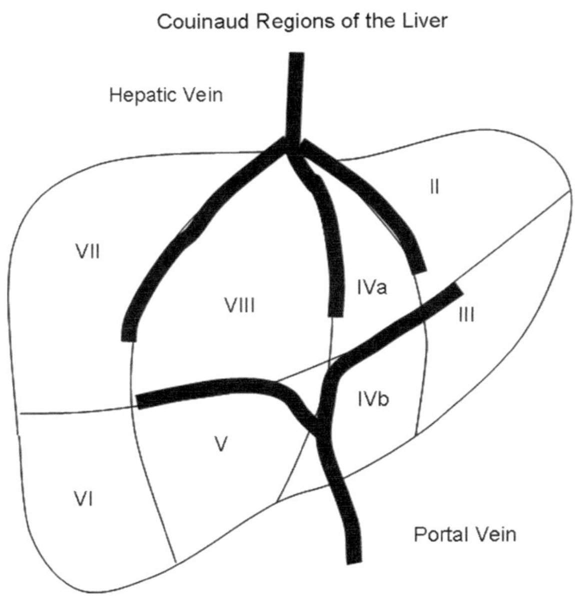

Fig. 8 Couinaud regions of the liver

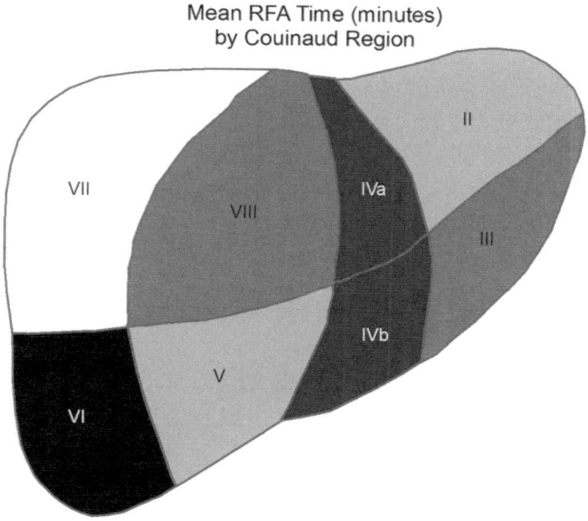

Fig. 9 Choropleth map showing mean RFA time by Couinaud region

liver lesions employing radio waves passing through a probe inserted into the liver. In this case we might choose a distance measure from the lesion center (centroid) to the hepatic artery bifurcation or to the nearest surface of the liver (anterior or posterior) where the probe could be inserted. Figure 9 gives a hypothetical example of how one might construct a choropleth map of RFA time according to Couinaud region. The distance measure we chose could be used in a spatial regression analysis (not shown) to determine if distance significantly affects RFA ablation time. On the other hand, in a study of chemo-therapeutic efficacy, more relevant distance measures might be proximity to the lymphatics and vasculature, as these would likely affect responsiveness to chemotherapy.

As another example of using features, side and orientation to create regional template maps, SEER codes (SEER 2004) provide a convenient and recognized classification of regions of the breast according to side, orientation (clock-face position) and presence of anatomic features (the nipple). Table 1 shows the SEER codes for the breast. Figure 10 shows a map of the breast based on the current regions (four quadrants plus the nipple and areolar areas). The four quadrants could be expanded to 12 regions based on clock face position (represented in the figure by light blue lines) plus the nipple and areolar regions. The nipple could be used as the reference point for measuring distance. The importance of lymph nodes in response to chemotherapy and spread of breast disease might call for addition of proximity to the nodes in an analysis of chemotherapeutic effectiveness.

Table 1 ICD quadrants of the breast

C50.0	Nipple
C50.1	Central portion of breast
C50.2	Upper-inner quadrant of breast (UIQ)
C50.3	Lower-inner quadrant of breast (LIQ)
C50.4	Upper-outer quadrant of breast (UOQ)
C50.5	Lower-outer quadrant of breast (LOQ)
C50.6	Axillary tail of breast
C50.8	Overlapping lesion of breast
C50.9	Breast, NOS (excludes skin of breast C44.5); multi-focal neoplasm in >1 quadrant

Source SEER (2004)

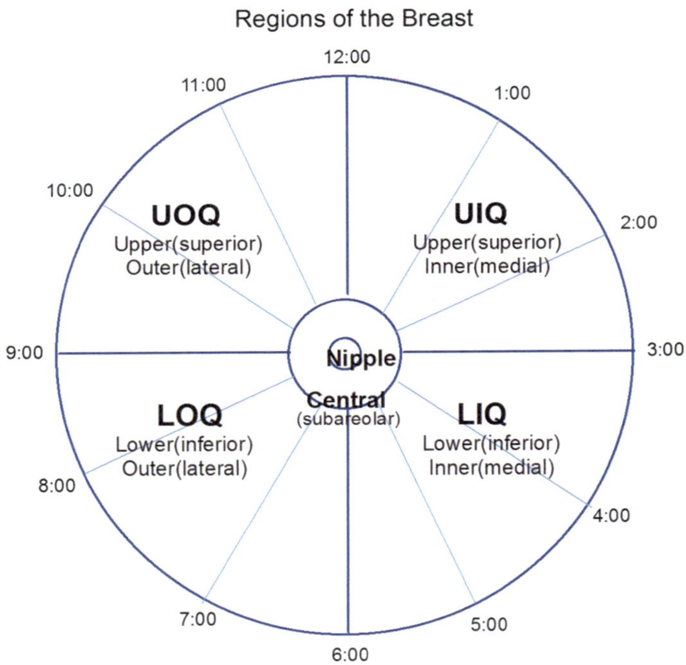

Fig. 10 Map of SEER regions of the breast

5 Conclusions

This chapter presents a framework for mapping and analyzing clinical data linked to spatial location within the human body. It is hoped that this will stimulate further mapping and analysis initiatives utilizing GIS and spatial statistical techniques in this context. There are limitless clinical questions that can be addressed with this methodology. Leveraging the collaboration of clinicians and GIS specialists, appropriate models for exploring the effect of human geography on disease and treatment can be established.

References

Bailey, T., & Gatrell, A. C. (1995). *Interactive spatial data analysis* (pp. 337–387). Lake Arbor: Crown Books.

Baldock, R. A., Bard, J. B., Burger, A., et al. (2003). EMAP and EMAGE: A framework for understanding spatially organized data. *Neuroinformatics, 1*, 309–325.

Brinkley, J. F., & Rosse, C. (2002). Imaging and the Human Brain Project: A review. *Methods of Information in Medicine, 41*(4), 245–260.

Bug, W., Wong, W. W., Gustafson, C., & Johnson, G. A., et al. (2007). Brain atlasing tool interoperation: NeuroTerrain-Smart Atlas synergistic visualization and analysis environment. In *Conference paper: Neural Engineering, 2007. CNE '07. 3rd international IEEE/EMBS conference,* Kohala Coast, HI, May 2–5 (pp. 280–283).

Ganai, S., Garb, J. L., Kanumuri, P., et al. (2006). Mapping the rectum: Spatial analysis of transanal endoscopic microsurgical outcomes using GIS technology. *Journal of Gastrointestinal Surgery, 10*, 22–31.

Garb, J. L., Ganai, S., Skinner, R., et al. (2007). Using GIS for spatial analysis of rectal lesions in the human body. *International Journal of Health Geographics, 6*, 1–14.

Kulldorff, M. (2010). *Software for the spatial and space-time scan statistics* (9.1st ed.). Calverton: Information Management Services Inc.

Moran, P. A. P. (1948). The interpretation of statistical maps. *Journal of the Royal Statistical Society B, 19*, 243–251.

Roth, N. M., & Kiani, M. F. (1999). A "geographic information systems" based technique for the study of microvascular networks. *Annals of Biomedical Engineering, 27*, 42–47.

SEER. (2004). SEER site specific coding guidelines: Breast. SEER Program Coding and Staging Manual 2004. Revision 1 edn., Appendix C, 470.

Smith, S. M., Jenkinson, M., Johansen-Berg, H., et al. (2006). Tract-based spatial statistics: Voxelwise analysis of multi-subject diffusion data. *Neuroimage, 31*, 1487–1505.

Standing, S., Borley, N. R., Collins, P., et al. (2008). *Gray's anatomy: The anatomical basis of clinical practice* (40th ed.). Edinburgh: Churchill Livingston.

Tobler, W. R. (1970). A computer movie simulating urban growth in the Detroit region. *Economic Geography, 46* (Article Type: research-article/Issue Title: Supplement: Proceedings. International Geographical Union. Commission on Quantitative Methods/Full publication date: Jun., 1970 / © 1970 Clark University), 234–240.

Uren, R. F., Howman-Giles, R., & Thompson, J. F. (2003). Patterns of lymphatic drainage from the skin in patients with melanoma. *Journal of Nuclear Medicine, 44*, 1–13.

Uren, R. F., Thompson, J. F., Howman-Giles, R., et al. (2006). The role of lymphoscintigraphy in the detection of lymph node drainage in melanoma. *Surgical Oncological Clinics of North America, 15*, 285–300.

Wade, T., & Sommer, S. (Eds.). (2006). *A to Z GIS: An illustrated dictionary of geographic informations systems. (Second ed.)*. Redlands, CA: ESRI Press.

Zaslavsky, I. (2006) *Smart atlas beta.* http://imhotep.ucsd.edu:7873/smart_atlas_annotation/Testing.htm

Zaslavsky, I., He, H., & Tran, J., et al. (2004). Integrating brain data spatially: Spatial data infrastructure and atlas environment for online federation and analysis of brain images. In *Biological data management workshop (BIDM 2004) in conjunction with 15th international workshop on database and expert systems applications (DEXA'04),* Zaragosa, Spain.

The Brain and Its Maps: An Illustrative History

Jordan P. Harp and Walter M. High

Abstract The human brain has been studied and mapped in a number of ways since antiquity. This chapter reviews investigative methods and the evolution of human brain maps from ancient times to the present. Special focus is given to the shift in modern history from purely structural mapping, through purely functional mapping, to neuropsychological maps displaying the relation between structure and function. The historical review that begins the chapter presents a number of key findings and investigators, major historical debates and detours in the field, and the more recently developed synergies between clinical treatment, neuropsychological measurement, and technological advancement. An introduction to current brain mapping techniques presents the mechanisms, strengths, and weaknesses of X-ray computed tomography (CT), single-photon emission computed tomography (SPECT), positron emission tomography (PET), electroencephalography (EEG), magnetoencephalography (MEG), transcranial magnetic stimulation (TMS) and transcranial electrical stimulation (TES), and the various forms of magnetic resonance imaging (MRI). A final section reviews promising current approaches and initiatives in human brain mapping and the value they may add to neuropsychological investigation and the refinement of human brain maps. These relatively new approaches include more advanced statistical analyses, better data management and informatics specialized to handle very large data sets, functional network analysis, nonlinear and chaos/complexity analysis, development of whole-brain connectome maps, and new advances in the imaging of the microstructure of brain cells. This chapter is organized in such a way as to highlight the shifts in preferred metaphors guiding different eras of human brain research–from inert lump of flesh, to a container for mental and spiritual functions, to mechanistic pump, to an information processing computer, to a hyper-connected and mutable network.

J.P. Harp (✉) · W.M. High
University of Kentucky College of Medicine, Lexington, KY, USA
e-mail: jordanharp@uky.edu

W.M. High
e-mail: walter.high@uky.edu

S.D. Brunn and M. Dodge (eds.), *Mapping Across Academia*,
DOI 10.1007/978-94-024-1011-2_7

Keywords Human brain mapping · History · Neuropsychology · Neuroimaging · Brain networks

Variously appearing in history as the seat of reason, an obscure lump of flesh, or a sophisticated cellular network for processing information and generating behavior, the human brain has been discussed and mapped since antiquity. The present chapter recounts and illustrates the evolution of the map in the study of the brain. The first section provides a short discussion of the scope of this chapter, introducing the reader to the defining features and various forms of map generally used by brain scientists. The second section is a historical review of maps of the brain as physical artifacts evidencing the evolution of brain science. This review specifically focuses on the role of the map in representing information about the structure and functions of the brain, as well as the interplay of the two. The third section describes current neuroimaging and neuro-stimulation technology and its role in brain mapping. The final section provides a speculative glimpse into the future of the field.

1 Maps and Maps

Throughout history, humans have attempted to amass and record knowledge of the world and of the human body. The means of transmission of recorded knowledge have included oral and written narratives, tabulated data, and representative images. A particular kind of representative image, the map, abstracts information from an object under study and distills it into an image formally similar, though not necessarily aesthetically similar, to that object. Thus, we have maps of roads, rivers, circuitry, sporting events, and empires, all of which bear little physical resemblance to their subjects, yet still convey useful information about them.

When neuroscientists discuss "brain mapping," the term usually refers to the use of tools to investigate the structural and functional features of the brain in such a way that the results of the investigation may be represented in the form of a map or map-like spatial representation. Examination of the various types of such maps in the course of history can provide a view of the development of human knowledge about the brain. The next section attempts to accomplish that task. A second, more specific use of the term "brain mapping" describes activities that fall within the broader definition but is limited to the use of technology currently *en vogue*, such as neuroimaging using magnetic resonance, positron emission, and X-ray, or elicitation of behavioral responses to sensory stimuli or electrical stimulation of the brain. This narrower set of brain mapping techniques will be discussed in the review, but in the interest of simplicity, our convention will be to use "brain mapping" in the first, broader sense.

The authors would be remiss in failing to point out the utility of the concept of "mapping" to an understanding of brain function. One of the great mysteries of the organ is the manner in which it assigns, or maps, representations of information to particular patterns of neural activity. Indeed, much of what the brain does is well described as creating, employing, and revising its own maps of the world—of concrete stimuli encountered, of formal qualities appreciated, of the emotional valence of objects and locations, of routes to navigate through space, and of the attractiveness of possibilities yet to be realized. Much as any mapmaker does, the brain simplifies information for adaptive use. To the extent that the map does not facilitate adaptive behavior, the mapmaker's job is ill done. The present chapter will not answer the question of how the brain makes and uses its maps, but it will touch on certain human investigations of that question.

2 Brain Mapping Through History

2.1 Getting Our Bearings—From Antiquity to Modernity

Pictorial representation of the brain began in antiquity. Perhaps the first illustrations of the brain have been preserved in what is called the "Alexandrian Series," based on medieval accounts of the work of Herophilus and Erasistratus in the 3rd century B.C.E. Almost all pre-scholastic figures represent the brain as a net-like structure. Galen, in his documentation of anatomy, provided some of the earliest descriptions of the structural features of the brain, including this net, or *rete mirabile*, which is, in fact, not to be found in the human brain at all. Because dissection was banned in ancient Rome, Galen extrapolated much of his description of human anatomy from dissection of other animals, such as the ox or pig, which do exhibit a *rete*. Due to the mistaken assumption that Galen had described human anatomy, the *rete mirabile* would figure prominently in illustrations of the human brain through Middle Ages, and it was often considered to be the interface of the intellect and the body (see Clarke and Dewhurst 1996 and Finger 1994 for fine, book-length historical reviews and additional illustrations).

Certain early Christian thinkers, such as St. Augustine (354–430 C.E.) and Nemesius, Bishop of Emesa (c. 390), drew on the work of Galen and the natural philosophy of Aristotle to locate mental function within the ventricles of the brain. This Cell Doctrine appeared in variable forms, but it generally held that the brain contained three cells (the lateral ventricles, the third ventricle, and the fourth ventricle) that contained the faculties of the mind, usually characterized as the *sensus communis*, or collection of impressions from the senses, and *imaginativa*, or image formation and fantasy in the first cell; *aestimativa* (judgment), *cogitativa* (thought), and *ratio* (reason) in the second; and *memorativa*, or memory, and occasionally *motiva*, the control of motion, in the third cell. Though later investigators of the brain would disprove the role of ventricles in the intellect, the identification of distinct intellectual functions and their assignment to different regions of the brain during this period established the project of brain mapping as a scholarly pursuit, and the Cell Doctrine persisted as the dominant view of the brain for more than a millennium. Figure 1 is a sixteenth century woodcut depicting both the Cell Doctrine and the *rete mirabile*.

With the rise of the Roman Empire, the works of Galen and the Alexandrians were largely inaccessible to Western Europeans, who generally did not read Greek. Latin translations of many of these works were not available until the 13th century C.E., spurring the authorship of manuscripts such as the one referenced in Fig. 1. In the meantime, the works of the Alexandrians and Galen were preserved in the Eastern Roman Empire, where Greek was the primary language, and from there became available to Islamic scholars and physicians of the Abbasid Caliphate in the 8th century C.E. Ibn Sina (Latinized to Avicenna; c. 980–1037) extended the tradition of Galen during the Golden Age of Islam, producing voluminous medical works among treatises in numerous other areas of scholarship. Ibn Sina's 14-volume *Canon of Medicine*, in which he adhered to the Cell Doctrine, though revising it to include five chambers, was widely used in European universities until the 17th century C.E. Eastern maps of brain anatomy and function, while exhibiting some advancement over time, continued to draw heavily on the medieval and classical models despite a Western rush of progress in the area during the 16th and 17th centuries (Clarke and Dewhurst 1996; Finger 1994).

As one of the first modern philosophers, Rene Descartes (1596–1650) bridged the gap between the medieval view of brain function and new developments based on medical science during the Scientific Revolution. Descartes elaborated on brain structure and function in his *De Homine* (1633), providing maps of the journey of sensation from the eyes and appendages to the brain. A mind-body dualist who believed the nonphysical substance of mind was in communication with, yet essentially different from, physical substance of body, Descartes located the mind-body interface in the pineal gland, to which sensations were conveyed for mental perception. Following Galen, he further posited that the mind controlled physical activity of the body by influencing the flow of cerebrospinal fluid, which he supposed flowed out to the body through the nerves, despite the demonstration by his contemporary, Vesalius, that nerves were not hollow.

During the Scientific Revolution in the West, Andreas Vesalius (1514–1564) is credited with reviving the scientific study of human anatomy. In his illustrations, Vesalius employed dissection of human cadavers to correct Galen's work (Fig. 2). Vesalius highlighted important structural elements of the brain, such as the corpus callosum, thalamus, basal ganglia, lenticular nucleus, pulvinar, cerebral peduncles, and optic chiasm. Along with other figures of the late Renaissance and Scientific Revolution, such as Leonardo da Vinci, Berengario da Carpi, and Dryander, Vesalius accomplished a major advance in maps of the brain, which from that point forward included cerebral convolutions, a sophisticated ventricular system rather than cells, and no *rete mirabile*. This shift paved the way for the influential English

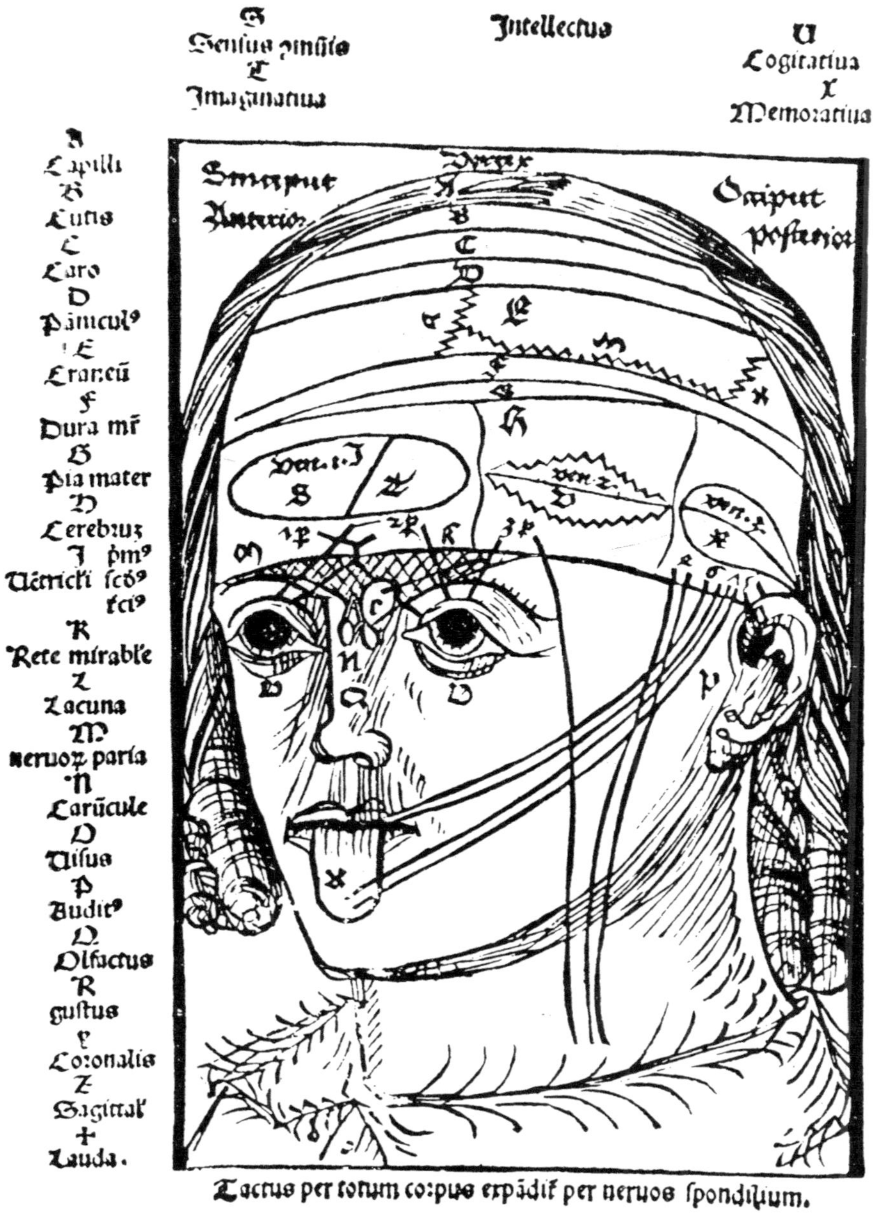

Fig. 1 This woodcut from the turn of the 16th century portrays the brain as conceived under the Cell Doctrine. The seat of each faculty is indicated as follows: (B) "Sensus communis," (C) "imaginatio," (D) "fantasia," (E) "estimativa," and (F) "memoria" (from the *Trilogium animae* of Ludovicus Pruthenus, published in 1498; image is in the public domain)

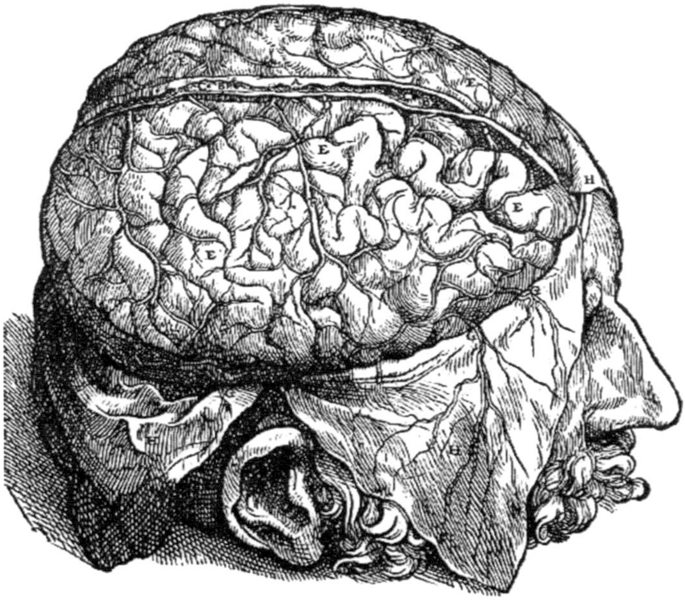

Fig. 2 This illustration by Andreas Vesalius is notable for depicting the human brain during dissection and for showing the gyri of the cortex in exquisite detail (Vesalius 1543; image is in the public domain)

anatomist Thomas Willis (1621–1675) to describe and illustrate both gray and white matter regions of the brain and to suggest that these regions may underlie the classic mental functions. Illustration of the cortex was later refined by the Danish scholar Nicolas Steno (1638–1686), and that of the white matter regions by the French anatomist Raymond de Vieussens (1644–1716).

It was not until the 19th century that the cerebral cortex began to figure substantively in models of brain function. German physiologist Franz Josef Gall (1758–1828) was the first to identify the cerebral gray matter as functionally active tissue and white matter as conductive. He also famously posited the doctrine of *cerebral localization*—that mental functions are subserved by discrete regions of the cortex. This belief led him to develop "cranioscopy," the inference of personality traits and mental ability based on the external shape of the skull. Later renamed "phrenology" by his disciple Johann Spurzheim, the practice of cranioscopy survived well into the nineteenth century but has since been wholly discredited. Despite his unfortunate tangent into pseudoscience, Gall's hypothesis of cerebral localization of function was later substantiated and brought to scientific acclaim by Broca and became a guiding (if embattled) doctrine in the development of modern neuroscience (Finger 1994).

2.2 *Covering New Ground—From the Modernity to the Millennium*

2.2.1 Mapping the Microstructure

The nineteenth and twentieth centuries brought technological advances that allowed for finer-grained description of nervous system structure. Italian pathologist Camillo Golgi (1843–1926) developed methods of staining cells to make them visible with a microscope and first identified particular cell types and organelles within them. Santiago Ramon y Cajal (1852–1934), a Spanish histologist, used Golgi's preparation to investigate the nervous system, producing hundreds of illustrations of the microscopic structure, providing evidence that the nervous system is made up of discrete neurons rather than a net. Based on cytoarchitecture, the structural properties of cells, German neurologist Korbinian Brodmann (1868–1918) divided the cerebral cortex into 52 discrete regions. Many of the original Brodmann areas have been associated with specific localized functions, such as primary auditory perception (areas 41 and 42), primary somatosensory perception (areas 1, 2, and 3), and primary visual perception (areas 17 and 18), though later investigators have found substantial variability across individuals in the location of Brodmann's cytoarchitectonic regions. Figure 3 is an image of Brodmann's map.

The tradition of structural mapping using novel techniques in basic neuroscience continued through the late 20th century to the present day. Marsel Mesulam developed maps in 1977 of the neural connectivity of the monkey brain, employing a novel means of revealing the paths of neural projections by following a tracer chemical introduced into the neuron's internal transport system normally used for nutrients and waste. An exciting contemporary development in the microstructural investigation of neurons was the invention of serial block-face electron microscopy (SBEM) in 2004, allowing automated, three-dimensional imaging of structures at the nanometer scale, such as the synaptic cleft and vesicles. This and similar techniques more closely resemble photography than mapping, but over time they

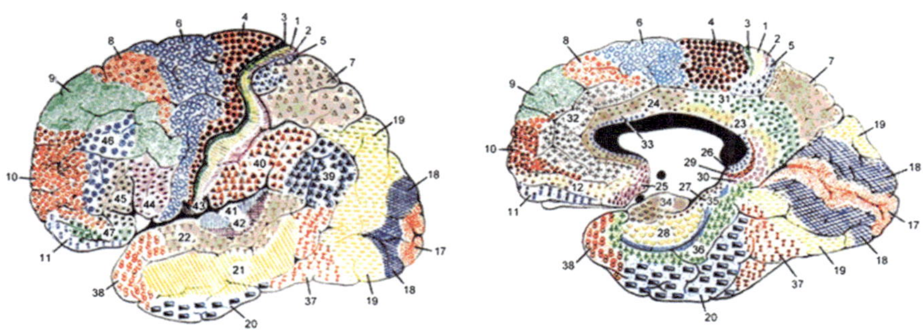

Fig. 3 This illustration displays the cytoarchitectonic regions of the brain identified by Brodmann through his study of cellular structure (*Source* Mark Dubin and the University of Colorado, http://spot.colorado.edu/~dubin/index.html)

allow the capture of information that may then be abstracted into useful maps—for example, of variation in synaptic cleft structure or vesicle density throughout the brain. An appropriate analogy might be the manner in which advances in satellite imaging have improved the construction of geographical maps of the planets.

2.2.2 Mapping Pure Function

Arising out of the interest in the brain kicked off by Gall and fomented by Broca and others, the study of brain function, agnostic to structure, arose parallel to the structural investigations described above and the project of cerebral localization described below. Ludwig Lichtheim (1845–1928), a German physician, made significant contributions to the study of disordered language, or *aphasia*. His primary early contributions included diagrams of functional models of language not specifically tied to brain regions. The models employed by later cognitive psychologists were often represented in the "boxology" diagram format as seen in Lichtheim's work, formal maps representing components of function and information processes as boxes, blobs, and arrows. In Lichtheim's case, these models were developed based on clinical investigation of individuals with brain lesions, often identified on autopsy, and as much on descriptions of other investigators' cases as on his own. As described below, Lichtheim's models would later be mapped to regions of the cortex. Figure 4 displays a comparison of Lichtheim's rendering of a model of which he was critical, his own revised model, and the later Wernicke-Lichtheim and Wernicke-Geschwind models that would associate Lichtheim's work with cerebral structures.

During the era of behaviorism, at least in American psychology, maps of functional processes gave way to stimulus-response flow charts and contingency management diagrams. With the rise of cybernetics and computer science in the second half of the 20th century, though, scholars of the brain rediscovered the conceptual metaphor of the brain as a machine. While preserving the focus on input and output from the behaviorist tradition, psychologists began to posit and test models of internal processing of information in the human mind. Donald Broadbent presented influential "mechanical models" of memory and selective attention, relying heavily on diagrams and visual metaphor to convey the workings of the model (Broadbent 1957). These models were refined and challenged during the 1960s and later by psychologists like Baddeley, Atkinson, Shiffrin, and Tulving, using behavioral tasks to test posited "control processes" and stimulus characteristics that affect information processing (see Driver 2001, for an historical review). Reviving the nearly century-old notion of mental chronometry, reaction time became an especially important measure in attempts by Sternberg (1969) and Posner (1978) to analyze complex cognitive tasks into more primitive functions and to identify cognitive processes that may compete or interfere with each other.

This conceptual turn resulted in the founding of cognitive science as a discipline, generally marked at a 1956 convention at the Massachusetts Institute of Technology, where psychologist George Miller presented his paper, "The Magical Number Seven, Plus or Minus Two," identifying the capacity of working memory.

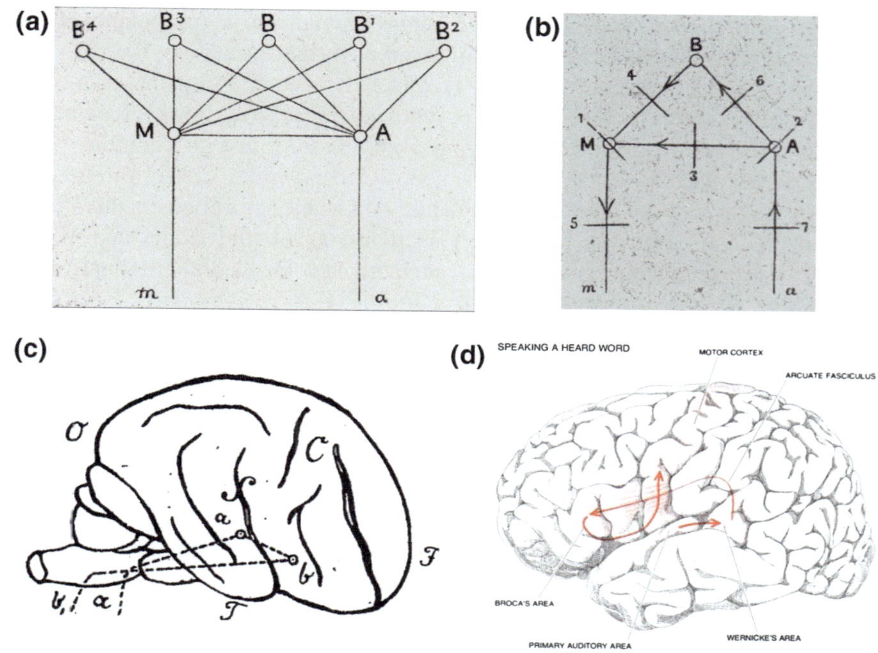

Fig. 4 Several iterations of the functional mapping of language to brain structure, showing the work of Lichtheim (**a**, **b**), Wernicke (**c**), and Geschwind (**d**) (*Sources* Lichtheim 1885; Wernicke 1874:19; Geschwind 1979)

Other presenters at the conference included Noam Chomsky, the founder of psycholinguistics, and computer scientists Allen Newell and Herbert Simon, who studied and developed artificial intelligence. Some psychologists, such as Ulric Neisser, employed computers to develop to model cognitive processes and to generate hypotheses for empirical testing. The conceptual center that held these investigations together was a focus on *representation* of information within the brain. Not only could brain function be depicted using external maps, it was found that the brain must generate internal representations of information, and the manner of that internal representation would affect the brain's future acquisition of information. The watershed example is in the 1962 work by Hubel and Wiesel, showing that cats exposed only to visual patterns of vertical lines would later only perceive and react to vertical lines, even when later placed in a richer visual environment. That is, the visual cortex was "cognitively tuned" to perceive vertical line information only, and the cat's experience of the world was impoverished because normal cognitive tuning was prevented. The brain maps information to meaning, and where that mapping does not occur, there is no map to which to refer when new information becomes available. Information without a referent is meaningless and, in effect, is never perceived at all.

Throughout the 1970s, semantic maps and memory retrieval models proliferated as cognitive psychologists sought to identify data structures and algorithms from computer science that best represented human thought as a computational and representational process. In studies following the cognitive tradition, the steps that make up cognitive processes and the factors that affect information processing became well characterized. Cognitive psychology in general, however, remained agnostic to the neural substrate underlying cognition until several traditions began to unite under the banner of cognitive neuroscience. Smith (2001) provides an impressive overview of these developments. A notable exception to this "black box" approach to cognitive psychology includes, again, the Hubel and Wiesel investigations noted above, which first characterized the neural mechanisms of visual information processing.

2.2.3 Mapping function to structure

Not long after Gall's 1818 proposition that intellectual functions may be localized to specific regions of cortex, Jean Pierre Flourens (1794–1867), a leading figure in the French scientific establishment sought to dispel the myth of phrenology by experimentally investigating the loss of function in animals after removal of parts of the brain. While Flourens did note that certain vital and motor functions may be disturbed by injury to the brain stem and cerebellum, he observed that large portions of the cortex may be ablated before loss of function becomes apparent, concluding that all parts of the cortex cooperatively and equally contributed to specific functions. Paul Broca (1824–1880), a French surgeon who initially adhered to Flourens' doctrine of *equipotentiality*, famously identified a cortical region essential for speech by associating the functional deficits of his patient "Tan" with the location of his lesion. This "lesion study" method contributed significantly to the field of behavioral neurology, which generated many hypothesized models of functional localization in the brain. German physician Carl Wernicke (1848–1905), prompted by observations that the use of speech may be disrupted by lesions outside of Broca's area, employed Broca's methods to identify the cortical locus underlying speech comprehension. The ultimate result of these efforts was the Wernicke-Lichtheim model, a map associating many cognitive functions subserving the use of language with particular brain structures (see Fig. 3c). Wernicke used this model to predict correctly the existence and character of a novel communication disorder called "conduction aphasia" that would be found if Broca's speech production area and Wernicke's speech comprehension area were disconnected from each other but otherwise left intact. The advent of this model represented a major transition into the modern era of brain science.

Debate raged for more than a century over the extent to which function could be localized to structure. The study of language function seemed to favor localization, though opponents pointed out that lesions in the appropriate areas were not strongly predictive of the kinds of functional deficit found. A major development was the use of electricity to directly stimulate cortical tissue to induce behavior in live participants. The principle was first demonstrated in 1870, when German physicians Hitzig and Fritsch applied an electric current to the exposed brains of live dogs, producing bodily movement that systematically related to the area stimulated. This technique allowed production of a somatotopic map of regions along the "motor strip" that, when stimulated, elicited a particular movement in the animal. Still, the scientific world was hesitant to accept cerebral localization after phrenology had failed so spectacularly. In 1881, German physiologist Friedrich Goltz independently demonstrated that lesion size, and not location, was predictive of level of functional deficit in animals, replicating Flourens' research. Kurt Goldstein and others of the German Gestalt School viewed this as evidence that the brain was a holistic unity that cannot be decomposed into regions that explain its function. Even as motor and sensory functions were found to be quite strongly localized, American psychologist Karl Lashley in 1950 demonstrated that higher functions, such as stored memories, did not appear to be localized to a specific region; even if cortical injury were to result in the loss of memorized information, the remaining uninjured cortical tissue could be recruited to relearn the information. English neurologist Hughlings Jackson (1835–1911) perhaps best typified a measured approach to the issue of localization, positing that hierarchies of function may indeed require particular regions for components of function, and at the same time, interplay among many regions of the brain is necessary for optimal function. Few brain scientists today would disagree with that proposition.

In the first half of the twentieth century, observation and assessment during and following clinical intervention played a primary role in the further refinement of the mapping of function to structure in the human brain. Following thorough documentation of projectile injury location and the positioning of blind spots (scotomae) in the visual fields of Russo-Japanese and World War I veterans, Japanese opthalmologist Tatsuji Inouye and British neurologist Gordon Holmes (Inouye 1909; Holmes 1918) demonstrated that the primary visual cortex of the occipital lobe shares the same physical layout as the image that strikes the retina (called "retinotopic" organization). Wilder Penfield (1891–1976), a famed American-Canadian neurosurgeon, developed with his colleague Herbert Jasper the Montreal procedure, wherein the cerebral cortex of an anesthetized human patient was stimulated with electrodes to evoke behavior and guide surgical incision. Through his use of this technique, Penfield created maps of cortical connections to limbs and organs in the body, which remain in use today. Some lasting contributions of these investigations include better outcomes for neurosurgery patients, a more sophisticated understanding of cerebral lateralization, and empirical demonstration that localized functions may overlap when mapped to physical brain regions. Figure 5 is a map summarizing the knowledge of functional localization as gathered by electrode stimulation. In this context of clinical investigation of brain function, contributions from the field of psychometric assessment allowed the use of reliable measures, standardized using a broad normative sample, to better characterize level of function. For example, Donald Hebb (1904–1985), a luminary figure in cognitive neuroscience and neuropsychology, administered the Stanford-Binet Intelligence Scales to patients with frontal lobe injuries and found that his sample had similar IQs to uninjured persons (Hebb and Penfield 1940).

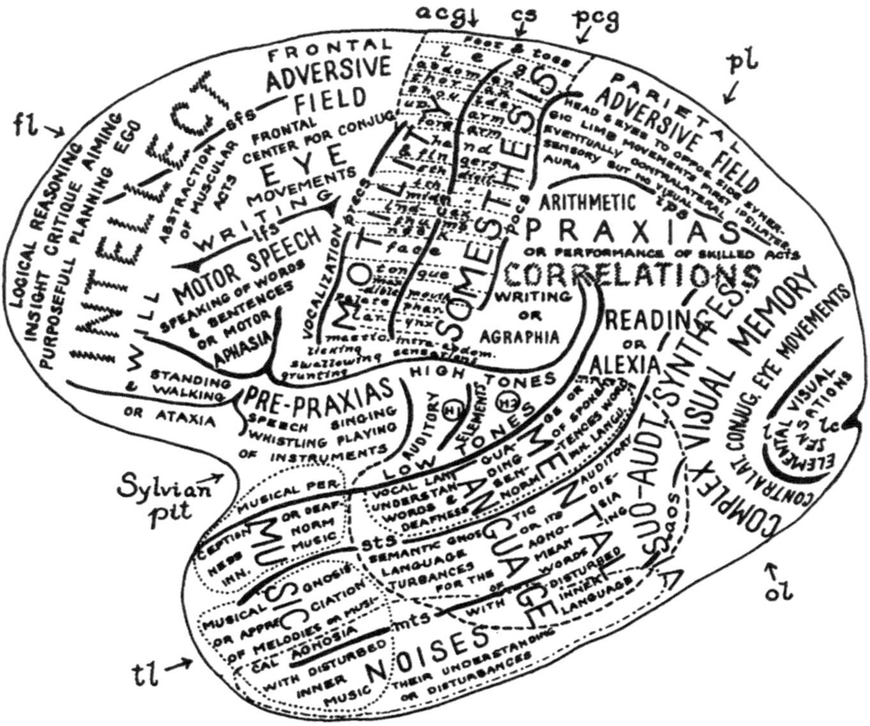

Fig. 5 A schematic summary of the state of knowledge of localisation of human functional brain in 1957. It was generated using data from lesion studies and experiments using direct cortical stimulation during neurosurgery (*Source* Savoy 2001:11, based on Stephen Polyak, The Vertebrate Visual System.)

The use of psychological tests in the clinical evaluation of neurological disorders became the foundation of the discipline of clinical neuropsychology. Through clinical contact with patients and systematic assessment of their abilities, neuropsychologists have contributed heftily to the description and localization of function in both intact and disordered brains. Many notable contributions arose from famous case studies. For example, in 1957, neurosurgeon William Beecher Scoville and neuropsychologist Brenda Milner published a case study of patient H. M. (Henry Molaison), now known as the most famous patient in the cognitive neurosciences (Scoville and Milner 1957). After Scoville removed the patient's temporal lobes to treat intractable epilepsy, Milner used a series of cognitive tasks to identify several dissociable components of memory function and to localize those functions in different regions of the cerebral cortex.

Starting in the 1950s, neuropsychologist Roger Sperry and, later, his student Michael Gazzaniga, demonstrated the striking results of functional disconnection between the two cerebral hemispheres and further advanced the understanding of cerebral lateralization of function in studies of "split-brain" individuals who had undergone resection of the corpus callosum, the large body of white matter responsible for transferring information between the two cerebral hemispheres (see Gazzaniga 2005, for a thorough review). The study of disordered language (aphasia) flourished at the Boston VA Hospital under behavioral neurologist Norman Geschwind and a coterie of influential colleagues including neuropsychologists Edith Kaplan, Harold Goodglass, and Nelson Butters. Geschwind revived and refined both the Wernicke-Lichtheim model of language (as the Wernicke-Geschwind model, see Fig. 4d) and the notion of disconnection syndromes—neurological presentations resulting from damage to association pathways between intact processing centers. More recent contributions to brain mapping from clinical neuropsychology include work from Goodale and Milner (1992) demonstrating the existence of two distinct systems for processing visual information (dorsal and ventral); Squire's (2004) investigations of the neuroanatomy underlying memory and frontal lobe functions; and right-hemisphere studies by Kenneth Heilman, a student of Geschwind, identifying roles of this previously opaque structure in spatial attention and planning, autonomic control, and emotional communication (Heilman and Van den Abell 1979, 1980; Tucker et al. 1977; Blonder et al. 1991).

Among the substantive contributions from clinical neuropsychology are an ever increasing array of cognitive testing instruments and a number of influential testing approaches. In 1919 at McLean Hospital in Boston, Shepherd Ivory Franz, a student of American psychologist James Cattell, was the first to administer a battery of objective psychological tests of sensation and perception in a clinical context (Colotla and Bach-Y-Rita 2002). Ward Halstead, in the 1940s at the University of Chicago, developed a battery of 21 tests for the purpose of classifying individuals as "normal" versus "brain injured," which was then adapted and refined by Ralph Reitan for clinical use as the 1974 Halstead-Reitan Battery. In addition to a primary index score allowing quantitative determination of brain injury, this standardized battery of tests was validated with one large sample of individuals, allowing examination of patterns among individual tests and, in turn, permitting quantitative determination of the location of brain lesions (Reitan and Wolfson 2009).

A complementary testing approach, termed the "flexible battery," arose out of the Gestalt tradition. Kurt Koffka (1886–1941), a founder of Gestalt psychology, made a foray into clinical neuropsychology later in his career, for a time assessing the judgment and insight of persons with neurologic disorders, and later creating what was likely the first flexible neuropsychological battery (Harrower-Erickson 1942). German Jewish neurologist Kurt Goldstein collaborated with psychologist Adhemar Gelb, using specific psychological tests to characterize effects of brain injury in World War I veterans (Goldstein and Gelb 1918); in 1941 Goldstein again collaborated with a psychologist, Martin Scheerer, to develop a battery for evaluating the effects of brain injury by capturing the procedure by which the individual achieves a result, rather than the presence or absence of a correct answer (Goldstein and Scheerer 1941). The flexible approach to neuropsychology has taken many forms, from Alexander Luria's systematic characterization of neurological syndromes through the use of focused bedside testing, to the analysis of qualitative changes in cognitive test profiles over time, pioneered by Heinz Werner. The result has been the refinement of measurement of neurologic disorders, the standardization of the "clinical signs" of classical neurology, and more specific understanding of the multiple effects that may follow a specific brain condition.

Extant "schools" of neuropsychological assessment marry the standardized and flexible approaches in distinctive ways (Barr 2008). The Boston School takes what Kaplan called a "process approach," administering a standardized battery of procedures to assess a syndrome while emphasizing attention to and quantification of testing behavior beyond the presence or absence of ability. The Iowa School, in the tradition of Arthur Benton, uses a flexible battery with well validated instruments to characterize neurological syndromes while taking into account effects of age and education. Hans Teuber took an approach that was similar to Benton's but focused specifically on "double dissociation," the identification of a lesion that focally disturbed one function while leaving another intact, combined with a complementary lesion that produces the opposite pattern. Each novel use of the flexible approach introduced a tool for better understanding the relationships between brain structure and cognitive function, and identification of a double dissociation allowed very strong inferences that more firmly grounded neuropsychological knowledge and allowed for improved brain mapping. Teuber, appropriately, is credited with introducing the term "neuropsychology" and arguing strongly for the combination of well validated tests with neurological practice, which would come to include the new and fast-developing methods of neuroimaging (Parlee 2012).

3 Current Brain Mapping Technology

As investigative methodology and behavior-based instruments became more refined in the behavioral sciences, neuroimaging and neurostimulation techniques developed in the context of medical intervention, drawing heavily on contributions from the physical sciences. Naturally, brain researchers began to attempt to localize function to cortical regions by combining these new technologies with their own behavioral tasks, experimental designs, and theoretical models. Major developments in neuroimaging technology are briefly described below, along with benefits, limitations, and illustrative examples in the field of brain mapping. See Savoy (2001) and Raichle (2008) for more thorough reviews. For in-depth study, see the *Brain Mapping* trilogy edited by Mazziotta et al. (2000).

The first form of in vivo neuroimaging widely used in medicine was pneumoencephalography, in which a bolus of air was invasively introduced into the cerebrospinal fluid. The patient would then be rotated in space according to a precise maneuver to position the air pockets within the cerebral ventricles, and an image would be collected using rudimentary X-ray technology. Though this method did allow the clinician to visualize brain tissue without opening the cranium, incision of the meninges presented the risk of infection, and even in the best cases, the patient was sure to have severe headache. When X-ray computed tomography (CT) was introduced in 1973, allowing non-invasive, three-dimensional visualization of the brain anatomy, incision and introduction of air became obsolete. CT is still very useful in contemporary medicine, but because it is better able to detect blood than brain tissue, it is of limited use in brain mapping. What CT and pneumoencephalography were able to offer in this arena was finer characterization of lesions in live patients than had before been available, improving the practice of behavioral neurology, neurosurgery, and clinical neuropsychology (Savoy 2001).

Imaging of brain function found its place in the mainstream with the development of single-photon emission computed tomography (SPECT) and positron emission topography (PET) in the 1970s. In SPECT/PET, radioactive isotopes of elements commonly found in biochemical compounds are introduced into the body, where they distribute, bind, and metabolize like their natural counterparts. The slight radioactivity is enough to be detected outside the body, and the signal varies with the concentration of the tracer element in a given tissue. Because different tracer elements decay at different rates, it is possible to introduce several agents at once and measure the processes that involve them simultaneously. The most widely used agents measure glucose metabolism and blood flow, allowing one to surmise which regions of the brain become more or less active during particular behavioral tasks. SPECT/PET represented a major step forward for brain mapping in that it allowed one to associate structure with function in live human subjects with uninjured brains. Its advantages over other current technologies are that it employs a wide array of chemical markers targeting numerous physiological processes and that undergoing a scan is not painful, time-consuming, loud, enclosing, or otherwise unpleasant for the participant. Of the two technologies, SPECT is more limited in spatial resolution, but it is overall less expensive to use. PET requires a significant financial investment in a cyclotron to manufacture the tracer elements on-site. Both present the risk of hazardous radiation exposure if used too frequently with one person (Raichle 2008).

Useful counterparts to SPECT/PET are electroencephalography (EEG) and magnetoencephalography (MEG), both of which non-invasively measure electrical activity in the brain with exceptional safety and temporal resolution. In EEG, electrodes placed on the scalp measure changes in electric fields within the cranium (Savoy 2001; Raichle 2008). MEG is similar but instead uses superconducting quantum interference devices (SQUIDs) to measure gradients in very low strength magnetic fields generated by electrical activity in the cortex. Either method allow continuous monitoring of brain activity for abnormalities in a clinical setting or may be used to gather localized changes in brain activity in response to structured behavioral and cognitive tasks. This latter stimulus-response approach, called the event-related potential (ERP) method in EEG and the event-related field (ERF) method in MEG, facilitates mapping function to cortical regions. For example, Posner mapped brain networks underlying visual attention using EEG and behavioral assays. A clear limitation of non-invasive measurement of electrical activity in the brain is that it can be difficult or impossible to measure activation of deeper tissue using surface measurement alone. Still, contributions from EEG and MEG continue to be considerable, and the very low cost of EEG equipment is an obvious benefit over other neuroimaging technologies.

Magnetic resonance imaging (MRI) is another live visualization technology that continues to contribute greatly to visualization of brain structure through various structural scan sequences and to mapping of brain function to structural activation through functional MRI sequences (fMRI). In MRI, a very strong magnetic field is applied to the body, which causes highly polar molecules, like water, to align along the axis of the field. Precisely placed coils emit radio waves to knock these molecules off of the axis, and receiver coils detect the time it takes for them to return to axial alignment. These principles had been applied in chemistry as nuclear magnetic resonance (NMR) since their discovery in 1946 at Harvard and Stanford Universities, but it was not until 1973 that Paul Lauterbur at Stony Brook University developed a method to recover cross-sectional MRI images similar to those of CT (Lauterbur 1973). In 2003, Lauterbur was awarded the Nobel Prize in Physiology or Medicine for his contributions.

MRI took off in both clinical and research settings, and in the past four decades, numerous specific signaling sequences and contrast agents have been developed to better visualize anatomical structure, functional activation, and markers of physiological processes (Raichle 2008). For example, diffusion tensor imaging (DTI) is a sequence that measures water diffusion in multiple directions, allowing the creation of maps of white matter tracts (tractography; Fig. 6). The diffusion weighted imaging (DWI) sequence can similarly detect barriers to diffusion, such as those that occur in ischemic stroke, in many cases within minutes rather than the hours it would take for the lesion to appear on CT. Magnetic resonance spectroscopy, another method adopted from chemistry and physics, allows one to measure the presence of particular metabolites in tissue. Structural sequences, like magnetization-prepared rapid acquisition with gradient echo (MP-RAGE) and several others, differentiate tissues according to fat and water content, allowing very high quality visualization of anatomical structure. Contrast agents include the more common gadolinium-based agents that improve visualization of blood vessels (angiography) and manganese-based agents that highlight neural activation (fMRI), among others. A major advance in fMRI, and hence in mapping function to structure, was the discovery of differences in the magnetic properties of hemoglobin in the blood that depend on whether it is oxygenated. Thus, deoxyhemoglobin, a naturally occurring by-product of oxygen use, was found to act as a contrast agent, resulting in the ability to compute a blood oxygen level dependent (BOLD) signal to index brain tissue activation. Since the advent of BOLD imaging, fMRI has become an extremely active area of brain mapping research. Figure 7 demonstrates the quality of brain tissue visualization afforded by MRI and an example of fMRI maps showing cortical activation.

Transcranial magnetic stimulation (TMS) and transcranial electrical stimulation (TES) are means of non-invasively manipulating electrical activity in cortical tissue, and they may be combined, to great effect, with neuroimaging methods (Raichle 2008). This manipulation is accomplished in TMS by running a strong electrical current through a coil to generate a magnetic field that, in turn, can induce changes in electrical current among neurons in the cortex when the coil is applied to the

Fig. 6 Visualization of a DTI measurement of a human brain. Depicted are reconstructed fiber tracts that run through the mid-sagittal plane. Especially prominent are the U-shaped fibers that connect the two hemispheres through the corpus callosum (the fibers come out of the image plane and consequently bend towards the *top*) and the fiber tracts that descend toward the spine (*Source* http://commons.wikimedia.org/wiki/Category:Diffusion_tensor_imaging#mediaviewer/File:DTI-sagittal-fibers.jpg), uploaded by the original author Thomas Schultz under a GNU license.)

scalp. In TES, electrodes are applied to the scalp, and an electric current is generated between them that runs directly through the cortex. TES causes some physical pain and is less often used. TMS, on the other hand, is generally painless and safe when used within established guidelines. A particularly useful feature of TMS is that it may be used to stimulate or inhibit activity, simulating either regional activation or lesion. Flexibility, safety, and relatively low cost have made TMS a fruitful method for brain mapping, allowing cortical stimulation in humans without a need for surgery. Its primary limitation is that its spatial resolution is fairly coarse, and it is more difficult to address deeper brain structures than those on the surface.

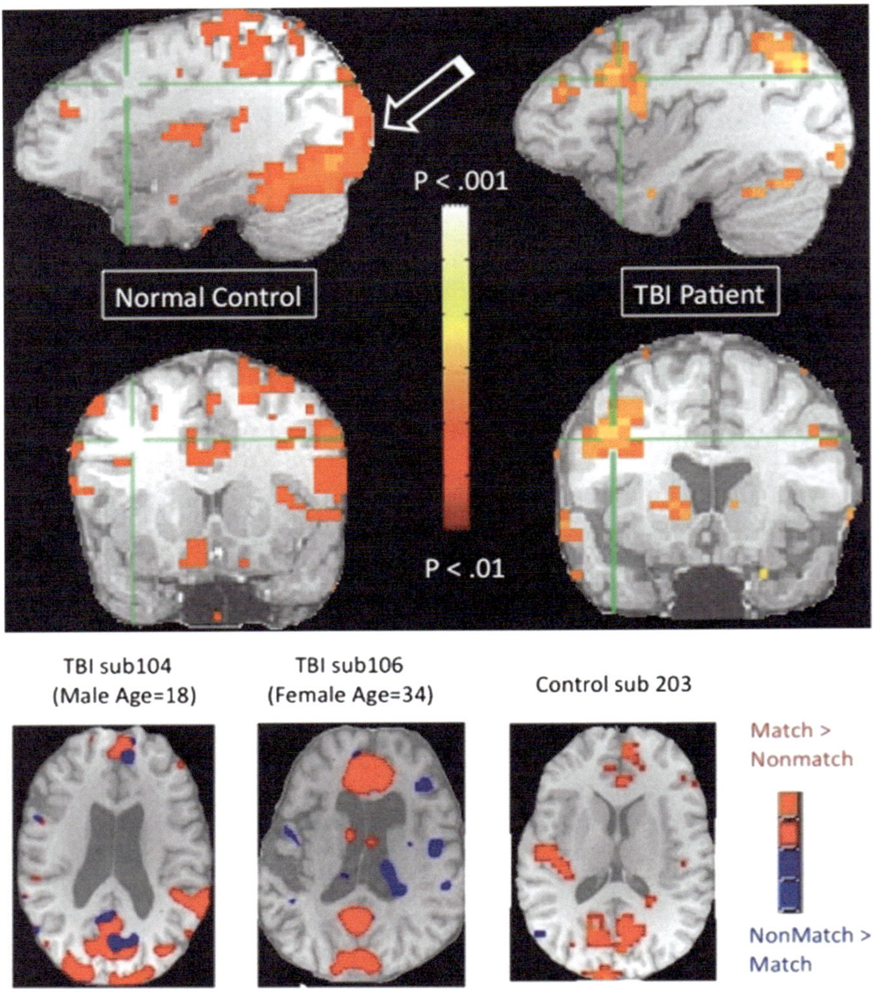

Fig. 7 Functional MRI imagery overlaid onto a structural MRI image. Note that the structural image, which takes much more time to obtain, has a significantly better spatial resolution than the functional image, which has a higher temporal resolution at the cost of spatial precision (*Source* Unpublished work from the authors)

4 What the Future May Hold

It is a difficult thing to correctly predict the course of progress in a field, especially when that field is so heavily dependent on innovative technology. Still, some facets of brain mapping are already under examination for further improvement, and others are the subject of promising, collaborative initiatives. Significant advances afforded by neuroimaging technology already include the ability to generate three-dimensional images, the development of neuroanatomical atlases that allow an individual's images to be standardized to a model brain for comparison across persons, and the ability to quantify structural and functional information in such a way that statistical analysis may be applied to clarify the reliability of findings. As in the past, attempts to further the field involve marshalling advances from multiple fields to solve problems in brain mapping. The physical sciences and engineering continue to develop novel imaging approaches, which are then evaluated in neuroscience laboratories for applicability to bench research and clinical use. One such technique currently under investigation for potential application is optical imaging of intrinsic signals, which maps activity by detecting changes in the light reflectance of tissue that arise from different physiological processes in the cell. In clinical neuropsychology, test development and validation continues apace, as does the fine-grained characterization of cognitive and behavioral changes proceeding from neurological disorders.

One challenge that will likely spur investigation in brain mapping in the near future is that neuroimaging technology allows collection of so many data points at once that it is difficult to analyze them sufficiently or to store them in a way that makes them accessible for future analysis. Classical techniques for data reduction, prediction, and control of inferential errors have in many cases been found wanting, and new approaches are flourishing. Network analysis allows researchers to combine more directly what is known of brain structure with functional imaging and behavioral data to investigate the circuitry underlying function. Structural equation modeling has also been useful in brain mapping, permitting combination of multiple measures and theoretical predictive relationships into complex, testable models. Nonlinear methods based on chaos and complexity theory examine cognition as an emergent property of self-organizing neural systems. Development of guidelines for the threshold of significance to be used in research and clinical settings is an active area of investigation as well. Using meta-analysis, researchers have sought to combine results from numerous studies to rigorously summarize the current state of knowledge, but variability in research methods, definitions, equipment, and database practices across research laboratories has complicated the task. Major initiatives currently focus on "neuroinformatics," the use of technology and research on best practices to encourage consistency across these domains and allow for easier database combination and meta-analysis (Jagaroo 2009). This development of infrastructure is a prerequisite for some of the more ambitious mapping initiatives currently underway, such as the multi-site Human Connectome Project, funded by the National Institutes of Health, which seeks to identify all structural and functional connections in the human brain in the interest of improving clinical intervention.

Maps have served key roles in the study of the brain throughout history. A map allows quick transmission of information for ready digestion, and like any technology, it is a double-edge sword, imparting a useful truth to its audience as readily as it purveys misunderstanding. For the ancients, the brain was a net with unspecified function, and in the end, the visible net itself was an artifact of overgeneralized animal research. Medieval scholars attempted to organize parts of the intellect within chambers of the brain, but their maps belied scant knowledge of anatomy, as well as a limited conception of what the brain might accomplish. Modern research has found that the brain's own internal maps of information, if impoverished experimentally, result in loss of adaptive behavior. With time, maps of the brain came to document physical features, now to the point of hyperrealistic nanoscale visualization in certain applications but strategically formal and abstract diagrams in others. Understanding of function developed such complexity that formal representation in maps became a necessary aid to thought. With the advent of cognitive science, formal representation of information in maps *within* the brain became central to scientific accounts of brain function. A review of brain maps uncovers a story of how the brain has been imagined over time, first as an inert net, then as the nexus of quickening spirit, next as an important but poorly understood lump of flesh, and eventually as an active, organized, and organizing collection of anatomical regions. Metaphors abound—vessel, pump, machine, computer—and lately the net has resurfaced, this time functionally active and more densely connected. It is a unique thrill to consider that the human investigation of the brain is indeed carried out by living brains, which are the storehouses of understanding and the sources of investigative activity and innovation. Across eras, the map has been a needful tool enabling humans to better conceive of the brain, to think about thinking, and to approach the profound questions of how we may be what we are and how we may better ourselves.

References

Barr, W. (2008). The historical development of the neuropsychological test battery. In J. Morgan & J. Ricker (Eds.), *Textbook of clinical neuropsychology*. New York: Taylor & Francis.

Blonder, L. X., Bowers, D., & Heilman, K. M. (1991). The role of the right hemisphere in emotional communication. *Brain, 114*(3), 1115–1127.

Broadbent, D. E. (1957). A mechanical model for human attention and immediate memory. *Psychological Review, 3*(64), 205–215.

Clarke, E., & Dewhurst, K. (1996). *An illustrated history of brain function: Imaging the brain from antiquity to the present*. Novato, CA: Norman.

Colotla, V. A., & Bach-Y-Rita, P. (2002). Shepherd Ivory Franz: His contributions to neuropsychology and rehabilitation. *Cognitive, Affective, & Behavioral Neuroscience, 2*(2), 141–148.

Driver, J. (2001). A selective review of selective attention research from the past century. *British Journal of Psychology, 92*(1), 53–78.

Finger, S. (1994). *Origins of neuroscience: A history of explorations into brain function*. New York: Oxford University.

Gazzaniga, M. S. (2005). Forty-five years of split-brain research and still going strong. *Nature Reviews Neuroscience, 6*(8), 653–659.

Geschwind, N. (1979). Specializations of the human brain. *Scientific American, 241*(3), 180–199.

Goldstein, K., & Gelb, A. (1918). Psychologische Analysen hirnpathologischer Fälle auf Grund von Untersuchungen Hirnverletzter. *Zeitschrift für die gesamte Neurologie und Psychiatrie, 41*(1), 1–142.

Goldstein, K., & Scheerer, M. (1941). Abstract and concrete behavior an experimental study with special tests. *Psychological Monographs, 53*(2), 1.

Goltz, F. (1881). Discussion on the localization of function in the cortex cerebri. In W. MacCormac (Ed.), *Transactions of the 7th international medical congress* (Vol. I, pp. 218–228). London: J. W. Kolkmann.

Goodale, M. A., & Milner, A. D. (1992). Separate visual pathways for perception and action. *Trends in Neurosciences, 15*(1), 20–25.

Harrower-Erickson, M. R. (1942). Kurt Koffka: 1886–1941. *The American Journal of Psychology, 55*(2), 278–281.

Hebb, D. O., & Penfield, W. (1940). Human behavior after extensive bilateral removal from the frontal lobes. *Archives of Neurology & Psychiatry, 44*(2), 421–438.

Heilman, K. M., & Van den Abell, T. (1979). Right hemispheric dominance for mediating cerebral activation. *Neuropsychologia, 17*(3), 315–321.

Heilman, K. M., & Van Den Abell, T. (1980). Right hemisphere dominance for attention: The mechanism underlying hemispheric asymmetries of inattention (neglect). *Neurology, 30*(3), 327.

Holmes, G. (1918). Disturbances of vision by cerebral lesions. *The British Journal of Ophthalmology, 2*(7), 353.

Hubel, D. H., & Wiesel, T. N. (1962). Receptive fields, binocular interaction and functional architecture in the cat's visual cortex. *The Journal of Physiology, 160*(1), 106.

Inouye, T. (1909). *Die Sehstörungen bei Schussverletzungen der kortikalen Sehsphäre: nach Beobachtungen an Verwundeten der letzten japanischen Kriege*. W. Engelmann.

Jagaroo, V. (2009). *Neuroinformatics for Neuropsychology*. New York: Springer.

Lashley, K. S. (1950). *In search of the engram*. Society of Experimental Biology Symposium No. 4: Physiological Mechanisms in Animal Behavior. Cambridge.

Lauterbur, P. C. (1973). Image formation by induced local interactions: Examples employing nuclear magnetic resonance. *Nature, 242*(5394), 190–191.

Lichtheim, L. (1885). On aphasia. *Brain,* 7, 433–484. [Reproduced in Lichtheim, L. 2006. Onaphasia. *Brain,* 129, 1347–1350.].

Mazziotta, J. C., Toga, A. W., & Frackowiak, R. S. (Eds.). (2000). *Brain mapping: The disorders*. San Diego, CA: Academic Press.

Mesulam, M. M., Van Hoesen, G. W., Pandya, D. N., & Geschwind, N. (1977). Limbic and sensory connections of the inferior parietal lobule (area PG) in the rhesus monkey: A study with a new method for horseradish peroxidase histochemistry. *Brain Research, 136*, 393–414.

Miller, G. A. (1956). The magical number seven, plus or minus two: some limits on our capacity for processing information. *Psychological Review, 63*(2), 81.

Parlee, M. B. (2012). Hans-Lukas Teuber: Envisioning neuropsychology. In A. Y. Stringer, E. L. Cooley, & A. L. Christensen (Eds.), *Pathways to prominence in neuropsychology: Reflections of twentieth-century pioneers*. Hove: Psychology Press.

Posner, M. I. (1978). *Chronometric explorations of mind*. Hillsdale, NJ: Lawrence Erlbaum Associates.

Pruthenus, L. (1498). *Trilogium animae*.

Raichle, M. (2008). A brief history of human brain mapping. *Trends in Neuroscience, 32*(2), 118–126.

Reitan, R. M., & Wolfson, D. (2009). The Halstead–Reitan neuropsychological test battery for adults—Theoretical, methodological, and validational bases. *Neuropsychological Assessment of Neuropsychiatric and Neuromedical Disorders,* 1.

Savoy, R. (2001). History and future directions of human brain mapping and functional neuroimaging. *Acta Psychologica, 27*, 9–42.

Scoville, W. B., & Milner, B. (1957). Loss of recent memory after bilateral hippocampal lesions. *Journal of Neurology, Neurosurgery and Psychiatry, 20*(1), 11.

Smith, E. (2001). Cognitive psychology: History. In N. Smelzer & P. Baltes (Eds.), *International encyclopedia of the social and behavioral sciences* (pp. 2140–2147). Oxford: Pergamon.

Squire, L. R. (2004). Memory systems of the brain: A brief history and current perspective. *Neurobiology of Learning and Memory, 82*(3), 171–177.

Sternberg, S. (1969). The discovery of processing stages. *Acta Psychologica, 30*, 276–315.

Toga, A., Clark, K., Thompson, P., Shattuck, D., & Van Horn, J. (2012). Mapping the human connectome. *Neurosurgery, 71*(1), 1–5.

Toga, A. W., & Mazziotta, J. C. (Eds.). (2000). *Brain mapping: The systems*. San Diego, CA: Academic Press.

Toga, A. W., & Mazziotta, J. C. (Eds.). (2002). *Brain mapping: The methods*. San Diego, CA: Academic Press.

Tucker, D. M., Watson, R. T., & Heilman, K. M. (1977). Discrimination and evocation of affectively intoned speech in patients with right parietal disease. *Neurology, 27*(10), 947.

Vesalius, A. (1543). *De humani corporis fabrica.*

Wernicke, C. (1874). *De Aphasische Symptomen-complex*. Breslau: Max Cohn & Weigert.

Weather Maps

James R. Carter

Abstract Data has to be observed, evaluated, aggregated, and distributed from many sites in a short period of time to create a weather map of any significance for forecasting. Telegraphy permitted the construction of the first meaningful weather maps. For a century international cooperation has facilitated the sharing of weather-related data for coverage beyond national boundaries. Globally weather data are collected regularly through many processes and at many heights and distributed widely. Many mapping systems have been built on this data-rich environment. Weather maps were some of the first graphics available for view online when the Internet went public. Television integrated weather maps into their programming early on. Weather maps are included in many newspapers. Maps are tools used in the study of weather and climate. Many maps will be created and studied in attempts to understand the patterns and behavior of atmospheric phenomena at many levels. A detailed map study of the 1974 Super-Outbreak of tornadoes led to terminology used today to characterize tornadoes. The world map of climatic patterns published at the start of the 20th century has been updated to portray the patterns of climate predicted at the close of the 21st century. Weather is a global phenomenon operating at many scales and weather maps of many types are employed to try to understand and forecast this part of our dynamic environment.

Keywords Global weather observation • Data rich mapping environment • Weather radar • Satellite imaging • Television weather • Climate models

Think of weather as the state of the atmosphere and climate being the expected range of weather. Thus, with weather we want to know what is happening now and predicted to happen in the near future. Climate takes the longer view based on what we have experienced. We want to understand how our atmosphere behalves at all spatial and temporal scales and maps are basic tools in gaining that understanding.

J.R. Carter (✉)
Department of Geography-Geology, Illinois State University, Normal, IL, USA
e-mail: jrcarter@ilstu.edu

S.D. Brunn and M. Dodge (eds.), *Mapping Across Academia*,
DOI 10.1007/978-94-024-1011-2_8

1 A World Map of Climates

Persons familiar with the world maps of climate are comfortable seeing red areas along the equator portraying tropical, moist zones; yellow and tan swaths indicating deserts at 25° North and South; and greens showing areas of dry summers and cool, moist winters around the Mediterranean Sea and in areas around Los Angeles, Santiago, Capetown and Perth. Such climate types and color symbolizations were developed by Wladimir Köppen in a quantitative study of world climates over many years. His first map of this nature appeared in 1900.

His climate classification was later updated by Rudolph Geiger and the current map is now known as the Köppen–Geiger map of World Climates (Kottek et al. 2006). Over the years variations in this system have been advanced but this map or something similar is found commonly in atlases and textbooks.

Sanderson (1999: 673) traced the evolution of classifying and mapping world patterns of climate, recognizing the dominance of the Köppen system. She notes many have explored and developed other climate classification systems but they have not been widely accepted. She suggested it is time to develop a new classification of world climates. However, Kottek et al. (2006) and Rubel and Kottek (2010) point to a number of recent studies using the Köppen–Geiger system as a reason to update the data on which the system is based. They created a new version of the map based on high resolution gridded data generated from digital models extending over the 50-year period 1951–2000 (Kottek et al. 2006).

Figure 1 is the current map based on their gridded datasets. With this detailed data they identified 31 classes of climates but note one class does not occur on the map and six classes occupy very small areas on the map at this scale. Note that with 31 classes it is difficult to define and portray this many colors such that each color is distinct from all others and can be identified with a particular climate type. Color choice and application are important components of mapping.

So, the world climate map that has been around for more than a century is still with us. Looking ahead, the authors engage in 'Climate Forecasting' by creating an animated map showing how the world patterns of climate will change through 2100 with their assumptions of global warming (Rubel and Kottek 2010). That animated map is the default image on the home web site (Institute for Veterinary Public Health 2015a).

2 A Global Map of Weather

Every day a great number of weather maps and weather related images are posted on the Internet. Among this mix is the relatively small Global Satellite and Surface Temperature Montage updated every 6 h (Space Science Engineering Center 2011). This map is a composite of satellite data, weather observations and model outputs. Created as an image of 640 by 480 pixels, each cell has a resolution of about

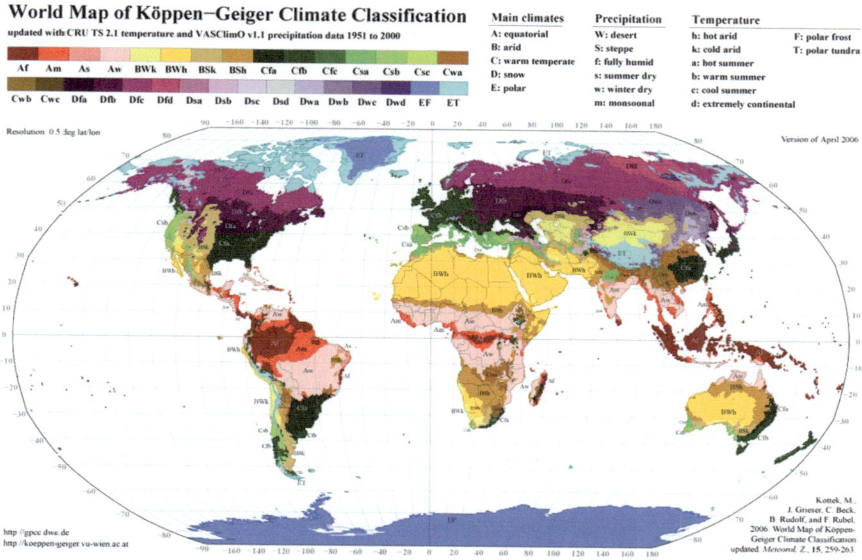

Fig. 1 Updated Koppen–Geiger map based on separate gridded datasets of temperature and precipitation for 1951–2000 (*Source* Institute for Veterinary Public Health 2015b)

35 km. As such it shows only larger weather features. Examining a single map in the series gives a broad perspective on global weather and is a complement to the climate map above. The Global Montage can be viewed as a sequence of 40 maps showing the weather evolving over a 10 day period. A single map is presented here for discussion, but many more await online viewers.

These maps are on a Mollweide equal area projection with lines of latitude parallel to the equator. Ocean temperatures are represented in a gradient of aqua colors ranging from 0 to about 32 °C. Land temperatures are portrayed in a color gradient ranging from purple through the color spectrum to dark red representing temperatures ranging from −60 to +40 °C. Note that the range of temperatures over land is much greater than over the oceans. These land/water temperature differences are drivers for much of our weather. Overlying all are images of clouds with the lightest being the coldest cloud tops (Fig. 2).

This map is for February 2, 2011, when the northern hemisphere was in winter. The patchy line of clouds extending east-west a little south of the equator shows the ITCZ (Inter-tropical Convergence Zone) where warm moist air converges to produce the rains of the humid tropics. At this time Indonesia, Malaysia and eastern Australia were covered with massive clouds. That inverted comma-like mass of clouds over NE Australia shows Tropical Cyclone Yasi, which was rated a category four storm.

A cloud-free zone north of the ITCZ extends from Central America to eastern China. Here the air is subsiding to produce the deserts and borderlands. The Indian

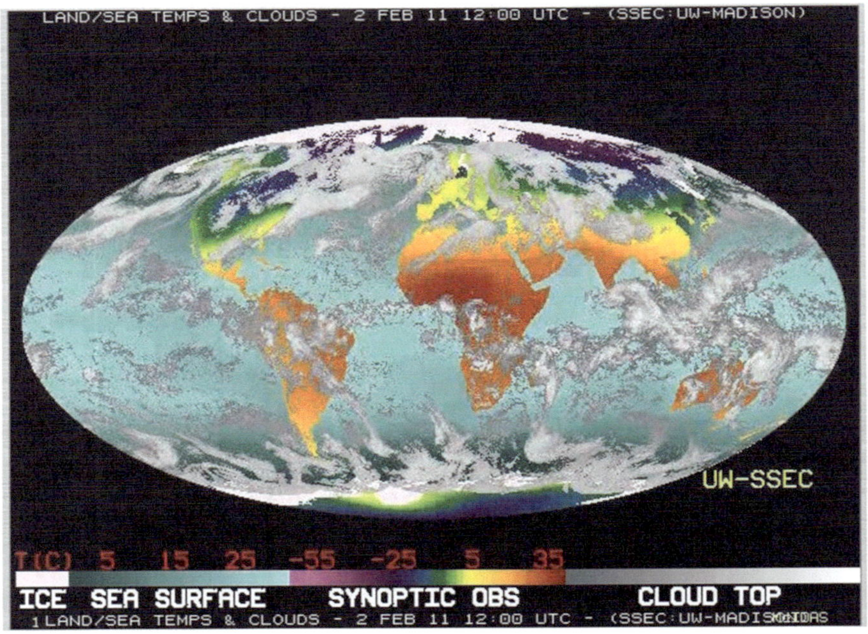

Fig. 2 An instantaneous global satellite and Surface temperature montage on equal-area map projection showing how the patterns of weather are consistent through broad latitude bands. This image represents noon at the Prime Meridian on February 2, 2011 (*Source* Space Science and Engineering Center 2011)

Sub-continent is cloud-free at this time but in 6 months that mass of clouds to the south will move north to bring the monsoonal rains (Carter and Kohrs 2011). Half of the Earth surface lies between 30°N and 30°S—the tropics and subtropics. Within this zone weather is dominated by the seasonal shift of the ITCZ with subsidence both to the north and south. Tropical cyclones, locally referred to as Hurricanes, Typhoons and Cyclones form in this area, move from east to west and normally move poleward.

At the time of this Global Montage much of Europe was cloud-free and relatively warm, especially compared to North America and Siberia, bathed in greens, blues and purples showing very cold air. Note this is at 12:00 UTC so it is noon in Europe and early AM in the Americas.

The cloud patterns in the middle latitudes in both hemispheres are more massive than in tropical areas and stretch out over hundreds of miles. In places the clouds spiral into a counter-clockwise rotations in the northern hemisphere and clockwise rotations in the southern hemisphere. Here cold air masses clash with warm air masses forming cold fronts and warm fronts, all moving west to east as is readily seen in a 10-day loop.

Viewing a 10-day loop of the Space Sciences Engineering Center Global Montage shows the distinction in the nature of the weather between the tropical areas and the middle latitudes. Because of these differences the nature of the weather maps created for the more tropical areas are frequently different from the weather maps of the middle latitudes, which have become the standard.

3 Evolution of the Weather Map

In the eighteenth and nineteenth centuries scientists in Europe and North America explored the nature of weather recognizing a relationship between atmospheric pressure, temperatures, winds, clouds and precipitation. Without an ability to collect and assemble observations from many sites in near real time, all scientific exploration was historical—reconstructing weather events based on observations from earlier times.

With the creation of the telegraph in the mid 1800s data were able to be assembled in near real time and soon weather maps were compiled in Europe and North America. Creating institutions and organizations to coordinate, construct and distribute weather maps was fitful at first. In 1873 the International Meteorological Organization was formed providing a mechanism to standardize the tools and procedures to gather weather data, code it, transmit it and map it in a systematic and consistent manner. No matter how attractive a map may be, it can be no better than the quality of the data that was used to make the map. The IMO evolved into the World Meteorological Organization under the United Nations which today still facilitates the systematic collection and sharing of weather data around the world. Such cooperation provides the rich source of quality information to produce weather maps.

Over the next few decades a basic weather map evolved as governmental organizations systematized the collection, dissemination, and display of data and maps to their own and international standards. As an example at the close of the 19th Century, the U.S. Weather Bureau was headquartered in Washington, D.C. Weather observers were set up across the country with the task of routinely making observations of temperature, pressure, wind speed and direction, and cloud cover at the same time twice a day. The collected data were coded and with priority were telegraphed to headquarters where the messages were decoded and plotted on a map. Then lines of equal pressure (isobars) and equal temperature (isotherms) were constructed to reveal centers of low and high pressure. Symbols indicating wind direction, speed and cloud cover were plotted at the location of the station on the map. Referencing preceding maps permitted plotting the tracks of lows. A team of persons constructed these maps and printed them in short order. They were then posted in public places and disseminated widely (Monmonier 1999: 7–9).

In many cases the observed data were also telegraphed to regional centers where local weather persons crafted maps integrating local conditions. To build and maintain a network of trained observers employing standardized equipment and procedures required a bureaucracy and a sizable budget. In the United States this became the Weather Bureau and similar organizations were established in other countries.

As this technology became more routinized and weather scientists became more familiar with the maps and the resulting weather they became more comfortable issuing weather advisories and warnings. In some cases scientists saw patterns in the many weather maps and identified types of weather situations and based predictions on what had happened in the past given this particular setup. They also compiled maps of preferred storm tracks based on patterns observed in the past (Monmonier 1999, 10–17).

Getting access to the appropriate information is always a problem, more so then when few stations existed off shore. Of particular note was the case in 1900 where a hurricane hit near Galveston, Texas, killing more than 8000 persons—the worst natural disaster in U.S. history. At that time there was no mechanism to exchange information with ships at sea and the U.S. Weather Bureau had discontinued accepting information from Cuba. The Weather Bureau predicted a tropical cyclone would move up the east coast, as shown on their maps. Thus, because the map did not show a storm in the area there was no basis to warn Galveston (Emanuel 2005, 83–90). This example illustrates the importance of timely, accurate and relevant maps.

Weather observation and mapping became rather standardized throughout the developed world in the late 1800s and the early 1900s. With the onset of the Great War in 1914 priorities prevented sharing information because knowledge of the present and forecast weather has strategic advantages (Fig. 3).

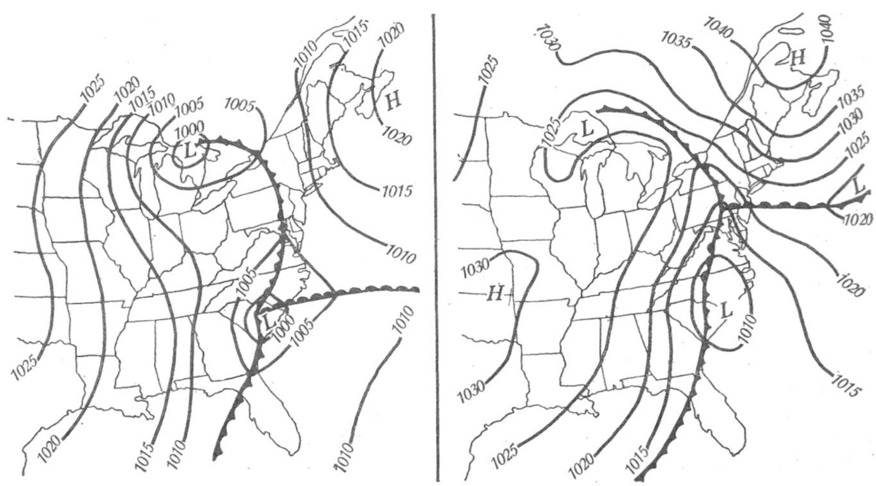

Fig. 3 This illustration demonstrates how the study of previous weather map patterns are useful in predicting the outcome of future weather events. The chart on the *left*, for 8th November 1913, was used as an analogue by forecasters to correctly predict a storm and heavy snow on the 25 November 1950 (*Source* Tannehill 1955: 88)

A good example of the surface weather maps in the first half of the twentieth century is seen in this pair of maps in *Water: the Yearbook of Agriculture 1955* to illustrate how weather patterns from the past may foretell the future (Tannehill 1955: 88). In this case the map of 1950 was similar to that of 1913 and in both cases the resulting weather was very similar.

These two maps were simplified for presentation in that book but they show the basic form of the surface weather map. The smooth black lines are isobars, showing surface pressure reduced to sea level. The numbers are pressure, expressed in hectopascals. Standard sea level pressure is 1013.25 hPa. (1 hPa is the same as 1 mb, or millibar, the term still used frequently in the U.S.) On the November 1913 map on the left there are two centers of Low pressure bounded by a small circle labeled 1000, on the Carolina coast and over the northern Great Lakes. In northeast Canada there is High Pressure greater than 1030. On these maps the isobars are at every 5 hPa.

The map for November 1950 shows two lows similarly located over the area and two Highs, one in almost the same area of northeast Canada. The other High is located in the U.S. west of the Mississippi River. On the 1913 map the pressure west of the Mississippi was quite high, but not labeled a High. Indeed, the two maps show similar pressure patterns that produced similar ‘destructive storm and heavy snow’.

The other features on these maps are the bold black lines with semi-circles and barbs, representing fronts. Today we are familiar with warm, cold and occluded fronts but fronts were not recognized until the late teens when a team of meteorologists in Bergen, Norway set up a dense network of observation stations and determined that at the center of the low pressure where warm and cold air come together are discontinuities, misrepresented by smooth curvy isolines (Friedman 1989). The Bergen group established the Polar Front Theory as a three-dimensional model where the cold air undercuts less dense warm air and warmer less dense air overrides the cooler (Monmonier 1999: 57–66). The concept of fronts inherent in the Polar Front Theory took time to be incorporated into weather maps. Thus, we must conclude that the fronts shown here on the 1913 map were not on the original weather map of that date.

The cold front is the black line with triangular barbs pointing in the direction the cold dense air is moving at ground level. The warm front employs semi-circles pointing in the direction the warm, less dense air is moving at ground level. In most cases the boundary between warm and cold air masses north of the center of Low pressure is above the surface. The trace of that boundary is called an occluded front because it is occluded from the ground. Today that is shown as alternating barbs and semi-circles pointing in the direction of motion. On these maps the symbolization is not clear.

These two surface weather maps are simplified versions of the maps meteorologist have been creating many times a day for decades. The weather scientists want details about what is observed at each weather station. This information is plotted as a complex code at each site. Over the years the weather maps have evolved as we learn more. The Government of Canada (2015) provides a web site where readers can bring up current and forecast surface weather maps for Canada and for the Northern Hemisphere.

The Government of Canadian (2015) statement at this site summarizes the process:

> Understanding current conditions is the starting point, and the most critical part, of any weather forecast.
>
> Hundreds of weather stations, ships, and aircraft across Canada, the US, and the rest of the world report readings of temperature, pressure, wind, moisture and precipitation. These reports are received at the CMC, and used to improve our picture of the current state of the atmosphere. The analyses made available four times daily on this page provide the latest snapshots of the state of the atmosphere over Canada and the Northern Hemisphere.

4 Weather Aloft and the 500 HPa Chart

In the early part of the twentieth century scientists sought to explore the atmosphere aloft. They employed many tools to capture information including kites, balloons and aircraft. With time we developed sounding devices and systematized the process so that we now regularly gather data from the surface to the upper reaches. With the development of computer models such observations are entered into routines which construct maps of the atmosphere aloft.

On surface weather maps we plot observed atmospheric pressure adjusted to sea level. We have to adjust pressure observations to a common elevation because pressure decreases rapidly with elevation. It is the small differences in atmospheric pressure that direct the winds.

The 500 hPa pressure surface has become a basic reference level to report and map upper-level pressure patterns. Sea level pressure is about 1000 hPa so the 500 hPa pressure level means about half of the atmosphere is above that level and half below. And, 500 is a nice round number. Currently, atmospheric pressures maps are created for many other pressure levels, such as 850, 700, 300, and 200 hPa.

We are interested in the height of that 500 hPa pressure level. On average that pressure occurs at roughly 3.5 miles, 5.5 km or 18,000 feet aloft. In the tropics where the air is warm the 500 hPa level will usually be well over 6 km high, while in the Arctic in winter that level is likely to be below 5 km. So, the 500 hPa chart uses contours to give shape to the pressure pattern at that level. (Note: the surface weather map uses isobars to show pressure patterns and the upper level pressure maps use contours to show the height of a pressure surface.)

The NOAA 500 mb (hPa) chart below is centered on the North Atlantic Ocean extending from the central U.S. to western Norway. It extends north from Cuba to the southern tip of Greenland. This is plotted on the Mercator Projection so that angles and directions are correct across the map (Fig. 4).

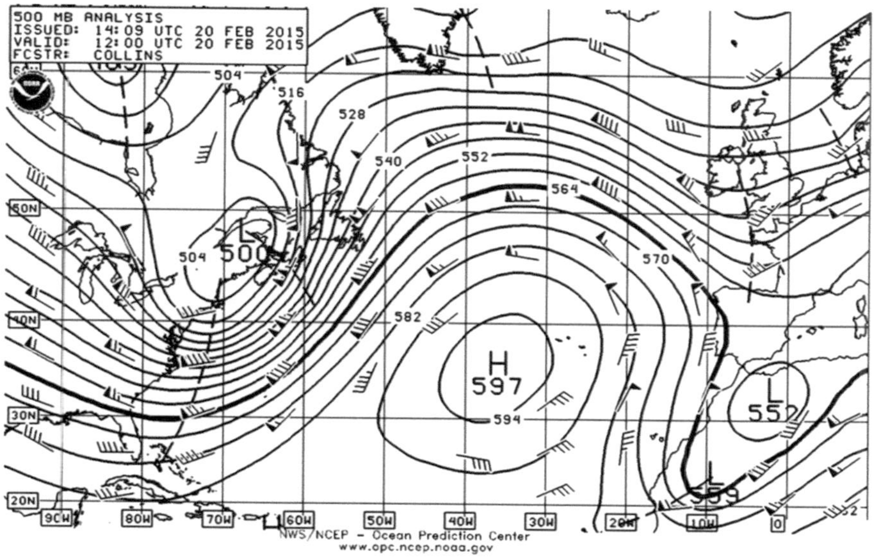

Fig. 4 Example of 500 mb (hPa) analysis for 14:09 UTC 20 February 2015 for the North Atlantic Ocean showing a ridge of high pressure at 35° West and troughs of low pressure off the North American coast and NW Africa (*Source* NOAA 2015a)

Those solid black lines running E–W are the contours in units of 10 s of meters. The darker index contour labeled 564 represents 5640 m. The contour interval is 60 m. The H-597 in the Atlantic is a high at 5970 m. The L-500 over Nova Scotia is a low at 5000 m. Between those two places there is a difference of 970 m in the height of that 500 hPa surface.

Air like all fluids flows downhill, thus it should flow from that high to the low along the pressure gradient. But, in the northern hemisphere the Coriolis Force diverts the air to the right so that at this level winds blow parallel to the contours. These winds, often called 'westerlies', are shown by arrows with barbs indicating wind speed.

Look at the height/pressure along 40°N moving west to east. Off the New England coast the contour is 510, it is 596 in the central Atlantic and about 556 in the Mediterranean Sea—representing heights of 5100, 5960 and 5,560 m respectively for the 500 hPa surface. Thus, moving west to east the 500 hPa surface dips off the east coast, crests over the Atlantic and dips again over the Mediterranean. Those dips are "troughs" and the crest is a "ridge", terms used commonly in weather discussions.

On this map dashed lines perpendicular to the contours delineate the center of the troughs. One such line extends from New England to Cuba. Ridges are not delineated. Moving W to E the contours and winds curve in a clockwise rotation going over the ridge, what is called anticyclonic flow. In general, anticyclonic flow represents good weather. Trough lines are located at the center of the counter-clockwise curve of the contours. This indicates cyclonic flow which indicates the convergence of air and storminess. The word ‘cyclone’ is applied to cyclonic storms of the mid-latitudes, tropical cyclones (hurricanes and typhoons) and cyclones (tornadoes).

On these upper level charts there are places where the winds are strong enough to be called Jet Streams. Labeling such ribbons of fast flowing air jet streams is more important for presentations to the public than for forecasters—they prefer to look at the numbers.

For more than a century the atmospheric communities around the world have been producing weather maps based on frequent observations at the surface and aloft. With new technologies computers construct the maps and generate the graphic products, but humans still study the maps and make interpretations. There are a variety of computer systems that employ different programs to create the models of the weather. Thus, there are often small variations in the forecasts. The public became aware of such variation in 2015 when the Nor’easter named Juno was predicted to move northeast off the Atlantic coast and three commonly used forecast models generated different predictions of the track of the low offshore. In one model New York City was predicted to get 8 inches of snow and in another 20 inches. In the past weathercasters made forecasts based on interpreting the map patterns. Now humans have to choose which computer model output to follow.

‘Ensemble’ is the term given to the production of a collection of maps (models) based on running the same program with slight variations in inputs or parameters (Wikipedia 2015). Creating ensembles is recognition that we are not all-knowing in mapping weather.

5 Mapping to Gain Understanding of the Atmosphere

While most weather maps are based on the systematically collected data, some persons use maps to discover aspects of weather not addressed by such standard data. Dr. T. Theodore Fujita produced two color maps as by-products of his research on tornadoes and intense storm events.

In an 18-h period in April 1974, 148 tornadoes swept across a large swath of the eastern U.S. in what was called Super Outbreak. Fujita assembled a team to identify and map the tornado paths by flying over sites and taking photos and using other sources of aerial photography and satellite imagery. In the process he identified the varying widths of each touchdown and classified each segment according to his F-scale. To disseminate knowledge of this event and to solicit feedback from the larger public he put together a map (55 × 41 cm) detailing the path and width of each touchdown, F-scale intensities along each segment, the underlying topography, a table listing each tornado and many graphs and statistics. Twelve thousand copies of this map were printed and distributed widely (Fujita 1974, 1975a, b; the map is available for download at National Weather Service 2015c).

In 1992 Hurricane Iniki proved very destructive as it crossed the Island of Kauai, Hawaii, from south to north, with the first winds largely from the east and the second winds dominantly from the west. Using more than 800 air photos and 2800 ground photos plus field inspections Fujita and his colleagues identified 27 downbursts and 2 mini-swirls accompanying the hurricane. The focus was the effects of the hurricane in this area of high and complex relief. A product of this investigation was a multi-color map for public information with contours shown in green and hypsometric tints for the higher elevations. Locations of Microbursts are shown with large, numbered dots and the two Mini-swirls are shown with open circles. In great quantity are vectors showing the direction of blow down material. Red dots and vectors represent the first wind products and blue those from the second wind. There is much information to be gained by integrating the vectors and downbursts with the topography. This is a map to be studied and enjoyed. The Iniki map was included as a separate in NOAA *Storm Data,* 1992 (the map is available from the University of Hawai'i at Manoa web site; Fujita 1992; NOAA 1992).

These two colorful maps are the products of a scientist who appreciated the value of sharing his findings with a larger audience through cartography. They deserve to be recognized for their cartographic value as well as their scientific insights.

6 Weather Maps in Newspapers

Weather maps in newspapers first debuted in the mid 1800s in various forms and were more occasional features than regular, reliable sources of information. In the early 1910s, a brief program by the U.S. Weather Bureau saw more widespread dissemination of a daily weather map, but this program, while popular, was terminated for many reasons with demands of the First World War (Monmonier 1999: 153–168). With time many national weather organizations came to work with the newspapers to make certain the maps were dispersed to inform the public and to build good will for the weather organization. But, in some cases and on some occasions there has been tension between the governmental weather organization and the media for a variety of reasons.

The modern era of weather maps in newspapers began in 1935 with the debut of the Associated Press's Wirephoto network. Over the next few decades, most newspapers added a daily weather map; Call (2005) found that 90 % of major newspapers had such maps by the 1960s. Call classified newspaper weather maps in three typologies: synoptic, which show spatial weather features such as fronts, highs, lows, and areas of precipitation; symbolic—essentially the map is spatially-arranged table of forecasts or observations; and "hybrid-symbolic" which is symbolic map with a minimum of weather data.

In 1982 the national paper *USA Today* came out with a full page dedicated to weather with a large color map showing predicted high temperatures, essentially a symbolic map with isotherms. This bold presentation of weather led to much greater use of color in the weather maps of other newspapers. While there is considerable variation in the nature of the weather maps in newspapers around the world most emphasize the national or regional weather, in part depending on the size of the country and the nature of the weather in that area (Monmonier 1999: 153–176).

As of 2015 newspapers are fighting for their niche in the news business but in most cases among the products they offer is a weather map. It may be in color with cold fronts shown in blue and warm in red if they show fronts. It may have golden suns, green rain patches and snow stars. Newspapers have been and will be a place to see weather maps in some form.

7 Television

During Second World War great advancements were made in science and technology as persons on all sides were trained and assigned to apply their knowledge and skills to gain strategic advantages. Radar became a functional tool. Aircraft became larger and flew higher, gaining new insights about the atmosphere aloft. Fleets and troops operated in tropical and polar environments and encountered storms and weather events they were not familiar with. Rocketry was developed which would have great impact on the atmospheric disciplines within years.

After that war the United States was relatively unscathed and had a great amount of technology, and persons who understood that technology. Henson (2010: 9) noted that in 1950 there were 9.7 million television sets in the U.S. broadcasting from stations scattered across the country. And, coming out of the War was a 'bumper crop' of persons trained in meteorology and looking for work. It was not long before television brought weathercasters with their weather maps into the homes of Americans.

In the U.S. television was little regulated and there was competition between broadcasters. (Note: this is not true in all countries, even today). The presentation of the weather became part of the competition between stations and weathercasters and maps competed for the eyes of viewers. In some cases clowns and skimpy clad girls were the attraction, but in many cases a knowledgeable weathercaster educated the viewers while developing the weather story on a series of maps. In those early years the weather person had to rely on numbers read from tabular forecasts or perhaps had access to crude maps not suitable to presentation to the public. So, they had to create the maps and graphs by sketching on large pieces of paper or standing behind a clear plastic wall and drawing maps backwards so it looked right for the viewers at home. The occasional doodle and a little humor created sizeable audiences to watch the weather locally and afar (Henson 2010).

With the development of color television, green screen technology (chromacolor) permitted a weather map to be shown on the TV screen while the actions of the weathercaster in front of a green screen were imposed on the map (Carter 1998). In many cases competent weather personalities walked viewers through the patterns on the maps and taught many viewers to read weather maps.

While television stations proliferated there were firms and research centers with meteorological expertise that specialized in assembling, customizing and packaging weather information and presentations. Still today, these firms provide maps and presentations to TV stations and industry. With many sources of expertise producing maps and forecasts there are variations and viewers switch channels to see competing forecasts with different maps. Today most television weather programming is made available for viewing on the web and mobile devices.

The weather community soon incorporated radar, a system that sends out a microwave signal and displays the reflected echo. As the technology developed meteorologists found they could detect specific weather events, including patterns that often lead to tornadoes. The U.S. government set up a system of weather radars that could provide national coverage. A few television stations purchased their own weather radar as a competitive advantage and so advertised that they had better weather forecasting for local concerns. With computer graphic developments, radar echoes became part of the dynamic image on the screen. Soon radar images were integrated with political boundaries and road networks to permit viewers to locate themselves relative to the weather event.

Within the United States the National Weather Service established systematic coverage of the conterminous country. The current system is known as WSR-88D, based on a prototype developed in 1988. With the Internet the output of this radar is made available as the images from a single weather radar site and as a composite for the 48 conterminous states. Users can click on the larger map which will take them to the radar display of the closest radar site.

Figure 5 shows the Composite Radar but viewers can toggle to base radar, two options of velocity showing how quickly things are moving toward or away from the radar site, and 1-h rainfall and total storm rainfall. In addition to these static maps, viewers have the option to show the map as a loop over the past hour. Each radar image is updated approximately every 5 min.

While this type of map image is available directly from the National Weather Service, many organizations take the basic data and repackage it into enhanced presentations. In many cases such presentations are presented at larger map scales on more detailed base maps on television and web based media. With their own Doppler radar systems television stations can provide their local viewers with more timely and more focused images and warnings. In some cases weathercasters can point to a specific neighborhood and warn residents in that area to take shelter. Large scale maps viewed in real time have become important tools in saving lives.

While weather has become an integral part of the news programming of most local television stations, the weathercaster has to compete for time on screen. It is not uncommon for a 3-min weather segment to be cut in half minutes before going on air. On the other hand, in the event of a severe weather event in the local viewing area, weather programming may displace the other forms of news (Carter 1998).

In 1982 The Weather Channel came into existence in the U.S. on cable television presenting weather and weather maps 24 h a day. In this environment, weather maps are the priority. The Weather Channel model was tried abroad but did not survive. At least one 24-h weather competitor was created in the U.S. but it was

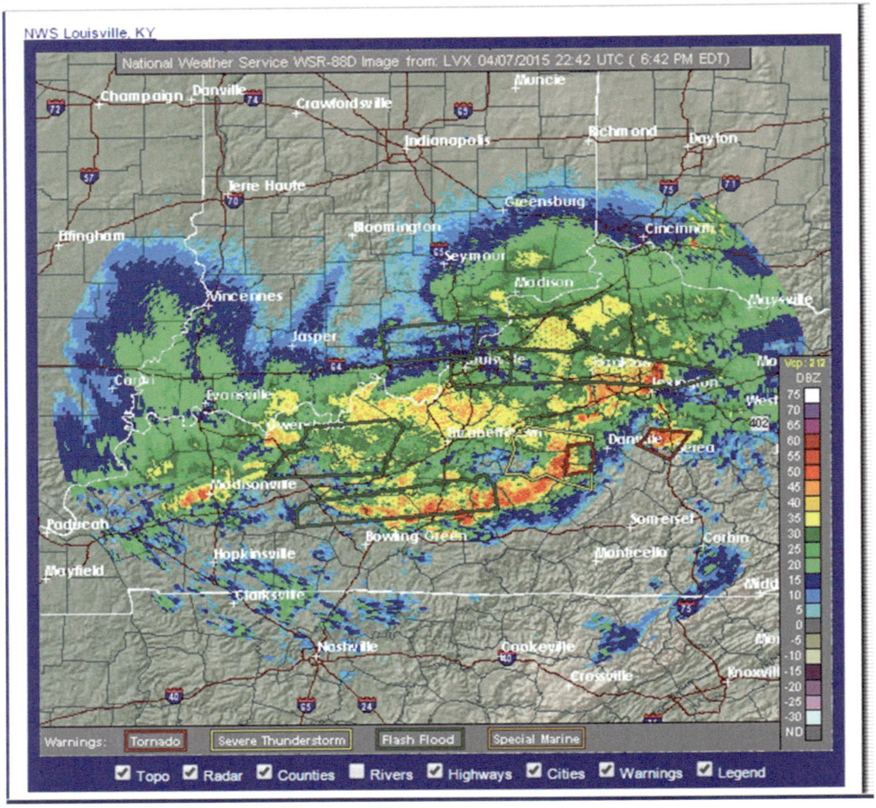

Fig. 5 The Short-range National weather service composite WSR-88D radar image map from the Louisville, Kentucky, radar station of 7 April 2015 at 22:42 UTC. Two red polygons show Tornado warnings, one yellow polygon forewarns of Severe Thunderstorms and many green polygons warn of Flash Flooding. On this map topo, county boundaries, highways and city symbols underlie the radar imagery. State boundaries and warning polygons overlay the radar imagery (*Source* National Weather Service 2015a)

unsuccessful. Thirty-three years later The Weather Channel continues showing animated isobars and fronts highlighting zones of rain, snow, ice, etc.

8 Satellite Imagery

In 1960 the first Tiros satellite was launched and within days was sending images of clouds. Meteorologists soon added these images to weather broadcasts to give perspectives unavailable before. In the half century since, satellites have provided a great amount of data which has been converted into images to add to weather maps.

The multitude of satellites has added greatly to our knowledge of the state and behavior of our atmosphere. The Space Science and Engineering Center (2015) at University of Wisconsin-Madison offers a diverse collection of satellite based images and maps relating to weather.

Satellite imagery has proven particularly effective in identifying and monitoring the tropical cyclones, which are known as hurricanes in the Americas and typhoons in East Asia. These systems form in the lower latitudes where clashing air masses and fronts are rare and mid-latitude weather maps have less meaning. So, satellite imagery provides a ready source of information about this large part of the world.

Many of us have seen loops of satellite images as weathercasters point out disturbances in West Africa moving west into the Atlantic to die out or to strengthen to become a tropical disturbance. Because there may be multiple tropical storms in existence at the same time and because such storms do not follow fixed paths we give them unique names when they reach tropical storm strength. With the potential nature of these storms we get to see many interesting weather maps showing the projected path of a particular storm, maps showing comparisons with the paths of previous large storms, and more. Then if the storm is predicted to move inland, we see maps of probable landfall, areas subject to flooding, potentials for interaction with other weather disturbances and warnings for travelers.

9 Computer Graphics, the Weather Map and the Internet

In the 1960s we developed the ability to store spatial information in the computer, analyze and manipulate the data and print out displays, including maps. Those line printer maps were crude but they were valid maps. By the 1970s we had graphic display terminals, where monochrome lines could be plotted on a screen. By the late 1970s the raster refresh display terminal emerged to bring forth what we now know as 'computer graphics.' On these screens combinations of red, green and blue pixels generated thousands of colors. Year by year the resolution of these displays increased and soon high quality graphic images were manipulated by scientists, artists and many others.

The atmospheric sciences were early and consistent users of this technology in large part because the there was an abundant flow of digital information ready to be processed, manipulated and displayed. Many university atmospheric science programs employed this technology for research and instruction.

Then in the early 1990s the World Wide Web became available for public access on something called the Internet. People could connect from home and office using their personal computers. In those early years there was little content on the web except the weather maps of those atmospheric science programs. They had the data, the technology and the freedom to share the displays. The maps were colorful, informative and the public took note. Likewise the university programs found they had viewers and took on the responsibility of providing maps and graphics for a larger audience. In 2015 a number of universities maintain web sites where they present images of our weather.

10 Lightning

Lightning is a weather phenomenon of lethal dimensions that historically has been difficult to monitor and map. Traditionally, when present, it is reported as observed at the station. In *Meteorology Today*, Ahrens (2003: 412) has a map showing "The average number of days each year on which thunderstorms are observed throughout the United States. (Due to scarcity of data, the number of thunderstorms is underestimated in the mountainous west.)"

With the emergence of electronic sensors and Internet connectivity we have gained the ability locate lightning strikes, assemble the data quickly and produce maps that can show the location of strikes in near real time. Holle and Cummins (2010) produced an atlas of monthly cloud to ground lightning flash density for conterminous United States and surrounding areas, expressed in flashes/km^2/month for the years 2004–2008. The data was collected as part of the National Lightning Detection Network (NLDN) which was first deployed for the entire USA in 1989. The maps reveal distinct patterns of lightning activity which vary month by month.

Lightning data collection, processing and distribution is largely a private/academic enterprise and proprietary systems are employed by different organizations. There are many ways these organizations package their systems and consumers can purchase the delivery of timely information to be delivered as contracted or set up their own sensors and become part of a network. With GPS equipped mobile devices and appropriate aps, maps and advisories of lightning threats will be sent when lightning storms are within a specific range of the device. It took the Internet to permit individual lightning strikes to be mapped and that same technology permits such maps to be delivered to individuals in near real time.

11 Examples of Lightning Maps

While aps can deliver lightning warnings to individuals and firms, there are systems that let us view larger patterns of lightning as maps. The United States and North America Precision Lightning Networks (USPLN/NAPLN 2015) provide an animated map of lightning strikes extending from Mexico and Cuba on the south to the southern half of Hudson Bay. The map employs dark blue for water and dark green for land with a lighter gradation for elevation. Lightning strikes are shown by + marks, the oldest in dark red are plotted first and then the red, orange, yellow and white are plotted in sequence. The legend is scaled in UTC time units showing when the strikes occurred. Movement of storms is evident by watching the sequence of the appearance of ever lighter + marks. Users can customize the spatial and temporal resolution of the display.

The World Wide Lightning Location Network (WWLLN 2015) headquartered at the University of Washington relies on a network of sferic sensors at universities and organizations around the world to create daily maps of lightning hour by hour. The maps at this site are presented in 4 panels on equator centered maps for the Americas, Europe-Africa, Asia-Australia, and the Pacific Ocean. Lightning is shown as blue dots on a black background. Each blue dot represents lightning within an hour period and the dots accumulate to show the total over 24 h. Satellite patterns of cloud motion complement the lightning patterns. Red circles show the location of the lightning sensors.

Using a similar technology LightningMaps.org (Blitzortung.org 2015) provides near real time maps of lightning for Europe, North America and Oceania, plotted on a variety of base maps that users can select between. Viewers can zoom into for considerable detail and by toggling appropriate switches watch red-yellow icons pop on the screen and fade out as newer flashes demand attention. According to the legend the capture and display of the strikes are delayed by less than 10 s. This lightning mapping system gives the user many options for viewing, including the ability to hear clicks with each strike and to display clouds and radar where available. This is a system in progress and shows where coverage is good and where more sensors are needed.

These lightning mapping systems are noted because they are suggestive of a direction efforts are moving to monitor and map this component of weather. Lightning has potential to damage and disrupt a world linked electronically and thus systems that can inform managers, decision makers and individuals with timely information are welcomed. In addition, being able to map lightning world-wide in near real time provides another perspective to help us understand weather. Such data and maps complement the traditional weather mapping systems.

The state of lightning mapping was tested when the author compared the way two mapping packages presented the evolution of thunderstorms in central Illinois. The Lightningmaps.org (Blitzortung.org 2015) images were viewed on a laptop computer and Spark (WeatherBug 2015) images were viewed on an IPad tablet. Both maps were zoomed into cover an area of about 60 miles/100 km on a side. Both systems employed base maps showing roads, hydrography, land cover and geographic names. The only substantive difference between the dynamic maps was the capture and display of lightning strikes.

The two mapping packages use different technology to detect lightning but the results were very similar. In only a few instances did a lightning symbol appear on one map and not seem to appear on the other and in most cases the capture of the lightning strike seemed to be quite synchronous. The Spark map employs a white dot which soon evolves into a yellow zig-zag icon to symbolize a lightning strike. That symbol stays in place for 30 min. Soon, a couple of areas appeared saturated with yellow icons while additional white dots added to the collection and showed the migration of the thunderstorm.

The lightningmaps.org display was viewed in 'real time' mode, which employs a larger red dot with a yellow center which evolves into a yellow dot which slowly grows darker to a brown color. It disappears after an hour. The change of color symbols over time made it easier to distinguish the total pattern of lightning strikes in areas with a high density of strikes.

In total, it is satisfying to have available near real time mapping systems of lightning so that individuals can monitor this weather threat on a number of personal devices that many people now have. While the functionality of such mapping systems is not world-wide, the trend is moving in that direction.

12 El Niño and La Niña

"The Year the Weather Went Wild" was the title for an article in *National Geographic* about winter 1976–1977 in the U.S. (Canby 1977) An annotated satellite image showed snow on the ground 1 day in all 48 states. According to long-range forecasters this weather related to surface temperatures of the Pacific Ocean. To illustrate they used a page-size perspective map showing the Jet stream, warm water and positions of highs and lows over North America, the Arctic and eastern Pacific Ocean in January 1977. A smaller map portrayed the normal January patterns.

Ecuador and Peru have long experienced an occasional warm coastal current which devastates their fishing industry. As it occurs at Christmas time they call this event 'El Niño' for the Christ Child. In 1982–1983 the rest of the world became familiar with this term to describe a world-wide weather phenomenon. In 1983 *National Geographic* employed a unique map projection to show the impacts of El Niño from India on the west to Africa on the east (Canby 1984). A two-page spread of five maps and a three-dimensional model was devoted to showing how warm currents move within the Pacific Ocean. That 1982–1983 El Niño event was reported with many maps in newspapers and magazines, including *Macleans, Newsweek*, *Readers Digest* and *Time* (Carter 2015).

The world was becoming aware that weather was global in its reach and that El Niño had a sister La Niña, the cold-water phase. ENSO (El Niño, Southern Oscillation) is the term now used for the larger complex incorporating both phases.

In our interconnected world great attention is devoted to forecasting the onset of El Niño and La Niña conditions. A rectangular area extending from 5°N to 5°S and from 120°W to 170°W in the central Pacific is designated as NINO3.4, the base area to evaluate departures and make predictions (NOAA 2015b). Periods of months of above average temperatures in this rectangle are designated El Niño events and periods below are La Niña events (Earth Systems Research Laboratory 2015).

The National Weather Service (2015b) provides a web page showing two animated maps of temperatures and temperature departures for the past 12 weeks of the Tropical Pacific Ocean. This gives information that people need to monitor trends relating to El Niño and La Niña.

There is much on the web relating to the effects of the ENSO weather phenomenon and many sites will have maps showing what has happened or is likely to happen with respect to El Niño or La Niña. The Australian Government (2015) has a collection of maps of Australia showing variations in rainfall patterns month by month for all El Niño and La Niña events back to 1902–1903.

The Royal Netherlands Meteorological Institute (2015) created world maps showing the correlation of precipitation and temperature departures by season around the world with the strength of the NINO3.4 index in the central Pacific. They also include a map of Tropical Cyclones relative to El Niño events. El Niño and La Niña are global weather phenomena and as such they provide much content to be mapped and studied (Fig. 6).

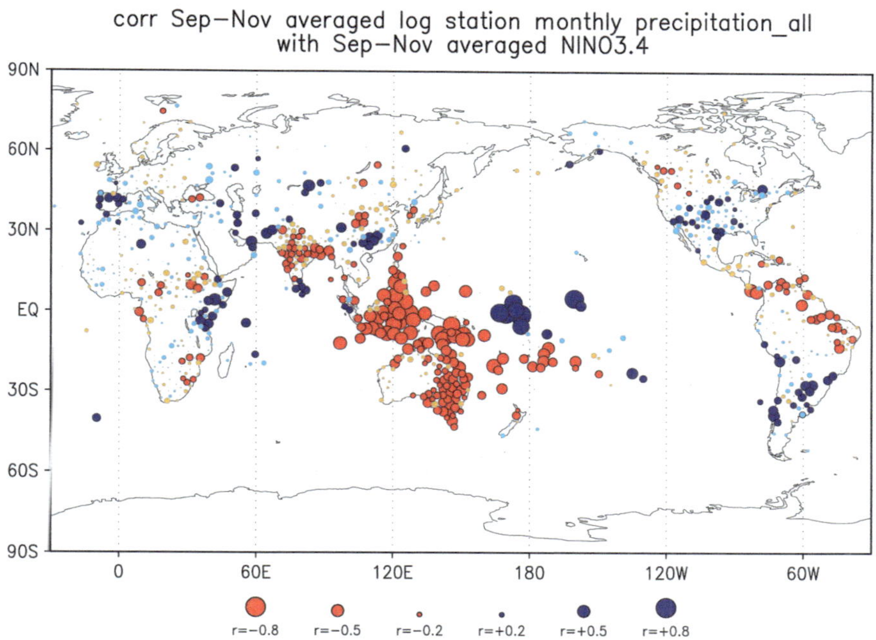

Fig. 6 September—November correlation of the strength of the NINO3.4 Index in the central Pacific with precipitation departures around the world. The Red circles over Indonesia, the Philippines, Australia, central India, and northern South America indicate relatively dry conditions during high El Niño events and relatively wet conditions during La Niña events. The *blue circles* over the Iberian Peninsula, east Africa, central North America, and the southern half of South America indicate the opposite response to the Index. The larger the circle, the stronger the correlation between the Index and precipitation departures (Royal Netherlands Meteorological Institute 2015)

13 Monitoring Global and Regional Patterns of Atmospheric Conditions

With the fleet of satellites from many countries now circulating the globe we are able to monitor many aspects of the atmosphere in addition to those factors that produce our day-to-day weather. NASA (2015) provides animated maps showing change month to month of 16 atmospheric variables over many years. Among the maps are land surface temperature anomalies, cloud fraction, snow cover, total rainfall, water vapor, net radiation, carbon monoxide, and aerial optical size.

A map of Yellow Dust warning (NHK World 2015) in East Asia was included among the many maps in the daily television presentation of weather on NHK World on April 16, 2015. The Korea Meteorological Administration (2015) produces a daily forecast chart of winds and particulate loadings at 1,500 m for all of south and East Asia—thus showing yellow dust incidents. On these daily maps particulate matter densities are symbolized with isolines and color tints while winds are portrayed with red arrows in proportion to speed and surface pressure is shown with fine blue isobars. This map is noted as one of many unique but relevant weather maps in the world mix of mapping the weather.

Another weather related variable regularly mapped is drought, a measure of the relative condition of the surface in terms of soil moisture, vegetative health, streamflow and water supplies. Drought as mapped is based on the relative dryness of an area. Thus, deserts can be in drought or can be wetter than normal, as can the humid tropics. Moisture conditions maps integrate the weather of the preceding weeks, months and even years. The U.S. Drought Monitor (2015) is updated weekly and employs five levels of colors of indicate drought intensity as contrasted to areas not in drought. At this site are links to other related map indices.

14 Map Products of Weather Model Outputs

As noted a great amount of weather observations are incorporated into computer systems which generate models of the atmosphere. Model data are now readily available for those with the appropriate technology to capture and display it. Three web sites are noted for the innovative and interesting map products available at the click of a button. Surely, there are and will be many more sites offering innovative displays of weather data in its many dimensions.

COLA (2015) provides innovative weather map presentations for the entire world broken into 8 viewing areas, i.e., Northern Hemisphere, Africa, etc. There are seven maps for each viewing area portraying unique perspectives on the weather, at the surface and aloft. An interpretation guide is included to help the viewer understand what is being shown. The maps at this site are updated twice a day. Two of the maps include streamlines, which are vectors showing the flow of the air at a particular level. Streamlines are the cartographic symbolization used at two other web sites of note.

For the conterminous U.S. the HINT.FM (2015) wind map presents animated streamlines of varied thickness pulsing in the direction of the air flow. The author notes, "The wind map is a personal art project, not associated with any company." The data come from the hourly posting of a government agency.

The Nullschool (2015) earth site portrays streamlines imposed on a variety of world map projections including a perspective globe, which can be rotated with the mouse. Viewers can select a wide variety of weather variables to portray as well as different map projections and scales. This is another personal project using model output data from a government agency.

15 Conclusion

We started by looking at global patterns of climate and then viewed a snapshot of global weather at a small scale. We saw how individual observations of temperature and pressure are integrated into weather maps and computer models that are used at many scales. Weather events range from local to global and weather maps are important tools helping us live with the vagaries of our atmospheric environment.

It is insightful to view these innovative map products in comparison to the current SSEC world weather map introduced at the start of the chapter and the current surface and upper air weather maps referenced in the chapter. Hopefully, viewers will be able to identify the same weather events represented in many different ways on these many maps. Indeed, weather is a data rich world and there is much to be seen and experienced with the weather maps now available on the Internet.

References

Ahrens, C. D. (2003). *Meteorology today: An introduction of weather, climate, and the environment* (7th ed.). Pacific Grove: Brooks/Cole.

Australia Government. (2015). *El Niño—Detailed Australian analysis*. http://www.bom.gov.au/climate/enso/enlist/. Accessed 14 Mar 2015.

Blitzortung.org. (2015). *Lightning maps*. http://www.lightningmaps.org/realtime. Accessed 2 May 2015.

Call, D. (2005). Fair weather ahead? Changes in newspaper weather maps: 1902–2005. *The Pennsylvania Geographer, 43*(2), 45–65.

Canby, T. Y. (1977). The year the weather went wild. *National Geographic, 152*, 798–829.

Canby, T. Y. (1984). El Niño's ill wind. *National Geographic, 165*, 144–183.

Carter, J. R. (1998). Uses, users, and use environments of television maps. *Cartographic Perspectives, 30*(Spring), 18–37.

Carter, J. R. (2015). Weather map. In M. Monmonier (Ed.), *History of Cartography, Vol. 6: Cartography in the twentieth century*. Chicago: The University of Chicago Press.

Carter, J. R., & Kohrs, R. A. (2011). Recognition of six-hourly global maps of ice, temperature and clouds—Real-time and 10-year archive. In *Proceedings, international cartographic conference*, Paris, France.

COLA. (2015). *Weather and climate data*. http://wxmaps.org/pix/analyses.html. Accessed 15 May 2015.

Earth Systems Research Laboratory. (2015). *Multivariate ENSO Index (MEI)*. http://www.esrl.noaa.gov/psd/enso/mei/. Accessed 21 Apr 2015.

Emanuel, K. (2005). *Divine wind: The history and science of hurricanes*. Oxford: Oxford University Press.

Friedman, R. M. (1989). *Appropriating the weather: Vilhelm Bjerknes and the construction of a modern meteorology*. Ithaca: Cornell University Press.

Fujita, T. T. (1974). Jumbo tornado outbreak of 3 April 1974. *Weatherwise, 27*(3), 116–124.

Fujita, T. T. (1975a). *Superoutbreak Tornadoes of April 3–4, 1974. University of Chicago*. http://www.erh.noaa.gov/iln/events/19740403/fujita_bigmap.png. Accessed 12 Apr 2015.

Fujita, T. T. (1975b). *New evidence from April 3–4, 1974 Tornadoes*. Department of the Geophysical Sciences, University of Chicago. SMRP research paper 127.

Fujita, T. T. (1992). Damage survey of hurricane Iniki in Kauai (map). *NOAA Storm Data, 34*, 9.

Government of Canada. (2015). *Analyses*. http://weather.gc.ca/analysis/index_e.html. Accessed 18 Mar 2015.

Henson, R. (2010). *Weather on the air: A history of broadcast meteorology*. Boston: American Meteorological Society.

HINT.FM. (2015). *Wind map*. http://hint.fm/wind. Accessed 18 Feb 2015.

Holle, R. L., & Cummins, K. L. (2010). Monthly distributions of U.S. NLDN cloud-to-ground lightning. In *Proceedings of the 21st international lightning detection conference*, Orlando, FL.

Institute for Veterinary Public Health. (2015a). *World maps of Köppen–Geiger climate classification*. http://koeppen-geiger.vu-wien.ac.at/. Accessed 14 Mar 2015.

Institute for Veterinary Public Health. (2015b). *World map the of Köppen–Geiger climate classification updated*. http://koeppen-geiger.vu-wien.ac.at/present.htm. Accessed 20 Apr 2015.

Korea Meteorological Administration. (2015). *Asian dust forecast chart: surface pressure, streamline and wind vector at 1500 m*. http://web.kma.go.kr/eng/weather/asiandust/forecastchart.jsp. Accessed 17 Apr 2015.

Kottek, M., Grieser, J., Beck, C., Rudolf, B., & Rubel, F. (2006). World map of the Köppen–Geiger climate classification updated. *Meteorologische Zeitschrift, 15*, 259–263.

Monmonier, M. (1999). *Air Apparent: How meteorologists learned to map, predict, and dramatize weather*. Chicago: The University of Chicago Press.

NASA. (2015). Earth observatory: Global maps. http://earthobservatory.nasa.gov/GlobalMaps/?eocn=topnav&eoci=globalmaps. Accessed 8 Apr 2015.

National Weather Service. (2015a). *Enhanced radar image, Louisville, KY radar, composite reflectivity, 04/07/2015 22:42 UTC*. http://radar.weather.gov/radar. Accessed 7 Apr 2015.

National Weather Service. (2015b). *Climate prediction center*. http://www.cpc.ncep.noaa.gov/products/analysis_monitoring/enso_update/sstanim.shtml. Accessed 21 Apr 2015.

National Weather Service. (2015c). *Map: Superoutbreak Tornadoes of April 3–4, 1974. Eastern regional headquarters*. http://www.erh.noaa.gov/iln/events/19740403/fujita_bigmap.png. Accessed 24 Feb 2014.

NHK World. (2015). *NHK world television*. (Program viewed 16 April 2015 on WILL-TV).

NOAA (National Oceanic and Atmospheric Administration) (1992). *Storm Data*. 34(9).

NOAA. (2015a). *Weather charts: North Atlantic—Eastern USA*. http://www.weathercharts.org/noaa-charts.htm.Accessed 20 Feb 2015.

NOAA. (2015b). Equatorial pacific sea surface temperatures. National Centers for Environmental Information. http://www.ncdc.noaa.gov/teleconnections/enso/indicators/sst.php. Accessed 28 May 2015.

Nullschool. (2015). *Earth*. http://earth.nullschool.net. Accessed 12 Mar 2015.

Royal Netherlands Meteorological Institute. (2015). *Effects of El Nino on world weather*. http://www.knmi.nl/research/global_climate/enso/effects/. Accessed 18 Feb 2015.

Rubel, F., & Kottek, M. (2010). Observed and projected climate shifts 1901-2100 depicted by world maps of the Köppen–Geiger climate classification. *Meteorologische Zeitschrift, 19*, 135–141.

Sanderson, M. (1999). The classification of climates from Pythagoras to Koeppen. *Bulletin of the American Meteorological Society, 80*(4), 669–673.

Space Science and Engineering Center. (2011). *Global satellite & surface temperature montage: Latest global montage*. University of Wisconsin-Madison. http://www.ssec.wisc.edu/data/comp/latest_cmoll.gif. Accessed 2 Feb 2011.

Space Science and Engineering Center. (2015). *Images and data*. University of Wisconsin-Madison. http://www.ssec.wisc.edu/data/. Accessed 14 Apr 2015.

Tannehill, I. R. (1955). *Is weather subject to cycles? Water: The yearbook of agriculture* (pp. 84–90). Washington, D.C.: U.S. Department of Agriculture.

US Drought Monitor. (2015). *The United States drought portal*. http://www.drought.gov/drought/. accessed 15 May 2015.

USPLN/NAPLN (United States Precision Lightning Network/North American Precision Lightning Network). (2015). *Real time lightning data*. http://www.uspln.com/product.html. Accessed 13 May 2015.

WeatherBug. (2015). *Spark 101*. http://www.earthnetworks.com/WeatherBug%C2%AE/Spark.aspx. Accessed 6 May 2015.

Wikipedia. (2015). *Ensemble forecasting*. http://en.wikipedia.org/wiki/Ensemble_forecasting. Accessed 15 May 2015.

WWLLN. (2015). World wide lightning location network. http://www.wwlln.net/new. Accessed 13 May 2013.

Mapping in Public Health

Lance A. Waller

Abstract The history and development of public health was motivated, illustrated, and implemented with the use of maps. In this chapter, we provide a brief review of the use of maps and related concepts in public health spanning from medieval plague quarantine plans to remotely sensed satellite measures of air pollution. We begin with early maps of infectious disease (e.g., plague, yellow fever, and cholera) including an in-depth review of Dr. John Snow's famed map of the 1854 London cholera epidemic. We also review the role of mapping in contemporary political debates regarding miasma versus contagion as underlying causes of disease and associated early public health responses such as sanitation and quarantine. We next highlight atlases of disease identifying, documenting, labeling, and mapping endemic areas of known diseases (e.g., yellow fever, cholera, and leprosy) across the globe. Finally, we outline the rise of quantification of observed patterns and the use of spatial statistical techniques to investigate epidemiologic hypotheses regarding geographic variations in disease risk and associations with potential local explanatory factors. Taken together, we find a rich history of mapping in the development, maturation, and modern implementation of public health science and practice.

Keywords Disease · Spot maps · Disease atlas · Spatial statistics

1 Introduction

Throughout human history, astute observers have noted that certain locations seem to be associated with particular outcomes. Western concepts linking health to location date at least to Hippocrates' famous treatise *On Airs, Waters, and Places*

L.A. Waller (✉)
Department of Biostatistics and Bioinformatics, Rollins School of Public Health, Emory University, Atlanta GA, USA
e-mail: lwaller@sph.emory.edu

S.D. Brunn and M. Dodge (eds.), *Mapping Across Academia*,
DOI 10.1007/978-94-024-1011-2_9

(Miller 1962). Even today we refer to tropical diseases, cancer clusters, and sick buildings to denote links between areas and (real or perceived) increased risk.

Broadly speaking, the study of public health can be viewed as the study of the incidence (new cases) and/or prevalence (existing cases) of disease or other health outcome, placed in context of the local environment where the term "environment" is very broadly construed and can include (among many other things) exposures to pollutants, pathogens, insect vectors, local variations in social and behavioral determinants of disease, regional health policies, and the impact of interventions. In many cases, these motivating questions inherently include a spatial component, i.e., many questions of recurring interest in public health begin with "where". As an illustration, consider the following:

- Where, when, and within which population subgroups are risks of disease highest?
- Where are environmental exposures the highest?
- Where are health interventions implemented, and how consistently are they applied?
- Where do we find concentrations of social determinants of disease (e.g., high poverty, low access to fresh fruits and vegetables, low access to health care, low access to transportation)?
- Where is the strongest association between a local exposure and local health outcomes?

The key concept is that a person's location links one to a surrounding local context of exposure(s) experienced by and impacting the health of the individual.

This notion that one's location impacts disease risk motivates the mapping of exposures, cases, potential risk factors (e.g., age, race, sex), and the numbers and composition of the at-risk population. As illustrated throughout this text, we consider mapping to be an active process of arranging, summarizing, and visualizing geographic data in order to reveal patterns driven by location with the goal of improving insight, enabling predictions, evaluating real or potential interventions, and, ultimately, gaining new understanding regarding potential causes.

The path from pattern to process is long and winding (and often not unique) with many potential missteps and challenges when it comes to inferring true causality, but the history of mapping in public health illustrates that, while we may not jump directly from observed pattern to true process, the geographic analysis of spatially-referenced public health data often provides important steps forward in our understanding of and steps toward control of the risk of adverse health outcomes.

In the sections below, we review the early history of medical cartography and the mapping of disease, the expansion to the development of atlases of disease, and the growth and development of quantitative spatial analysis tools for public health. In each case, we find contributions of mapping, sometimes in the forefront, sometimes behind the scenes, involved in the study and practice of public health.

2 Early Examples: Miasma and Contagion, Quarantine and Sanitation

As summarized in Koch's (2005) *Cartographies of Disease*, the earliest extant maps of disease on record include Arrieta's 1694 maps of quarantine plans for plague in the province of Bari in the then-kingdom of Naples. The impact and fear of plague drove the local government to plan for and map a quarantined area, an extreme acknowledgement of the locality of disease risk, i.e., a strict interpretation that disease occurs *here* and must be kept from moving *there*. Note that quarantine does not necessarily acknowledge the cause(s) of the local increases, but presumes that by limiting interaction with the location and/or the individuals within it, one can limit risk in those outside the impacted area. The practical aspects of quarantine can involve severe consequences by condemning those inside to high risk as dramatized in Camus' *The Plague* (1948) and observed quite recently in the response to the 2014 Ebola outbreak within poor, high-incidence neighborhoods of Liberia's capital of Monrovia, which revealed that quarantine as a concept often fuels the polemics of political pundits more than providing any positive impact on the practice of public health.

Moving beyond the identification of high local risk for the purpose of containment, the 1700s and 1800s saw a use of disease maps in the debate between *miasma* (bad air) and *contagion* (transmission by infection with an infectious agent) as the cause of epidemic diseases. The debate not only involved the *causes* of disease but also the proper public health response to them, with rapid advances in both areas due to maps. Maps provided insight into the role of bacteria and other pathogens, e.g., in drinking water, in the cause of disease. The period saw the initiation of the "sanitation movement," with an increased focus on clean water and the development of citywide sewer systems, a movement also clearly influenced by maps. However, as well known and documented examples illustrate, in neither case were maps free of misunderstanding and misinterpretation.

As a first example, New York public health official (and avowed sanitation supporter) Valentine Seaman's 1798 "spot map" of yellow fever provides the first known map of resident locations of individual cases of disease (Stevenson 1965; Garfield 2013), revealing a concentration of cases near the docks where ships often brought cases (sick individuals infected in endemic areas and moved to New York) and where swampy ground and the build-up of sewage effluvia from "areas of convenience" led to foul odors and, as noted by Seaman, concentrations of mosquitos (Fig. 1). As a proponent of sanitation, Seaman interpreted the concentration of mosquitos as further evidence of local miasma, not as related to the transmission of disease. The recommended responses of cleaning up the areas under the docks were effective, even if the primary route of transmission was missed through focus on preconceptions and assumptions.

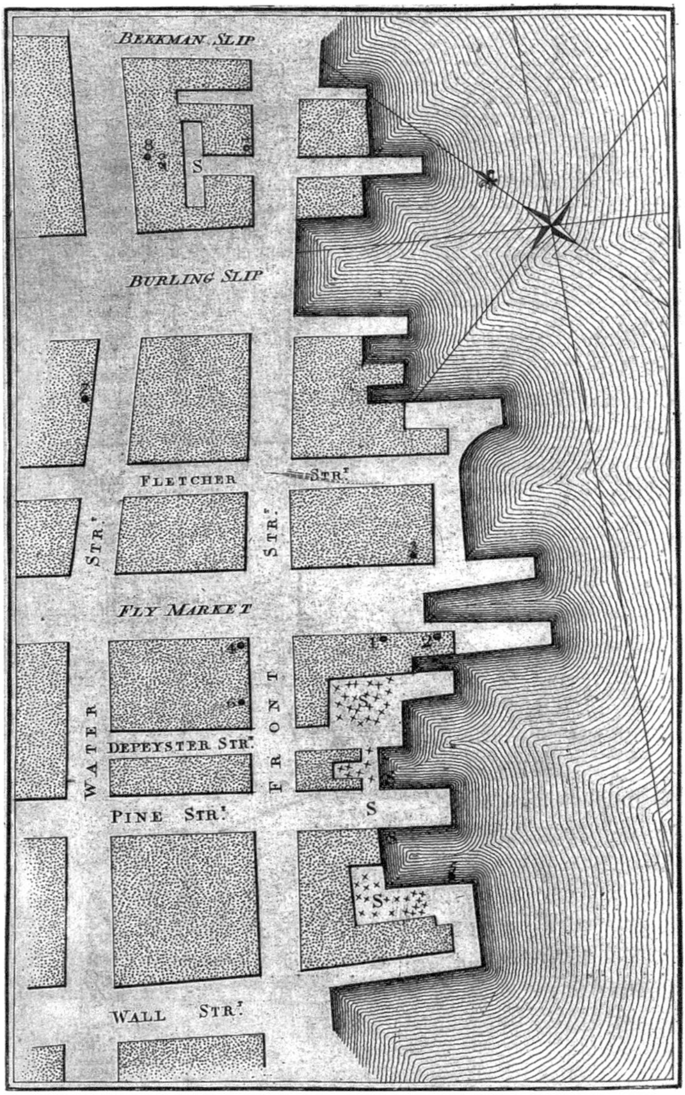

Fig. 1 Valentine Seaman's 1798 map of yellow fever residences in New York City (*Source* courtesy of Wellcome Library, London, http://wellcomeimages.org.)

Following, but better known than Seaman's work, is the famed investigation of the 1854 London cholera outbreak by Dr. John Snow. Snow's detailed, data based, and highly influential work (Snow 1855) results in his listing among the founders of modern epidemiology. A recent London travel guidebook, in a note regarding the John Snow pub (a landmark for epidemiologists worldwide) summarizes the story as follows:

> In 1854, Londoners were dropping like flies from cholera until Dr. Snow figured out that the bacteria were carried by water. The water pump he turned off, thereby saving countless lives, was near the site of this pub. (Wurman 2006)

Snow's innovative use of maps, and his theory of a water-borne cause of cholera make for a compelling story that is often serves as the introduction to discussions of the role of maps in public health. Naturally, the real story is a bit more nuanced but ultimately even more interesting than the brief summary above. Brody et al. (2000), Koch (2005), Johnson (2006), and Garfield (2013) provide readable, detailed, and fascinating summaries of Snow's work and the role and impact of the maps involved. Briefly, Dr. Snow, as both a physician with a suspicion that water was involved in transmission and a resident of the Soho neighborhood strongly impacted by the 1854 outbreak, interviewed the families of, and mapped the residence of, each cholera death in the neighborhood. He also mapped the location of public water pumps and noted which individuals accessed each pump. Brody et al. (2000) and the others note that Snow's map was not the first map of the Soho cases of the 1854 outbreak, nor the only one considered by the London Board of Health, a group largely in the miasma camp (noting that the disease appeared concentrated in poorer sections of town where one could *smell* the filth).

The first spot map of cholera cases for the outbreak was provided by Edmund Cooper, an engineer for the Metropolitan Commissioner of Sewers, who was investigating a public concern that the sewer system had somehow connected to a site of a mass burial of victims of the 1665 plague outbreak in London. Cooper's map presented deaths by address with respect to "gully-holes" where the sewer vented to the surface. Comparing Cooper's gully-hole map to Snow's map with public water pumps reveals that the gully-holes and pumps are largely in the same locations, particularly around the Broad Street pump which was the focus of Snow's investigation and, ironically, which was slightly misplaced (two doors from its actual location) in Snow's original map. Both maps show a high concentration of deaths in the vicinity of the Broad Street pump on Snow's map and near the Broad Street gully-hole in Cooper's. Both Snow and Cooper supplemented their maps with additional field work. Cooper, arguing against the plague burial theory, noted that sewer workers within the tunnels near the gully-hole had not become ill. Snow, arguing for the impact of the pump, noted that (a) some individuals far from Broad Street who used water from the Broad Street pump had fallen ill and (b) that no residents of a workhouse nearby with its own well had fallen ill. Snow (and others) effectively lobbied the London Board of Health to remove the handle from the Broad Street pump, a dramatic public health response that occurred as the epidemic was abating.

The fact that Snow's theory was correct makes it tempting to suggest that the spot maps of disease and potential causes provided overwhelming and irrefutable evidence indicating the underlying cause of disease, but it is important to note that proponents of the miasmist theory of disease viewed the same maps and saw clear evidence for their own argument. In Seaman's case, Seaman himself sees a clear miasmist interpretation of the observed concentration of cases near the docks and sewage, noting the concentration of mosquitos as further evidence of his theory noting "circumstances favoring the rise of putrid miasmata, equally favor the generation of these insects" (quoted in Stevenson 1965). In Snow's case, Brody et al. (2000) note both (a) that the Committee on Scientific Inquiries of the General Board of Health rejected Snow's theory in favor of a miasmist interpretation, and (b) that contemporaneous cholera expert Edmund A Parkes in his review of Snow's monograph noted:

> On examining map given by Dr Snow, it would clearly appear that the centre of the outburst was a spot in Broad-street, close to which is the accused pump; and that cases were scattered all round this nearly in a circle, becoming less numerous as the exterior of the circle is approached. This certainly looks more like the effect of an atmospheric cause than any other; if it were owing to the water, why should not the cholera have prevailed equally everywhere where the water was drunk? (Parkes 1855)

In both cases, maps provided important clues but not necessarily incontrovertible evidence alone, and map-readers (Seaman, Snow, and Parkes) clearly continued to bring their own assumptions, presupposition, and bias into their interpretation of the maps. However, while the maps did not provide definitive proof of the underlying cause, they certainly framed both the discussion and the public health response in a manner leading to subsequent discoveries, a more complete understanding, and best response practices.

3 Documenting Spatial Variation in Disease Risk: Atlases of Disease

While Seaman and Snow set the stage for the use of detailed maps of case residences in neighborhoods to identify factors related to local outbreaks of a disease, another branch of medical geographers aimed to define the reach of disease across districts, countries, continents, and the globe. The earliest examples of global maps of the prevalence of various diseases were modeled after similar maps of species of plants and animals and include Friedrich Schnurrer's 1827 "Charte über die geographische Ausbreitung der Krankheiten" (Brömer 2000) and Heinrich Berghaus's 1848 map of human diseases in his first major atlas of global thematic maps *Physikalischer Atlas* (Camerini 2000; Fig. 2). Barrett (2000). provides evidence that a German physician, L.L. Finke, may have generated a single copy of a global disease map as early as 1792. These early efforts spawned an ongoing effort to map the incidence and prevalence of disease across administrative areas in order to identify areas at increased risk for disease.

In the 20th century, Walter (2000) notes a transition of interest from mapping infectious disease to mapping chronic disease with a particular emphasis on cancer, beginning in the United Kingdom as early as the late 1920s and into the 1930s with Stocks' series of individual maps of regional cancer mortality rates adjusted for local differences in the age and sex distributions of the at-risk population, an

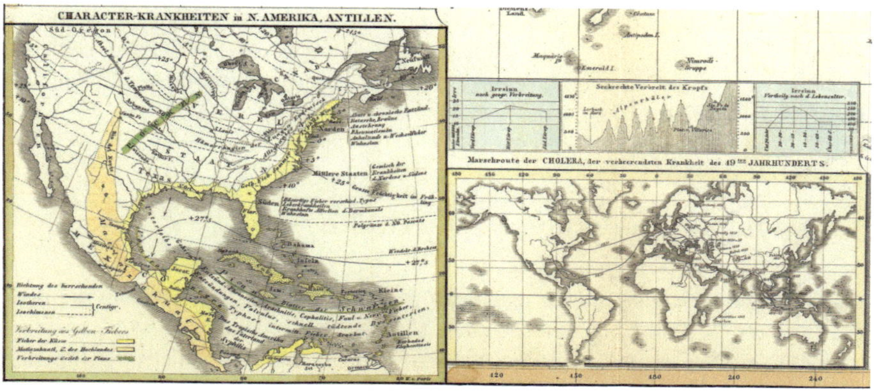

Fig. 2 Selection from Heinrich Berghaus's 1848 map of human diseases, illustrating areas of Yellow Fever in North America and a world map illustrating the spread of cholera (*Source* courtesy of the David Rumsey map collection, www.davidrumsey.com)

important adjustment allowing better comparability between rates from regions with different population compositions (Walter 2000). Howe (1977) extended these efforts to produce an atlas of thirteen major causes of death, leading the way to a proliferation of printed disease atlases, particularly cancer atlases, in the 1970s, 1980s, and 1990s (Walter and Birnie 1991). Updated atlases of chronic and infectious disease continue to appear, with an increasing number of on-line resources available with custom basic mapping, query, and data download capabilities.

The development of disease atlases in the late 20th century corresponded to the rapid increase of three contributing factors: (1) the availability and portability of administrative health data, (2) the development of geographic information systems (GIS) to manage, link, and display large quantities of georeferenced data, and (3) the development of spatial statistical methods to address analytic challenges to the analysis of local disease rates. These three areas of development provided access to increasing amounts of information, the ability to link data from disparate sources (e.g., health outcomes from medical records, local exposures from air monitors, and demographics from the census) onto the same geography, and a growing set of analytic tools to address increasingly spatial questions of interest including:

- Are there "clusters" or "hot spots" of disease incidence?
- Do high local rates correspond to high local exposures?
- What local demographic groups are impacted the most for particular health outcomes?
- What areas are good habitats for vectors that transmit disease?

While atlases provide relevant data summaries and displays, questions such as those above provided the motivation for development of specialized statistical approaches representing a toolbox for the spatial analysis of public health data, our next topic of review.

4 Quantifying Pattern in Public Health

In order to provide quantitative responses to the questions motivated by the maps described above, public health researchers focused on the development, assessment, and application of spatial statistical tools to address a growing list of geographic questions in public health (Waller and Gotway 2004; Lawson 2008). Cressie's (1993) landmark *Statistics for Spatial Data* brought together theory and applications of point process models (is an observed pattern of events more clustered or more regular than we would expect under random allocation of events to possible locations?), geostatistical prediction (using observed values at a fixed set of locations, what is the best prediction of the value associated with a new location?), and regression models for regional data (what is the estimated association between an outcome and covariates measured on the same set of zones?). While not focused specifically on public health, each element of Cressie's (1993) typology of methods has application in quantifying geographic distributions in public health research as illustrated by linking to the questions above.

Point process methods can address whether an observed pattern appears more clustered than one might expect. A *spatial point process* defines a probabilistic model of data patterns, and a spatial Poisson process often defines the pattern we would observe under a null hypothesis of no local variation in disease risk. The general mathematics of spatial point patterns tend to compare observed patterns to a uniform distribution in space (wherein an event is equally likely to be observed at any location), but in application to the distribution of disease, it is more common to compare observed patterns to an *equal risk* setting where a case is equally like to occur in any individual. Since individuals aggregate spatially into villages, towns, and cities, we not only seek spatial concentrations of cases, but we are looking for clusters of cases above and beyond the level of spatial aggregation of individuals observed in the population at risk. Waller and Gotway (2004: Chapters 5 and 6) review general principles for approaches to detect spatial clustering of disease in spatially heterogeneous populations, and outline two families of approaches. The first compares the spatial distribution of cases to that of controls (or non-cases), while the second compares the level of clustering observed in cases (as measured by the number of additional cases expected within a given distance of each case) to the level of clustering observed in controls. The first approach can identify the location of specific *clusters*, while the second provides a summary of the spatial scale of *clustering*. Both concepts are of interest and both can detect deviations from the null hypothesis, but the two approaches cannot distinguish between underlying *causes* of the pattern. More specifically, Bartlett (1964) notes that, without additional information, it is mathematically impossible to distinguish between a pattern generated by local increases in risk for otherwise independent cases (e.g., a cluster due to a shared local environmental exposure), and a pattern generated by interactions between cases (e.g., a cluster due to a set of cases infected by nearby cases) from a single observed pattern of point locations. Interestingly, this challenge is exactly a mathematical manifestation of the public health debates regarding miasma and contagion during the time of Snow and Seaman, and a part of the reason that mapping alone could not distinguish between the two theories.

In order to move past this challenge, two primary types of additional information are of interest in public health. The first is time, since a similar cluster occurring in the same location across multiple time periods suggests a local source of increased risk rather than spatially non-specific contagion, while similar sized clusters at different locations over time provides the opposite conclusion. Exposure data provide a second source of additional information, wherein one links local exposures to local observed disease risk, and assesses associations, often through the use of linear or, more commonly for count-based outcomes, logistic or Poisson regression.

Spatial statisticians addressed two challenges in the application of regression techniques to disease rates observed in small administrative areas. The first, referred to as *small area estimation*, addresses the tension between geographical and statistical precision in presenting maps of regional rates of disease. Geographically, one prefers small areas to provide a map with local detail, while, statistically, rates based on data from smaller numbers of individuals become increasingly unstable. This is a particular problem for maps of local rates of rare diseases. For example, if public health records suggest a rate citywide of one case per 100,000 individuals, local estimates from a city of 100,000 individuals divided into neighborhoods of 100 individuals will include, on average, one neighborhood with a local rate of 1 per 100, even with no local increase in risk. To address this situation, statisticians developed techniques to “borrow information” from neighboring regions through the use of local weights, typically based on a hierarchical random effects models (Clayton and Kaldor 1987; Besag et al. 1991).

The inclusion of a spatial random intercept within these models also provides a mechanism to address a second challenging analytic problem, namely, adjusting regression estimates to account for residual spatial autocorrelation, the often-observed feature of spatial data that nearby observations tend to be positively correlated, perhaps due to a shared unmeasured (or even unmeasureable) local feature. Allowing residuals to be correlated complicates statistical inference, again due to Bartlett’s result: in this case, from a single data set, it is very difficult to know if an observed pattern is due to correlation or to covariate effects. The hierarchical models mentioned above provide a specific component of the model defining spatial covariance and identifies the combination of covariate values and spatial correlation that provides the best fit to the model.

Collectively, these methods for stabilizing rates and fitting spatial regression models define a class of methods referred to in the public health literature as *disease mapping* and have seen wide application in the public health literature to provide stabilized estimates of local disease rates and to assess associations with local covariate values such as demographics and exposures (Waller and Gotway 2004; Lawson 2008; Banerjee et al. 2014).

The next area of spatial statistical developments relating to mapping in public health includes the geostatistical prediction of exposure values. To illustrate, consider the measurement of air pollutant levels at fixed monitors located at particular locations across a city. These values offer detailed measurement of pollutant levels but *only for those locations*. The field of geostatistics grew out of exploratory mining analysis seeking to predict the yield of mines in new locations, and works as follows: Given measurements at a set of fixed locations, we predict a measurement at a new location by calculating a weighted average of the observed data with weights relating to how similar we expect our prediction to be to each of the observations. Assuming positive spatial correlation, i.e., nearby observations are more similar than those far apart, we give more weight to observations near to the prediction location. Geostatistical methods formalize the definition of optimal weights in order to minimize uncertainty in the prediction (typically by minimizing the mean square prediction error or other summary of prediction error). By predicting a measurement value for each of a set of points across the study area, we can now construct a contour map of predicted pollutant levels across the study area and the estimated prediction error associated with each location.

Predicted exposures and estimated associations between exposure and outcomes allow the analyst to make maps of predicted risk for any location. Such maps provide powerful communication tools in public health, by pooling information from multiple sources, linking them geographically, and analyzing them statistically. Placing the results on a map places predicted risk in context for residents and policymakers alike, and clearly identifies the impact of living or working *here* versus *there*, a strong motivating question for the first maps of disease. Such risk maps can be basic regression predictions based on exposures and demographics, or more complex, e.g., incorporating satellite images to define locations of favorable habitats for disease vectors and identifying areas where potentially infected vectors and people coincide (Kitron 2000), or identifying areas at high risk for violent crime based on alcohol outlet density (Gruenewald et al. 2006).

The methods above provide a means to quantify spatial patterns for statistical analysis, but this need not mean the map has been subsumed by quantitative, statistical analysis. In fact, Waller (2014) suggests quite the opposite, identifying and calling for new uses of maps to keep a spatial perspective within spatial statistics. As an example, *geographically weighted regression* (Fotheringham et al. 2002) and *spatial varying coefficient* (Gelfand et al. 2003) models allow the associations between exposures and health outcomes to vary by location, e.g., the association between local socioeconomic status and birthweight could be stronger in some neighborhoods than others. Such methods use maps to reveal where associations are stronger and where associations are weaker, offering insight into areas where targeted interventions may be more effective (Fig. 3). Waller (2014) also suggests maps of model residuals to reveal *where* a statistical model fits well and where it does not. Such application require spatial thinking, statistical thinking,

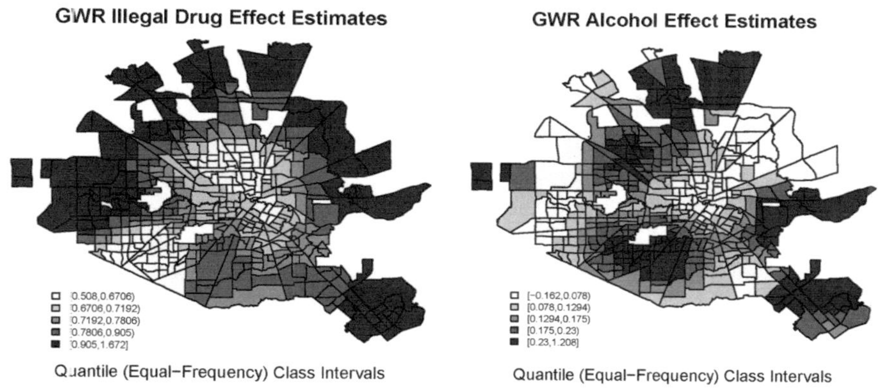

Fig. 3 Geographically weighted regression example showing maps of local association with violent crime reports for illegal drug arrests and alcohol sales. Different neighborhoods show different effects (*Source* adapted from Waller et al. 2007)

and, increasingly, spatial statistical thinking to gain the full public health benefit of mapped data.

5 Conclusion

Our whirlwind tour of mapping in public health ranges from medieval plague quarantines to GIS-linked satellite and health record data, all enabling and addressing spatial questions in public health. As illustrated above, the act of mapping is a critical component of public health research, but rarely a stand-alone solution. Maps identify public health issues, suggest quantitative associations, evaluate the impact of interventions, and communicate results to the neighborhood residents, public health researchers, and policy makers. The range of applications is as broad as the field of public health itself. While our review focuses on chronic and infectious disease, similar examples of the impact of mapping exist in other areas of public health. For example, health services research contains a considerable literature exploring geographic variations in health care delivery and marketplaces (Wennberg and Gittelsohn 1973; Wennberg and Cooper 1998), one goal of which is to enable location-specific valuation of services for Medicare reimbursements in the United States (IOM 2012a, b). Clearly, mapping is essential to public health research, practice, and policy, and the mapping process begins anytime a neighbor, reporter, student, investigator, or government official raises a public health question that begins "Where….".

References

Banerjee, S., Carlin, B. P., & Gelfand, A. E. (2014). *Hierarchical modeling and analysis for spatial data* (2nd ed.). Boca Raton: Chapman & Hall/CRC Press.

Barrett, F. A. (2000). Finke's 1792 map of human diseases: The first world disease map? *Social Science and Medicine, 50*, 915–921.

Bartlett, M. S. (1964). The spectral analysis of two-dimensional point processes. *Biometrika, 51*, 299–311.

Besag, J., York, J., & Mollié, A. (1991). Bayesian image restoration, with two applications in spatial statistics (with discussion). *Annals of the Institute of Statistical Mathematics, 43*, 1–59.

Brody, F., Rip, M. R., Vinten-Johansen, P., Paneth, N., & Rachman, S. (2000). Map-making and myth-making in Broad Street: The London cholera epidemic, 1854. *Lancet, 356*, 64–68.

Brömer, R. (2000). The first global map of the distribution of human diseases: Friedrich Schnurrer's 1827 'Charte über die geographische Ausbreitung der Krankheiten'. *Medical History, 44*(20), 176–185.

Camerini, J. R. (2000). Heinrich Berghaus's map of human diseases. *Medical History, 44*(20), 186–208.

Camus, A. (1948). *The plague*. New York: Alfred A. Knopf.

Clayton, D. G., & Kaldor, J. (1987). Empirical Bayes estimates of age-standardized relative risks for use in disease mapping. *Biometrics, 43*, 671–682.

Cressie, N. A. C. (1993). *Statistics for spatial data* (2nd ed.). New York: Wiley.

Fotheringham, A. S., Brunsdon, C., & Charlton, M. (2002). *Geographically weighted regression: The analysis of spatially varying relationships*. Chichester: Wiley.

Garfield, S. (2013). *On the map: A mind-expanding exploration of the way the world looks*. New York: Gotham Books. Chapter 13.

Gelfand, A. E., Kim, H. J., Sirmans, C. F., & Banerjee, S. (2003). Spatial modeling with spatially varying coefficient processes. *Journal of the American Statistical Association, 98*(462), 387–396.

Gruenewald, P. J., Freisthler, B., Remer, L., LaScala, E. A., & Treno, A. (2006). Ecological models of alcohol outlets and violent assaults: Crime potentials and geospatial analysis. *Addiction, 101*(5), 666–677.

Howe, G. M. (Ed.). (1977). *A world geography of human diseases*. London: Academic Press.

IOM (Institute of Medicine). (2012a). *Geographic adjustment in medicare payment: Phase I: Improving accuracy*. Washington, DC: The National Academies Press.

IOM (Institute of Medicine). (2012b). *Geographic adjustment in medicare payment: Phase II: Implications for access, quality, and efficiency*. Washington, DC: The National Academies Press.

Johnson, S. (2006). *The ghost map*. New York: Riverhead Books.

Kitron, U. (2000). Risk maps: Transmission and burden of vector-borne diseases. *Parasitology Today, 16*(8), 324–325.

Koch, T. (2005). *Cartographies of disease: Maps, mapping, and medicine*. Redlands: ESRI Press.

Lawson, A. B. (2008). *Statistical methods for spatial epidemiology* (2nd ed.). New York: Wiley.

Miller, G. (1962). "Airs, waters, and places" in history. *Journal of the History of Medicine and Allied Sciences, 17*(1), 129–140.

Parkes, E. A. (1855). Review: Mode of communication of cholera by John Snow. *British and Foreign Medico-Chirurgical Review, 15*, 458.

Snow, J. (1855). *On the mode of communication of cholera* (2nd ed.). London: John Churchill.

Stevenson, L. G. (1965). Putting disease on the map: The early use of spot maps in the study of Yellow Fever. *Journal of the History of Medicine, 20*, 226–261.

Waller, L. A. (2014). Putting spatial statistics (back) on the map. *Spatial Statistics, 9*, 4–19.

Waller, L. A., & Gotway, C. A. (2004). *Applied spatial statistics for public health data*. Hoboken: Wiley.

Waller, L. A., Zhu, L., Gotway, C. A., Gorman, D. M., & Gruenewald, P. J. (2007). Quantifying geographic variations in associations between alcohol distribution and violence: A comparison of geographically weighted regression and spatially varying coefficient models. *Stochastic Environmental Research and Risk Assessment, 21*, 573–588.

Walter, S. D. (2000). Disease mapping: A historical perspective. In P. Elliott, J. C. Wakefield, N. Best, & D. J. Briggs (Eds.), *Spatial epidemiology: Methods and applications* (pp. 223–239). Oxford: Oxford University Press.

Walter, S. D., & Birnie, S. E. (1991). Mapping mortality and morbidity patterns: An international comparison. *International Journal of Epidemiology, 20*, 678–689.

Wennberg, J. E., & Cooper, M. M. (Eds.). (1998). *The Dartmouth Atlas of health care 1996*. Chicago: American Hospital Association.

Wennberg, J., & Gittelsohn, A. (1973). Small area variations in health care delivery: A population-based health information system can guide planning and regulatory decision-making. *Science, 182*, 1102–1108.

Wurman, R. S. (2006). *Access London* (10th ed.). New York: Harper Collins.

Languages

Roland Kehrein

Abstract This chapter offers an overview of the different ways in which languages, speech varieties or single linguistic variants have been represented on maps, i.e., with reference to geographic space. It discusses historical aspects of language mapping, which initially referred to the mapping of languages (resulting in "language maps") and further developed into the representation of linguistic phenomena on maps (resulting in "linguistic maps"). Moreover, former and current theoretical and methodological issues of variational linguistics and their specific impact on map topics in a linguistic atlas will be laid out. In this respect, the academic field of language cartography has fundamentally changed over the last 130 years from the mere depiction of linguistic variants in certain geographic locations or areas to the systematic consideration of additional (above all social) factors influencing linguistic variation. In summary, linguistic geography has developed from a monodimensional to a pluridimensional approach to maps and atlases. Finally, as for future prospects—which the following still had been when this chapter was originally submitted in 2011—the application of Geographical Information Systems (GIS) will be introduced, allowing for statistic correlations of linguistic data and other, e.g., socially or economically relevant data in geographical space. Such investigations offer completely new insights into the overall interrelations of such aspects providing the opportunity to learn more about the functioning of societies.

Keywords Linguistics · Dialectology · Linguistic cartography · Linguistic variation · Languages · Dialects

---

R. Kehrein (✉)
Research Centre Deutscher Sprachatlas, Marburg University, Marburg, Germany
e-mail: kehrein@uni-marburg.de

S.D. Brunn and M. Dodge (eds.), *Mapping Across Academia*,
DOI 10.1007/978-94-024-1011-2_10

1 Introduction

Language is a device that serves a huge number of purposes. It not only helps us manage our everyday communicative life, it also enables us to access information about nearly everything that exists in and outside the world. On the other hand we may also be seduced, deceived or betrayed by language when people use it thus. In general, accidental misunderstandings can arise in communicative interaction, especially when information is translated from one language into another, since ways to express complex phenomena differ in different languages—difficulties like these have even caused wars in the history of mankind.

Conflicts caused by linguistic differences may also result from other factors: aside from its function as a means of communication, a language or a speech variety serves as a means of constructing identity on several levels. This function ranges from the level of individual identity—there are no two human beings who are completely identical with respect to their voice, their way of speaking or their linguistic knowledge (competence)—to that of a common ethnic/cultural identity. With the latter, linguistic consistency is accompanied by traditional costumes and other visible forms of appearance (e.g., hairstyle), religious beliefs and customs, cultural assets such as songs or tales and so on. Many of these historical habits are now no longer recognized by non-ethnologists in the so-called Western civilization. They have been replaced by different beliefs and customs which still form groups that are often also characterised by special ways of using language or even by a language variety of their own. In the pre-industrial era, ethnic/cultural and national identities were much more tightly bound to geographical or regional identity. Language played a major role in constituting such bonds by uniting the members of the respective groups and at the same time excluding non-members. One of the oldest examples of language serving as an indicator of ethnic identity can be found in the Old Testament book of Judges (12, 5–6) where the victorious tribe of the Gileadites identified and then killed Ephraimite fugitives on the basis of their pronunciation of the word "Shibboleth." Hence, in this biblical episode the fate of 42,000 human beings was decisively influenced by a single linguistic feature.

Even if the Babylonian confusion, another famous Old Testament passage in which language plays a major role, might not be the true explanation for the multitude of 4000–7000 different languages found around the globe, the conclusion remains the same: language may be described and differentiated in relation to geographic space. In other words: language forms an excellent subject for thematic cartography.

Basically, we can distinguish two ways in which "language" can be represented on maps. These I will call "language maps" (or "maps of languages," *Sprachenkarten*) on the one hand and "linguistic maps" (or "maps of linguistic forms," *Sprachkarten*) on the other (cf. Thun 2000; Girnth 2010). In Romance linguistics terms, Swiggers (2010) differentiates here between a *géographie des langues* 'geography of languages' (he refers to the reference manual by Breton 1976 as an example) and *géographie linguistique* 'linguistic geography' and emphasises that in "the history of linguistic practice and thought [...] the geography of languages preceded the emergence of linguistic geography" (Swiggers 2010: 272–273). This article follows this historical development and starts with the description of language maps.

2 Language Maps

On language maps, the geographical distribution of languages—in the sense of 'means of communication primarily utilised by the community domiciled in the respective area'—is depicted. This means that from a language map or even an atlas of the world's languages you can identify "the locations in which a given language is spoken and also which languages are spoken in a particular region" (cf. Asher and Moseley 2006:1). The first known map containing information about languages is the map entitled "Lusatia superior" by Bartholomäus Scultetus which was initially produced as a woodcarving in 1593 (it was later reproduced in a slightly modified way in several atlases, for example the one edited by Guiljelmus and Johannes Blaeu in 1645; Fig. 1). This map, commissioned and financed by the

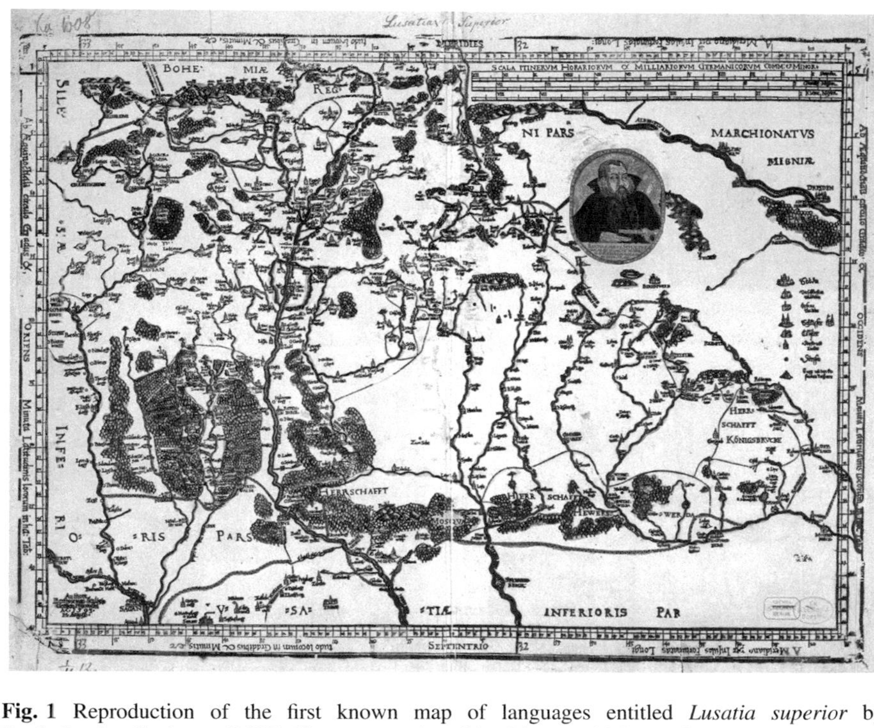

Fig. 1 Reproduction of the first known map of languages entitled *Lusatia superior* by Bartholomäus Scultetus (1593) (*Source* reproduced by Blaeu and Blaeu 1645.)

political leaders of Upper Lusatia, depicts the topography of the province of Upper Lusatia at the end of the 16th century including information about settlements, rivers, woods and (groups of) mountains. These data were collected by the scholar of mathematics and astronomy who surveyed the countryside in eleven so-called "Lusatian journeys." As an additional piece of cartographical information, Scultetus drew lines labeled "Deutsch" (German) or "Wendisch" (Sorbian). These indicated a borderline existing between the two languages in the Upper Lusatian territory. As mentioned, "languages" in this context are proxies for the ethnic groups who use the respective Germanic or Slavic languages. Historians surmise that the commissioners of the map may have asked for this information to be depicted since the question of national identity and dominance was raised in this region in the second half of the 16th century (cf. Reuther 1957: 64). From a cartographic point of view it has to be mentioned that, while we know from his notes that Scultetus used measurement devices to collect the topographical information, it remains unclear how he determined the exact course of the borders between the German and Sorbian regions (cf. Harms 1979: 50). This alleged technical shortcoming, however, cannot be regarded simply as a consequence of the historical epoch. Rather, it is a problem which remains unresolved to the present day, as we will recognize in a few paragraphs.

In 1723, some 130 years after Scultetus produced his map, Lambert ten Kate published a voluminous work entitled "Aenleiding Tot de Kennisse van het Verhevene Deel der Nederduitsche Sprake," a comparative description and grammar of the Germanic languages in Europe. The compendium starts off with a section on the development of the different languages and peoples and their distribution across Europe. This part concludes with a map providing a cartographic overview of these topics, entitled "Volk- en Tael-Verspreiding over Europa." On it, the cardinal branches of the "Europische Tael-Boom" (the 'European languages' tree'), viz., the "Kimbrische Tak," the "Theutonische Tak" and the "Keltische Tak," are marked out by horizontal coloured lines labelled in capitals while the individual language groups within these families and the Slavic language are merely represented by lower-case letters and uncoloured areas without borderlines. Following the map of the languages and peoples in Europe, ten Kate offers samples of the languages depicted in the form of translations of the first line of The Lord's Prayer.

Also based on translations of The Lord's Prayer are the early maps of Gottfried Hensel from 1741 which depict the languages of Europe, Africa and Asia by reproducing the first verses in the respective alphabet systems of the languages considered. Furthermore, Hensel classifies the languages using coloured areas to differentiate the three ethnic and hence language groups along the lines of the genealogies of the Book of Genesis: Noah's sons Japheth (yellow), Shem (red) and Ham (green). For the European languages Hensel even adds a subclassification indicated by different shades of yellow (also see Robinson 1982: 130–132). The three maps containing translations of The Lord's Prayer are supplemented by a map of North and South America in which the language groups are merely specified without translations (Fig. 2 shows Hensel's collection of the four maps).

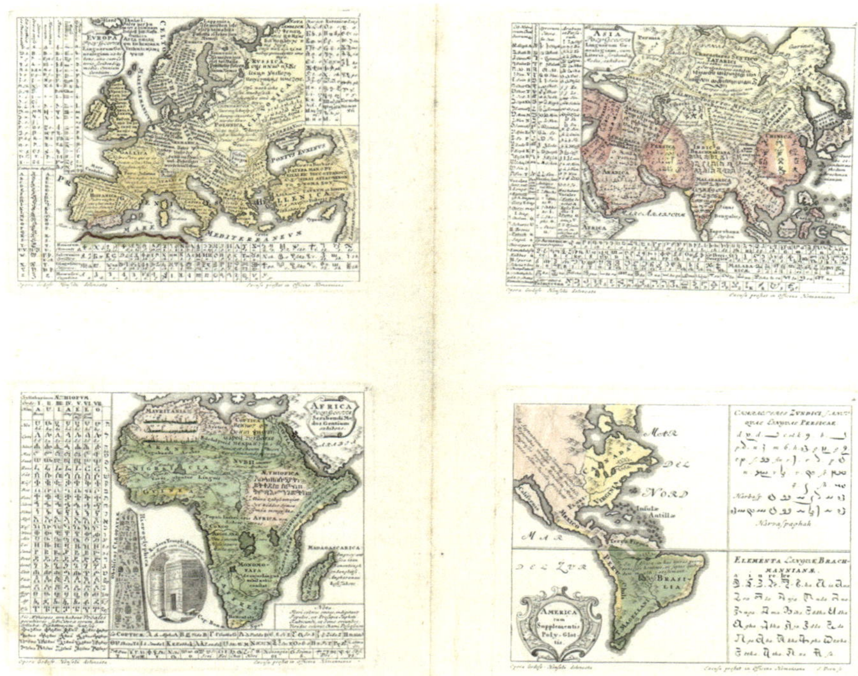

Fig. 2 Classification of the world's languages along the lines of the genealogies of the Book of Genesis: Noah's sons Japheth (*yellow*), Shem (*red*) and Ham (*green*) based on translations of The Lord's Prayer (Hensel 1741)

These three examples of early maps (in principle) outline all forms of cartographic representations of the distribution of languages or language varieties and already clearly reveal the most important problem. Where exactly and on what basis are cartographers able to draw a boundary separating two varieties/languages? Taking a look at the most recent collections of maps depicting the world's languages it is obvious that this central question has still not been solved (and most probably never will be, as the section on linguistic maps will suggest). For example, the editors of the "Atlas of the World's Languages," R.E. Asher and Christopher Moseley, point out that "section editors have used informed judgement to determine where the boundaries between languages should be placed, but such boundaries are rarely definitive" (Asher and Moseley 2006: 2). Similar statements can be found in the introductory parts of the sixteenth edition of the "Ethnologue" (cf. Lewis 2006: 16–17). The main reason for this uncertainty lies in the dynamic nature of languages and language varieties and the resultant perpetual change. Moreover, language is a phenomenon which is inescapably bound to its users. Hence, language data in the sense of "locations in which a given language is spoken" (Asher and Moseley 2006: 1) can only be collected in settled areas (nomadic languages aside).

This results in white spaces of varying sizes for (relatively) unpopulated areas on language maps (e.g., desert or high mountain areas; Fig. 3). A final problem confronting the editors of language maps is whether an observable variety ought to be considered a language or a dialect of a language. (Throughout this text the term dialect is always used to refer to regional dialect.) The most important criteria for differentiating between the two are “mutual intelligibility—the degree to which in a pair of languages each is intelligible to speakers of the other—and […] the percentage of basic vocabulary shared by a pair or a group of speech varieties” (Asher and Moseley 2006: 1). Mutual intelligibility is assumed for dialects, which is not necessarily borne out, as studies of German dialects in the Middle Rhine area prove. In perception experiments it can be shown that dialect speakers from neighbouring dialect formations, namely Rhenish Franconian and Moselle Franconian, barely understand one another when they use their deepest dialects (cf. Schmitt 1992). The decision to classify a speech variety as a language or a dialect is by no means an easy one and in many cases remains contestable.

In the time between the early maps of languages and the two recent compendia of the world’s languages just mentioned, numerous cartographic representations of languages and speech varieties have been published in every populated part of the world. In these, geographical or political areas in which two or more languages or varieties meet are of particular interest. Such areas can be global or continental in scale (e.g., Klaproth’s 1823 map of Asia or the Pacific area mapped in Wurm et al. 1996), or nation-states (e.g., Bernhardi 1844), the border region between two states,

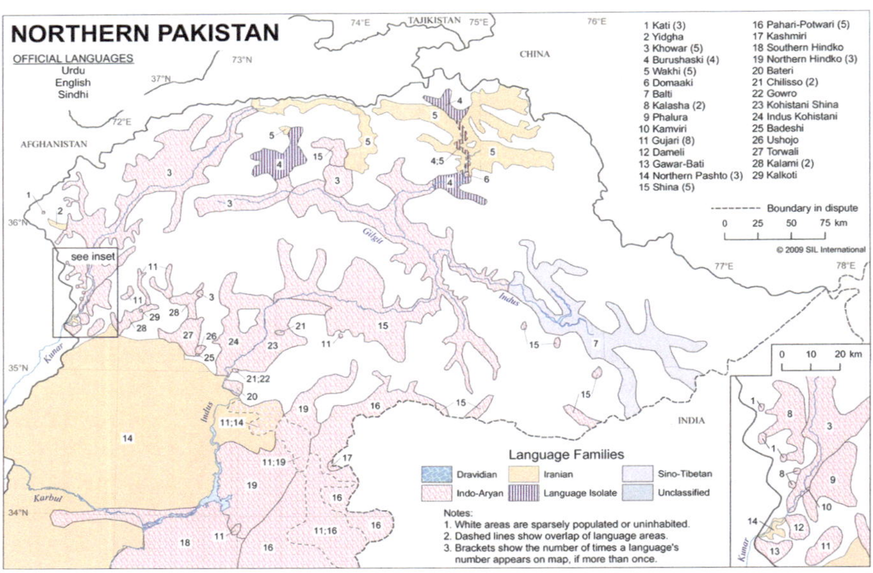

Fig. 3 The distribution of language families in Northern Pakistan where uninhabited regions remain white (Lewis 2006: 816)

or urban regions (e.g., Jupp and McRobbie 1989). Languages can be treated as map themes but so can other regional or social varieties (e.g., maps of multilingualism as a result of multi-ethnicity in urban areas, cf. the maps of the ethnolinguistic situation in Cape Town in van der Merwe and van der Merwe 2006 depicted as Fig. 4). A special issue is the UNESCO-funded cartographic representation of languages threatened with extinction. The "Atlas of the World's Languages in Danger" edited by Christopher Moseley is available on the internet at <http://www.unesco.org/culture/languages-atlas>.

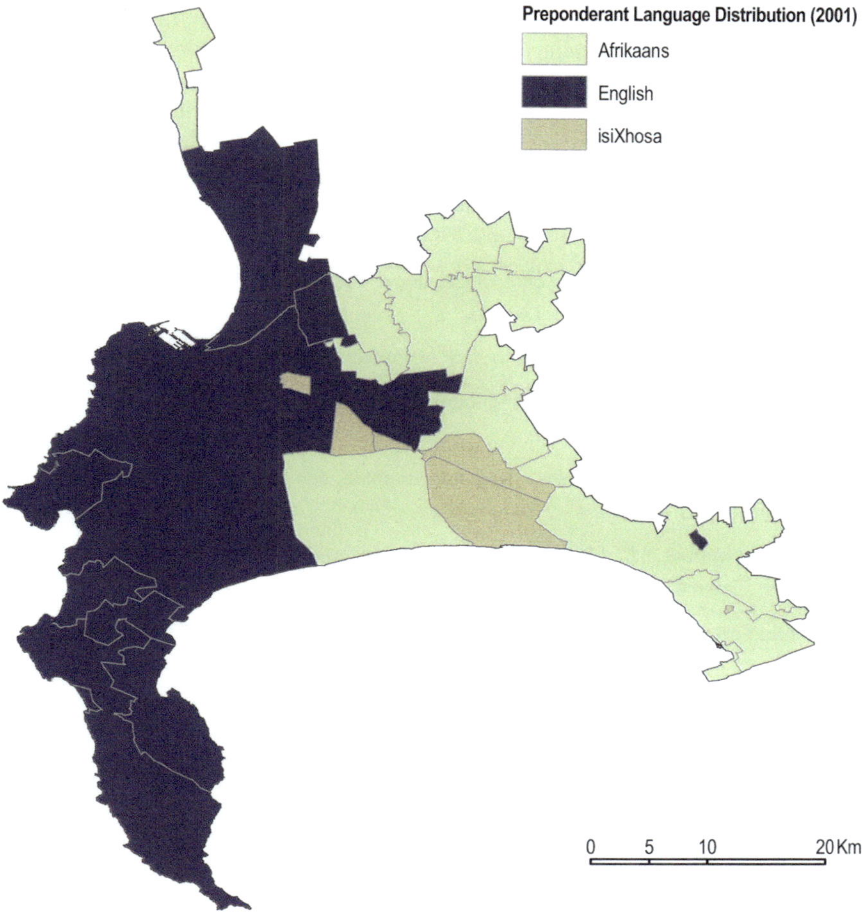

Fig. 4 Multi-ethnic and multilingual urban areas: the distribution of Afrikaans, English and isiXhosa in Cape Town (van der Merwe and van der Merwe 2006: 101)

A look at Ethnologue's overview map of the world's languages reveals that the linguistically most heterogeneous regions in the world are to be found on the Pacific island of New Guinea, in Middle America (i.e., southern Mexico and Guatemala) and eastern Africa (above all Nigeria and Cameroon). Figure 5 shows many dots within a relatively narrow geographic space (each dot representing the geographical centre of the region in which an individual language is spoken). In contrast, far fewer languages are registered across all of Europe, which suggests that although the European states are generally quite densely populated, they form linguistically rather homogeneous areas. Of course, this impression shifts radically if dialects or other regional varieties are taken into consideration.

A completely different and independent wing of variation linguistic studies involves the linguistic "blending" that results when speakers from diverse cultural and linguistic background settle in the same location. Many of the world's metropolises can be regarded as linguistic melting pots. The resultant constellations of extreme cultural and linguistic heterogeneity trigger very specific language dynamic processes studied in sociolinguistic or language-contact investigations. Maps documenting such multilingual situations in urban settings have already been referred to (see for example Jupp and McRobbie 1989 or van der Merwe and van der Merwe 2006; a collection of essays on the linguistic heterogeneity of London was edited by Baker and Eversley 2000).

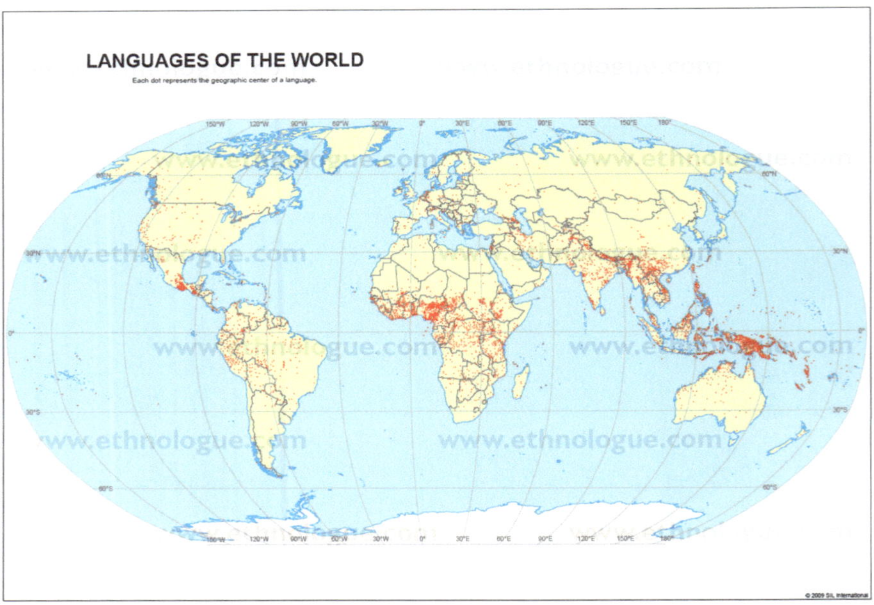

Fig. 5 The world's languages, each dot representing the primary location of a living language listed in the *Ethnologue* (Lewis 2006; NB. due to Ehnologue's terms of use the copyright and "watermark" must not be removed)

3 Linguistic Maps

The origin and development of linguistic cartography may be regarded as an accomplishment of the linguistic subdiscipline of dialectology (also called linguistic geography or dialect geography; cf. Crystal 1998: 26; Chambers and Trudgill 1998: 14) and theoretical and methodological progress is thus neatly connected with developments in dialectology as such. An intensive scientific investigation of regional varieties or dialects as substantial parts of language systems such as German, English, French and so forth was essentially initiated and promoted by German and Romance scholars beginning in the mid 19th century. Earlier works primarily collected and documented dialect vocabulary because the extinction of dialects was generally expected (the first collection of this type was Richey's "Idioticon Hamburgense" published in 1755). This special interest in the regional varieties can also be inferred from the production of language maps depicting the areal distribution of dialects in many parts of Europe (e.g., Šafařík 1842 for Slavic-speaking territories; Bernhardi 1844 for the dialects of Germany; or Fuchs 1849 for dialects of the Romance languages) which, as mentioned, also occasionally served the political purpose of marking the geographical extension of particular national languages. (On Bernhardi, see for example Dingeldein 2001 or Scheuringer 2010: 159.) As pointed out, delimiting dialect areas can be controversial, especially given the lack of clear criteria on which to base borders. Hence, "within this field, a major debate took place [in the mid 19th century] concerning the (im)possibility of drawing boundaries between dialects [… opposing] 'typophile' linguists such as Graziadio Isaia Ascoli (1829–1907) and 'typophobic' linguists […] such as Paul Meyer (1840–1917) and Gaston Paris (1839–1903)" (Swiggers 2010: 273). Out of these debates or, more generally, the examination of such issues—in which the Neogrammarian assumption of the exceptionlessness of sound laws played a major role – linguistic cartography emerged (a recent overview on the development of linguistic maps and atlases is also given in Lameli 2010). Before examining this theoretical issue in more detail it is useful to enumerate the most important methodological issues that have arisen around the production of linguistic maps over the past 130 years. These, to be discussed further later in this chapter, include:

1. Choice of the geographical area from which linguistic data are collected.
2. Number of locations at which data are collected, i.e., the density of the network of locations.
3. Methods of data collection and data processing.
4. Criteria for the choice of informants.
5. Choice of map types and mapping techniques.

The map themes of most linguistic maps are individual linguistic phenomena, the linguistic variables and their dialectal variants depicted on the map face. Hence, comprehensive descriptions of regional varieties in an area of interest are usually presented as an atlas, which Ormeling defines as an "intended combination of maps, based on an objective or narrative" (Ormeling 2010: 37). The following paragraphs thus deal primarily with linguistic atlases. Since the early linguistic atlases of the dialects of the German (by Georg Wenker) and Romance language territories (by Jules Gilliéron) are regarded as most influential in terms of theory and methodology, I will deal with these in a little more detail, and since theoretical progress in both traditions is generally comparable (cf. the recent overviews by Scheuringer 2010 and Swiggers 2010), I will focus on the earlier projects conducted by Georg Wenker.

4 Theoretical Developments in the Mapping of Linguistic Phenomena

In their introductory book on dialectology, J.K. Chambers and Peter Trudgill indicate that the "first dialect survey that can properly be called dialect geography was begun in Germany by Georg Wenker in 1876" (Chambers and Trudgill 1998: 15). Until his death in 1911, Georg Wenker worked on three linguistic atlas projects, the second and third of which may be regarded as extensions of their predecessors:

1. "Sprach-Atlas der Rheinprovinz nördlich der Mosel sowie des Kreises Siegen" (1878)
2. "Sprach-Atlas von Nord- und Mitteldeutschland" (1881)
3. "Sprachatlas des Deutschen Reichs" (1889–1923)

The initial aim of his first atlas project was to areally classify the dialects of his home province, the *Rheinprovinz*, on the basis of a precise investigation of the regional distribution of linguistic features regarded as typical of the respective dialects ("in allgemeinem Ansehen stehender Eigenthümlichkeiten der dortigen Dialekte" (Wenker 1886: 189)). In pursuing this objective he was driven by the conviction that the dialectal forms of each linguistic variable had to differ in a consistent and comparable manner—a necessary (for sounds at least) implication of the Neogrammarian hypothesis that sound laws were exceptionless. This would result in clear cut boundaries between individual regional dialects (cf. Wenker 1886, 189; only in later publications does he explicitly mention the Neogrammarian concept of sound laws, cf. Wenker 1895a: 41). The dialect data collected from 1500 locations in the *Rheinprovinz*, however, presented a completely different picture. If Wenker were to describe his impression of the courses of the borderlines today, he might reach for the analogy of a "bowl of spaghetti." These findings forced him to incorporate transitional zones (he calls these "Mischdialekte" 'mixed dialects') in his summarising (language) map of the dialects of his home province (the "Sprach-Karte der Rheinprovinz nördlich der Mosel," cf. Wenker 1877 and Fig. 6). In order to establish a broader base for a replacement theoretical perspective, Wenker extended his area of investigation, first to the northern and middle sections of the German Empire and later to cover the entire Empire. On the basis of further analyses of the pictures offered by his maps, Wenker concluded that the regional

Fig. 6 Georg Wenker's "Sprach-Karte der Rheinprovinz nördlich der Mosel" (Wenker 1877) in which dialect regions are delimited and indicated on the basis of analyses of the world's first linguistic atlas (Wenker 1878)

classification of dialects is merely a by-product of the creation of linguistic atlases and is more of interest to historically inclined ethnographers than to linguists. Wenker considered dialect surveys and the linguistic maps they produced as research tools, which serve linguistic analyses by providing a reliable and exact database ("eine sichere und […] mikroskopisch genaue Grundlage," Wenker 1886: 190; cf. also Wenker 1895a: 42–43), especially in relation to the investigation of language change. Against the background of knowledge from historical linguistics, the geographical coexistence of synchronically observable linguistic facts may be interpreted as hints about the diachronic unfolding of a development at a particular location ("ein geographisches Nebeneinander gleichzeitiger Erscheinungen in ein historisches Nacheinander gleichörtlicher Entwicklungen" (Wenker 1886: 191). This means maps should be interpreted as the results of, or rather intermediate stages in, processes of ongoing change, i.e., the interpretation of diachronic developments in apparent time. Wenker mentions the theoretical advance he was able to derive from his maps only in passing, when responding to severe criticism from Bremer, namely, that sound change does in fact affect all of the sounds in a variety that are similarly distributed (as the Neogrammarian sound laws would have it) but that these changes progress lexeme by lexeme (cf. Wenker 1895b: 20–26). This theoretical implication, however, only became popular when Labov (unaware of Wenker's findings) published his article on "Resolving the Neogrammarian Controversy" in 1981.

5 Theoretical and Methodological Developments Concerning Data Collection

In methodological respects, Gilliéron's and Wenker's linguistic atlases may be described as "monodimensional atlases," in that they relate the variation of linguistic features exclusively to the areal dimension—in line with their theoretical goals, of course, of being able to trace the diachronic "life" of the respective varieties by interpreting the map pictures. This type of monodimensional linguistic map or atlas is still widespread, and both Gilliéron and Wenker set enduring standards for data collection and representation. Georg Wenker applied a so-called indirect method, utilising questionnaires that contained a set of standardised sentences which informants were asked to transfer into the local dialect in written form. The questionnaires were sent to virtually every location with a school in the late 19th century German Empire and the teachers were asked to fill out the forms together with the local pupils. The most important advantage of this method of data collection is that a huge number of locations and informants can be addressed in a relatively narrow time frame. Wenker and his collaborators were thus able to collect questionnaires from more than 45,000 locations. All these are reflected in the maps of the "Sprachatlas des Deutschen Reichs," making this atlas with its nearly 1650 map sheets the most comprehensive cartographic illustration of the local and regional dialects of a language worldwide (the maps of Wenker's linguistic atlases are available at http://regionalsprache.de/SprachGis/Map.aspx). Admittedly, Wenker's data collection method has one major disadvantage, discussed in linguistic publications from the outset. The data about the oral varieties, i.e., the local dialects, were written down using the letters of the alphabet and—which many detractors regarded as even worse—by linguistic laypersons. Wenker was thoroughly aware of this problem (cf. Wenker 1886) and thus validated the informants' responses wherever possible by consulting other dialect surveys such as Fischer (1895) or the phonetically exact dialect grammars of Neogrammarian scholars (cf. Wenker 1895b). Moreover, recent analyses have shown that Wenker's informants turn out to be very reliable informants. In some cases even fine-grained phonetic differences were expressed using diacritic symbols. Nevertheless, aware of Wenker's method of data collection and its (presumed and actual) shortcomings, Jules Gilliéron produced the "Atlas linguistique de la France" (ALF; published together with Edmond Edmont 1902–1910) using a completely different approach to data collection, the so-called direct method. This entailed Edmond Edmont, a linguistically trained interviewer, riding through France on a bicycle and making on-the-spot phonetic transcriptions of informants' answers. The advantage of this method is, of course, that all phonetic transcriptions satisfied a particular scientific standard: they were done by the same interviewer and were thus directly comparable. The disadvantage of this way of collecting data is that only a rather small number of locations can be included in an atlas. Hence, for the whole of France only 639 locations are represented in the ALF.

Many subsequent projects combined the advantages of Wenker's and Gilliéron's methods sending trained interviewers to informants in as many locations in the area under investigation as possible. This became feasible because conductors no longer explored large areas but rather concentrated on smaller regions. This regional focus was inspired by the *Wörter-und-Sachen* approach which pleaded for a combined investigation of linguistic (in the first instance lexical) features and ethnographic aspects which required a confinement of the atlases "to more narrowly circumscribed regions for which a specifically adapted questionnaire had to be constructed" (Swiggers 2010: 276). Two of Gilliéron's students, Karl Jaberg and Jakob Jud, were the first to combine linguistic (i.e., lexical) information and pictures of the respective objects on the maps of their "Sprach- und Sachatlas Italiens und der Südschweiz" (AIS published 1928–1940; Fig. 7). Their introduction to this atlas was of considerable relevance for theoretical and methodological progress in linguistic cartography, especially data collection. Jaberg and Jud also expanded the scope of the dialect variants collected to include not just the oldest but also the most common forms. In order to achieve their goal, they not only investigated some locations twice but also took account of information added by the informants' wives whom they for several reasons considered to be better informants (cf. Jaberg and Jud 1928: 189–193). Hence, on the maps of the AIS—or rather in the maps' commentaries—linguistic variation at individual places of interest is considered for the first time. Gauchat (1905) had already confuted the idea that speakers from a single location all spoke an homogeneous dialect. In a study of the Swiss location

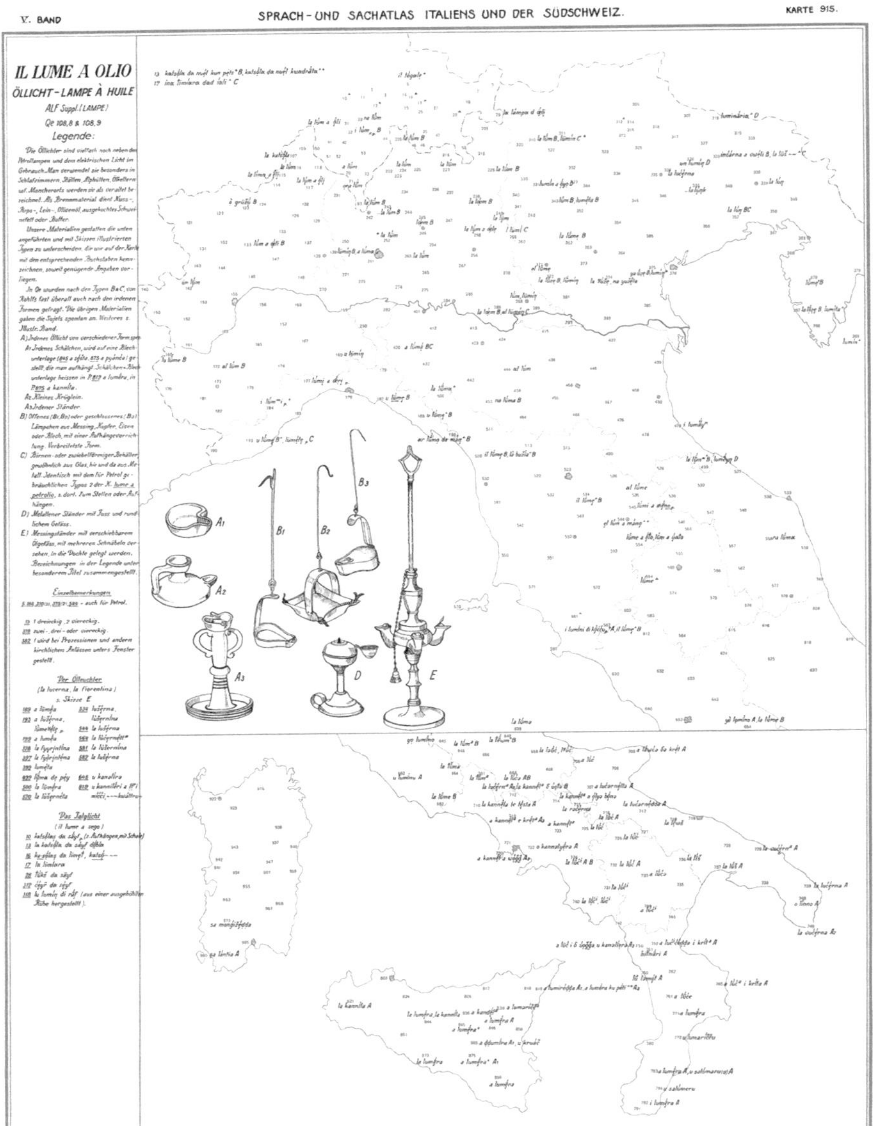

Fig. 7 Map of the “Sprach- und Sachatlas Italiens und der Südschweiz” displaying the different types of oil lamps together with the lexical variants referring to these (Jaberg and Jud 1928–1940: map 915)

Charmey, Gauchat was able to prove that “there is a very complex mix of features correlating with age, sex and social class distinctions within what we tend to regard as a single ‘dialect’” (Swiggers 2010: 274). This study crucially inspired Labov’s work on language variation in the US (cf. Labov 1981: 267).

The methodological advance made with Jaberg and Jud's AIS is of considerable relevance given that there is a direct connection between their work and Hans Kurath's "Linguistic Atlas of New England" (1939–1943) (cf. Chambers and Trudgill 1998: 17); Kurath also explicitly names the ALF and the AIS as models for his own atlas. However, he too introduces some "notable innovations, suggested in part by the peculiar linguistic situation in America" (Kurath 1938: 3), where linguistic differences between ethnic groups and social classes are much more salient than in the rural societies of the early European atlas projects. Hence, in each location at least two informants were considered, "one aged and unschooled (illiterate, if possible) and the other middle-aged and possessing a grammar school or even a high education" (Kurath 1938: 3–4). Additionally, in 38 locations so-called cultured speakers have been included. This means that for the first time in a linguistic cartography project different groups of speakers, selected by well-defined social criteria, served as informants. Another notable innovation was the production of phonographic records of associated speech to supplement the phonetic transcriptions done by the fieldworkers. The maps included all informants' responses on each topic. Kurath's Linguistic Atlas of New England thus represents a first step towards the modern pluridimensional linguistic map or atlas. This approach was further developed in the survey for the influential "Linguistic Atlas of the Gulf States" (1986–1992) conducted by Lee Pederson and his colleagues from the late 1960s on. In this project informants were chosen by applying the following criteria (cf. Pederson 1977: 6–8):

- age (two to four generations, age 15–75)
- sex (balanced)
- ethnic group (four different groups, i.e., "Anglo, Afro, Cajun, Latino")
- social class (five levels, i.e., "aristocratic, higher-than-median, median, lower-than-median, indigent")
- education (three levels, i.e., "folk, common, cultivated")
- social experience (two levels, i.e., "provincial, worldly")

In the meantime, a considerable number of pluridimensional linguistic atlases covering many parts of the world have been published, always focussing on areas of particular linguistic heterogeneity. Above all, the atlases covering speech communities in Latin America created by Harald Thun and colleagues have to be mentioned, since the Pluridimensional Geolinguistics (PG) approach has been finalised in these. Thun defines the approach as follows:

> "Pluridimensional Geolinguistics operates with an expanded model of linguistic space. It combines the traditional geolinguistic superficies ('areality') with the axis formed by socially distinct groups (the dimension of 'sociality'), creating in this way a conceptual model of a tridimensional space of superposed areal levels. Unlike monodimensional or areal geolinguistics, Pluridimensional Geolinguistics does not focus exclusively on 'old rural men' […] but successively on 'old women', 'young men', etc., in order to identify and compare the variation connected with these extralinguistic parameters." (Thun 2010: 507)

Thun and colleagues examine communities in which several varieties are utilised in everyday communication. Hence, the "atlases aim to document the consequences of contact between languages in space [..., because] it seems virtually impossible to analyze the real linguistic situation in any Latin American country (and of other regions in the world) without taking into account linguistic contact" (Thun 2010: 509). Situations of language contact can also be found in many urban regions and thus we also find linguistic atlases dealing with such areas (e.g., Mang 2004 or Stör 2005).

In a preliminary summary we can say that the most important aspects of the first four methodological issues listed above have now been mentioned:

1. With regard to the geographical areas chosen as the focus of linguistic atlases and maps we have encountered political areas (nation-states or federal states), areas of special linguistic relevance (border regions between several languages or dialects) and urban regions. To this list we can add multinational projects which have been conducted in Europe (the "Atlas Linguarum Europae," Alinei et al. 1983)—and several parts of the Slavic language area (e.g., the "Общеславянский лингвистический атлас" [Obščeslavjanskij lingvističeskij atlas] (Avanesov et al. 1965–2008), i.e., the 'Slavic Linguistic Atlas' in which eleven national committees are involved, or the "Малый диалектологический атлас балканских языков. [Malyj dialektologičeskij atlas balkanskich jazykov] (Sobolev 2003–2005) dealing with the languages of the Balkan states; for a detailed description of linguistic cartography of the Slavic languages cf. Kloferová 2010). Finally, "The World Atlas of Language Structures" (2005) covers the whole world.
2. Concerning the density of the network of locations investigated for a linguistic map we have seen that this aspect is closely associated with the size of the area under investigation: the smaller the region the higher the density of the network can be. At the extremes of the scale we have Georg Wenker's "Sprachatlas des Deutschen Reichs" with more than 45,000 locations across the 19th century German Empire and the "The World Atlas of Language Structures" where each language is represented by just one symbol.
3. Also closely related to the number of locations is the question of the method used to collect the linguistic data. We can roughly differentiate between a direct and an indirect method. With the advantages and disadvantages of these methods taken into consideration, they are both still applied nowadays, each for specific types of data. While in most cases phonetic data are collected in direct interviews that are recorded and later phonetically transcribed, studies dealing with lexical or syntactic phenomena often employ the indirect method, i.e., using questionnaires and orthographic transcriptions.
4. As regards the choice of informants for linguistic atlas projects we have tried to show that considerable theoretical progress has taken place. Even though the first atlases did not control for social language variation, Swiggers' view that "early dialectological work entertained the mythical idea of the 'pure' dialect speaker in a rural environment" (Swiggers 2010; 270) and Chambers and

Trudgill's introduction of the acronym NORM (for nonmobile, older, rural males, whom they held the majority of informants for linguistic atlases to be; cf. Chambers and Trudgill 1998: 29) both have to be regarded as a considerable oversimplification of the facts. The controlled variation of different groups of informants was introduced by Hans Kurath and has since been further developed, culminating in modern pluridimensional approaches.

6 Map Types, Mapping Techniques and Map Themes in Linguistic Mapping

As a final issue, the choice of map types and mapping techniques was listed above. In linguistic cartography we can distinguish between four basic map types based upon how the linguistic information is represented and upon the type of geographical reference (for a discussion of these issues also see Kirk 2001 or Girnth 2010):

1. Area maps, in which regions are delimited by lines (so-called isoglosses) and the linguistic information is represented
   a. by the relevant transcription, giving us an area text map, or
   b. by a symbol or (colour) shading (sometimes graphs or diagrams), known as an area symbol map.
2. Point maps, in which the linguistic information is represented at the relevant locations
   a. by the transcription itself, resulting in a point text map, or
   b. by one or more symbols (sometimes graphs or diagrams), resulting in a point symbol map.

These basic map types were all introduced by the early atlas projects described in some detail above. In line with his initial aims, Wenker started off with area symbol maps depicting a combination of several related features on a single map. In his huge "Sprachatlas des Deutschen Reichs" project, however, he switched to a combination of area text and point symbol maps in which the dominant variant of a whole region delimited by coloured isoglosses is orthographically transcribed as a so-called leading form and deviations from this variant are indicated by smaller symbols at the respective locations. Thus, his maps provide information for all of the locations investigated (Fig. 8). Gilliéron chose a different type of map in which the phonetic transcription of a surveyed phenomenon is specified for each location (Fig. 9). While in most cases, in line with common thematic cartography practice, the base map contains physical and/or topographical information, we sometimes find linguistic maps which completely lack such background information and merely indicate the topological relation between competing variants (e.g., the first linguistic map of lexical features of American English dialects by Hempl 1896: 440

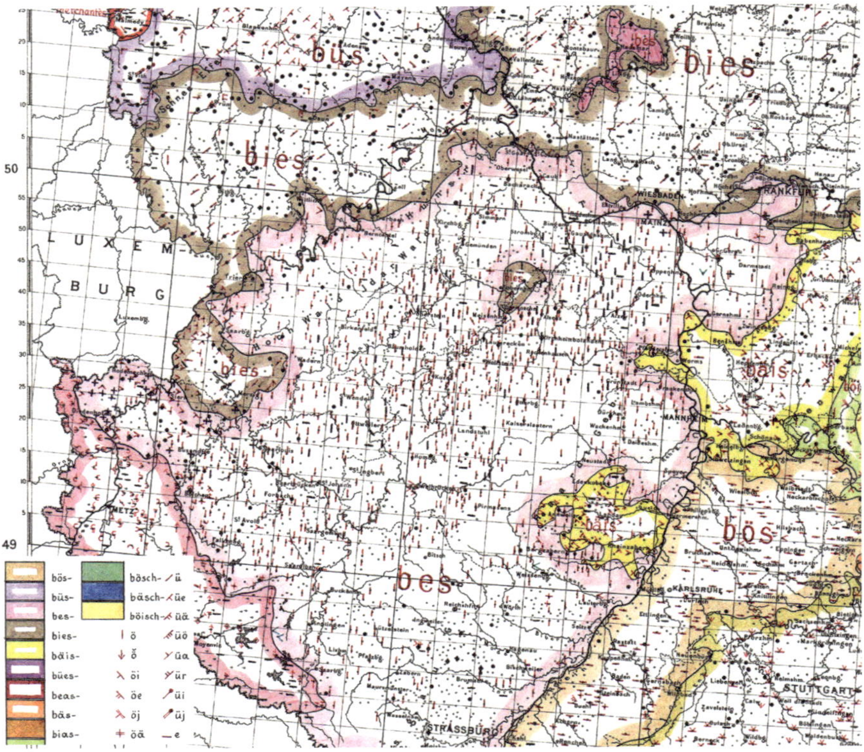

Fig. 8 Enlarged detail of a map of the "Sprachatlas des Deutschen Reichs" (Wenker 1889–1923) displaying areas in which one predominant variant occurs delimited by isoglosses and deviations from this variant at single locations indicated by symbols, thus combining the mapping techniques *area text map* and *point symbol map*

which is reproduced as Fig. 10, or some maps from the "Linguistic Atlas of the Gulf States"). In recent times, digital editions of atlases have been gaining relevance, not least because they offer a means of presenting the raw material in the form of sound recordings.

The themes of linguistic maps can be features from any level of linguistic description and many different degrees of complexity can be observed. The simplest and most widespread case is the depiction of the distribution of regionally distinct variants of a defined linguistic variable. This may be the expression of a certain concept (i.e., the lexical level), the local pronunciation of a specific phoneme from a reference system (i.e., the phonetic level) or, for example, the way in which certain past tense forms (i.e., the morphological level) or relative clauses (i.e., the syntactical level) are constructed. Individual phenomena are sometimes combined into more complex data representations, as seen in Wenker's early work. A very complex case in point is provided by a number of structuralist publications in which complete phoneme systems are compared with respect to their geographical

Fig. 9 Cutaway view of a map of the "Atlas linguistique de la France" (Gilliéron and Edmont 1902–1910) in which the *point text* mapping technique was applied (phonetic transcription at each survey location)

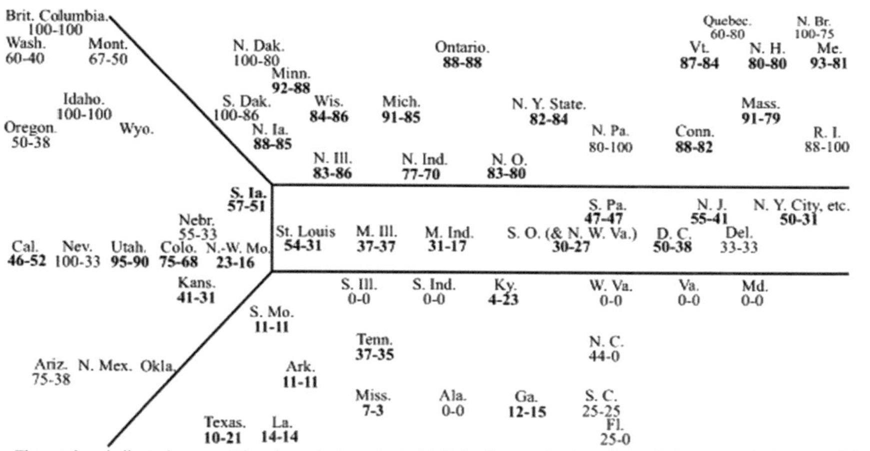

The numbers indicate the percent favoring voiceless *s* (as in 'sin'); the first number in each case is the percent for 'to grease,' the second for 'greasy.' Where the percents are based upon but few replies, fainter type is used.

Fig. 10 The first linguistic map of American English dialects on the use voiceless *s* in *to grease* and *greasy* lacking a base map with physical information (Hempl 1896: 440)

distribution under application of the notation of diasystems (cf. Weinreich 1954 or Moulton 1961). Most of the themes mentioned are mapped using isoglosses (if the network of investigated locations is dense enough) or point symbol or point text maps. This also holds true for more complex themes, where diagrams, tables or graphs are mapped to specific locations or whole areas. Since full-colour printed maps have become affordable and increasingly since the emergence of digital publication media, map themes have been able to become more complex and still remain legible, for example by simultaneously varying the shape, colour and size of the relevant symbols. These mapping techniques are also applied in the combination of linguistic information with two or more nonlinguistic factors relating to the choice of informants in the pluridimensional studies mentioned above (e.g. the "Atlas lingüístico Diatópico y Diastrático del Uruguay" 2000).

So-called "mental maps" represent a completely different view of language and the distribution of regional varieties. Maps of this type are drawn by linguistic laypersons and depict their perception of the varieties they know about, i.e. their number, their names, and their topological relationship to one another and/or to a given frame (in most cases the borders of one or more adjacent states). These aspects of perceptual dialectology are often neatly connected to further issues such as the individual (and common) evaluation of certain speech varieties (for an overview cf. Preston 2010).

Only in recent times and only in some regions of the world has an analysis of real-time processes of linguistic change become achievable. It can be attempted where early atlases (from the late 19th or early 20th century) coexist with younger compendia of maps covering the same regions (for more detailed information cf. Schmidt 2010; Kehrein 2012). In order to be able to draw plausible and reliable comparisons, Schmidt (2010) argues for the strict observance of certain methodological principles in the construction of so-called "dynamic linguistic maps" (Fig. 11):

> The primary principle is to maximize the comparability of the different sets of data. With reference to the linguistic topic to be analyzed in a dynamic linguistic map, those results from the various underlying surveys that are directly comparable need to be unambiguously highlighted and missing information made clearly visible. [… The] secondary principle […] says that a minimization of information loss, strictly related to the linguistic topic of the dynamic linguistic map, should be sought. (Schmidt 2010: 387)

In the final paragraphs of this chapter I return to early language and linguistic cartography, both initiated in order to demarcate the geographical range of certain languages or dialects. Although dialectologists—as described above—very soon discovered that the borders of different linguistic features only coincide in exceptional cases, dialectology did not completely abandon the goal of establishing linguistically based geographical dialect classifications. Instead, they started to search for weighted borders between regional varieties via a quantification of differences. The first to represent such weighted differences on a map was Carl Haag (1898), who developed what later became known as a honeycomb map. On this map (Fig. 12), coinciding isoglosses are bundled, forming stronger borders with each additional line. Ultimately, Haag was able to deduce a regional classification of the dialects of a region and to identify the features relevant for the differentiation. Bundled isoglosses aside, we often find quantified dialect differences depicted using

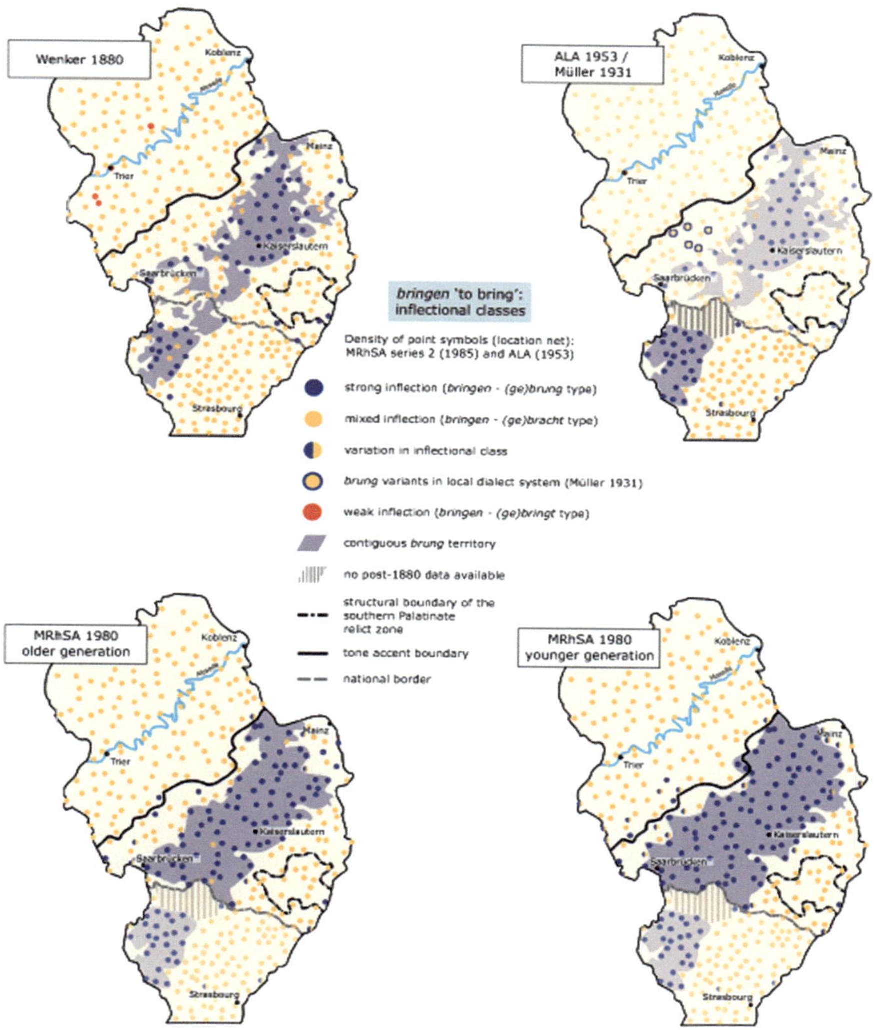

Fig. 11 Dynamic linguistic map displaying linguistic data (i.e., inflectional classes of the verb "bringen" 'to bring' in the Middle Rhine and Alsace areas) which have been collected at identical locations at different points (between 1880 and 1985) in time making processes of language change observable (cf. Schmidt 2010: map 1905)

choropleth maps in which (colour) shadings signify the degree of similarity or dissimilarity between respective geographical areas (cf. Ushiyama 1969; Bailey et al. 1993; Lameli 2013; for a recent overview see Goebl 2010). Additional map types applied in this context include joint-count maps (e.g., Kretschmar 1997) and network maps (e.g., Nerbonne and Siedle 2005).

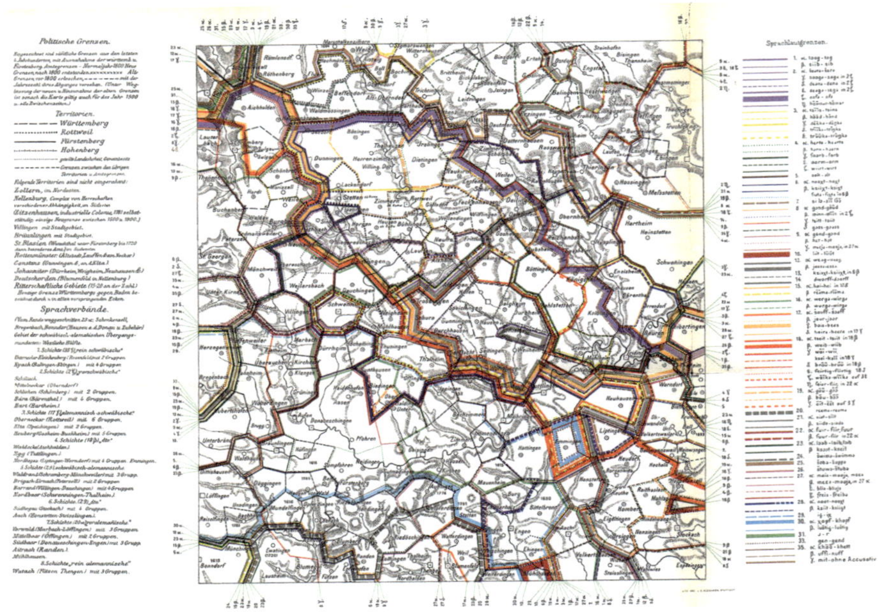

Fig. 12 Early honeycomb map in which single isoglosses (i.e., "Sprachlautgrenzen" in the legend on the right) are bundled resulting in a regional structure of the surveyed dialect areas (in this case the transition zone between Swabian and Alemannic dialects, cf. Haag 1898)

7 Prospects

As in so many other fields of modern life, the use of computers plays an important role in linguistic cartography: in the processing of language data, the construction of maps and, finally, also in their interpretation. This development started with the use of the computer simply as a printing control device but now permits the processing of large amounts of data and extraordinarily large numbers of data sets. For example, between 2001 and 2003 the Marburg Research Centre *Deutscher Sprachatlas* created a digital edition of the multicoloured maps of Georg Wenker's linguistic atlas mentioned earlier. This edition is not merely an electronic publication; the original maps have been digitised and simultaneously "enriched" with relevant geographical information. The geocoded maps may be precisely compared with any other geocoded types of data (for a summary of this project see Kehrein et al. 2006 and Rabanus et al. 2010). Other linguistic atlas projects offer internet users the ability to create maps on the basis of database queries concerning any of the phenomena included during the collection of data, whereas large parts of the data collected for printed atlases are never published. Finally, the use of GIS software and digital media opens up a wide range of possibilities for data processing and the cartographic representation of language or linguistic data. For an overview, the reader is referred to the handbook on "Language Mapping" published in 2010. In this final part, we take just one very recent study which closes a circle first opened by early "linguistic mappers" and raised in the introductory part of this chapter: the notion that language and dialect play a major role in the constitution of people's cultural identities. In the study published by Falck et al., "the effect of dialect similarity on gross regional migration flows" (Falck et al. 2010: 2) as an indicator of economic exchange is investigated. This investigation was conducted in Germany, correlating data based on Wenker's "Sprachatlas des Deutschen Reichs" project (after a quantification of dialect similarity) with data on cross-migration flows between 2000 and 2006. Essentially, the study shows "that current regional migration is significantly positively affected by similarity of the dialects prevalent in the source and destination areas in the late twentieth century. This result remains robust even after controlling for physical distance and travel time across regions and for origin and destination fixed effects, as well as for host of region-pair-specific characteristics" (Falck et al. 2010: 3; maps depicting such relations have been published in Lameli 2013). The interpretation offered for the fact that language data appear to serve as an almost ideal indicator of cultural characteristics is that dialects and their specific structures at a certain point in time "were shaped by past interactions, prior mass migration waves, religious and political divisions, ancient routes and transportation networks, and so forth" (Falck et al. 2010: 30). Given that linguistic data are quite easily accessible, they are well placed to act as a strong indicator of some sort of cultural memory and hence to serve as a reliable benchmark for comparisons to all kinds of cultural dimension.

In the future, more studies like that just sketched will need to be conducted in variational linguistics. One special challenge will be the search for extralinguistic variables that correlate with linguistic phenomena with respect to geographical distribution. An additional future objective will be to supplement the existing and exhaustive phonetic-phonological linguistic atlases with atlases dealing with additional linguistic levels, particularly syntax and morphology but probably prosody or textual and pragmatic issues as well. Finally, some developmental work still needs to be done concerning the cartographic representation of pluridimensional approaches and the depiction of processes of language change.

References

Alinei, M., et al. (Eds.). (1983). *Atlas Linguarum Europae (ALE)*. Assen/Maastricht/Rome: van Gorcum/Istituto Poligrafico e Zecca dello Stato.

Asher, R. E., & Moseley, C. (2006). *Atlas of the world's languages* (2nd ed.). London/New York: Routledge.

Avanesov, R. I., et al. (Eds.). (1965–2008). *Slavic Linguistic Atlas*. Moscow/Wrocław/Warsaw/Kraków/Zagreb: Nauka/Ossolineum/Instytut Języka Polskiego PAN/Hrvatska akademija znanosti i umjetnosti/IRJA RAN.

Bailey, G., et al. (1993). Using apparent time data to chart linguistic diffusion. *Language Variation and Change, 5*(3), 359–390.

Baker, P., & Eversley, J. (Eds.). (2000). *Multilingual capital: The languages of london's schoolchildren and their relevance to economic, social and educational policies*. London: Battlebridge Publications.

Bernhardi, K. (1844). *Sprachkarte von Deutschland. Als Versuch entworfen und erläutert von Karl Bernhardi*. Kassel: Bohné.

Blaeu, G., & Blaeu, J. (1645). *Terrarum orbis terrarum sive atlas novus in quo tabulae et descriptiones omnium regionum*. Amsterdam.

Breton, R. J.-L. (1976). *Géographie des langues*. Paris: Presses Universitaires des France.

Chambers, J. K., & Trudgill, P. (1998). *Dialectology* (2nd ed.). Cambridge: CUP.

Crystal, D. (1998). *The Cambridge encyclopedia of language* (2nd ed.). Cambridge: CUP.

Dingeldein, H. J. (2001). "Karl Bernhardi und die Sprachgrenzen im Deutschen." In A. Braun (Ed.), *Beiträge zu Linguistik und Phonetik. Festschrift für Joachim Göschel zum 70. Geburtstag*. Stuttgart: Steiner (ZDL Beihefte. 118, 161–175)

Falck, O., et al. (2010). Dialects, cultural identity, and economic exchange. *IZA DP* 4743.

Fischer, H. (1895). *Geographie der schwäbischen Mundart. Mit einem Atlas von achtundzwanzig Karten*. Tübingen: Laupp.

Fuchs, A. (1849). *Die romanischen Sprachen in ihrem Verhältnisse zum Lateinischen. Nebst einer Karte des romanischen Sprachgebiets in Europa*. Halle: H. W. Schmidt.

Gauchat, L. (1905). L'unité phonétique dans le patois d'une commune. In E. Bovet (Ed.), *Festschrift Heinrich Morf zur Feier seiner fünfundzwanzigjährigen Lehrtätigkeit von seinen Schülern dargebracht* (pp. 175–232). Halle: Niemeyer.

Gilliéron, J., & Edmont, E. (1902–1910). *Atlas Linguistique de la France*. Paris. [Reprint 1968. Bologna: Forni].

Girnth, H. (2010). Mapping language data. In A. Lameli, R. Kehrein & S. Rabanus (Eds.), *Language and space: Volume 2: Language mapping* (pp. 98–121). Berlin/New York: De Gruyter Mouton.

Goebl, H. (2010). Dialectometry and quantitative mapping. In A. Lameli, R. Kehrein & S. Rabanus (Eds.), *Language and space: Volume 2: Language mapping* (pp. 433–457). Berlin/New York: De Gruyter Mouton.

Haag, C. (1898). *Die Mundarten des oberen Neckar- und Donaulandes (schwäbisch-alemannisches Grenzgebiet: Baarmundarten)*. Reutlingen: Hutzler.

Harms, H. (1979). *Themen alter Karten*. Oldenburg: Ernst Völker.

Hempl, G. (1896). Grease and greasy. *Dialect Notes, 1*, 438–444.

Hensel, G. (1741). Synopsis universae philologiae, in qua: Miranda unitas et harmonia linguarum totius orbis terrarum occulta, e literarum, syllabarum, vocumque natura & recessibus, eruitur [...]. Nürnberg: Homann.

Jaberg, K., & Jud, J. (1928a). *Der Sprachatlas als Forschungsinstrument. Kritische Grundlegung und Einführung in den Sprach- und Sachatlas Italiens und der Südschweiz*. Halle (Saale): Niemeyer.

Jaberg, K., & Jud, J. (1928–1940). *Sprach- und Sachatlas Italiens und der Südschweiz*. Zofingen: Ringier.

Jupp, J., & McRobbie, A. (1989). *Australian languages—An introductory Atlas*. Canberra: Centre for Immigration and Multicultural Studies, Research School of Social Sciences, Australian National University.

Kehrein, R. (2012). Linguistic atlases: Empirical evidence for dialect change in the history of languages. In J. M. Hernández-Campoy & J. C. Conde-Silvestre (Eds.), *Handbook of historical sociolinguistics* (pp. 480–500). Oxford: Blackwell.

Kehrein, R., Lameli, A., & Nickel, J. (2006). Möglichkeiten der computergestützten Regionalsprachenforschung am Beispiel des Digitalen Wenker-Atlas (DiWA). In G. Braungart, et al. (Eds.), *Jahrbuch für Computerphilologie 7* (pp. 149–170). Paderborn: Mentis.

Kirk, J. M. (2001). Maps: Dialect and language. In R. Mesthrie (Ed.), *Concise encyclopedia of sociolinguistics* (pp. 350–362). Amsterdam: Elsevier.

Klaproth, J. (1823). *Asia Polyglotta Nebst Sprachatlas*. Paris: J. M. Eberhart.

Kloverová, S. (2010). Mapping the Slavic languages. In A. Lameli, R. Kehrein, & S. Rabanus (Eds.), *Language and space: Volume 2: Language mapping* (pp. 317–333). Berlin/New York: De Gruyter Mouton.

Kretzschmar, W. A. (1997). Generating linguistic feature maps with statistics. In C. Berstein, T. Nunnally, & R. Sabino (Eds.), *Language variety in the south revisited* (pp. 392–416). Tuscaloosa: University of Alabama Press.

Kurath, H. (1938). *Linguistic Atlas of New England*. Providence: Brown University for the American Council of Learned Societies.

Kurath, H., et al. (1939–1943). *Linguistic Atlas of New England*. Providence: Brown University for the American Council of Learned Societies.

Labov, W. (1981). Resolving the Neogrammarian controversy. *Language, 57*(2), 267–308.

Lameli, A. (2010). Linguistic atlases—Traditional and modern. In P. Auer & J. E. Schmidt (Eds.), *Language and space: Volume 1: Theories and methods* (pp. 567–592). Berlin/New York: De Gruyter Mouton.

Lameli, A. (2013). *Strukturen im Sprachraum. Analysen zur arealtypologischen Kompexität der Dialekte in Deutschland*. Berlin/Boston: De Gruyter.

Lewis, P. M. (Ed.) (2006). *Ethnologue. Languages of the world. sixteenth edition*. Dallas: SIL International. Online version (2009). http://archive.ethnologue.com/16

Mang, A. (2004). *Sprachregion Nürnberg. (Bayerischer Sprachatlas, Regionalteil 2, SMF 6.)* Heidelberg: Winter.

Moulton, W. G. (1961). Lautwandel durch innere Kausalität: die ostschweizerische Vokalspaltung. *Zeitschrift für Mundartforschung, 28*(3), 227–251.

Nerbonne, J., & Siedle, C. (2005). Dialektklassifikation auf der Grundlage aggregierter Ausspracheunterschiede. *Zeitschrift für Dialektologie und Linguistik, 72*(2), 129–147.

Ormeling, F. (2010). Visualizing geographic space and the nature of maps. In A. Lameli, R. Kehrein & S. Rabanus (Eds.), *Language and space: Language mapping* (pp. 21–40). Berlin/New York: De Gruyter Mouton.

Pederson, L. (1977). *A compositional guide to the LAGS project*. Atlanta: Emory University.

Preston, D. R. (2010). Mapping the geolinguistic spaces of the brain. In A. Lameli, R. Kehrein & S. Rabanus (Eds.), *Language and space: Language mapping* (pp. 121–141). Berlin/New York: De Gruyter Mouton.

Rabanus, S., Kehrein, R., & Lameli, A. (2010). Creating digital editions of historical maps. In A. Lameli, R. Kehrein & S. Rabanus (Eds.), *Language and space: Volume 2: Language mapping* (pp. 375–385). Berlin/New York: De Gruyter Mouton.

Reuther, M. (1957). Bartholomäus Scultetus (1540–1614), der Schöpfer der ersten deutschen Sprachgrenzenkarte. *Petermanns Geographische Mitteilungen, 101*, 61–65.

Richey, M. (1755). Idiotikon Hamburgense oder Wörterbuch zur Erklärung der eigenen, in und um Hamburg gebräuchlichen, Nieder-Sächsischen Mund-Art. Jetzo vielfältig vermehret, und mit Anmerckungen und Zusätzen Zweener berühmten Männer, nebst einem Vierfachen Anhange, ausgefertiget. Hamburg: Conrad König.

Robinson, A. H. (1982). *Early thematic mapping in the history of cartography*. Chicago/London: The University of Chicago Press.

Šafařík, P. J. (1842). *Slovanský zeměvid [Slavic Map]*. Prague: Vincenc Paseka.

Scheuringer, H. (2010). Mapping the German language. In A. Lameli, R. Kehrein & S. Rabanus (Eds.), *Language and space: Volume 2: Language mapping* (pp. 158–179). Berlin/New York: De Gruyter Mouton.

Schmidt, J. E. (2010). Dynamic linguistic maps and validation. In A. Lameli, R. Kehrein & S. Rabanus (Eds.), *Language and space: Volume 2: Language mapping* (pp. 385–401). Berlin/New York: De Gruyter Mouton.

Schmitt, E. H. (1992). Interdialektale Verstehbarkeit. Eine Untersuchung in Rhein- und Moselfränkischen. Stuttgart: Steiner (Mainzer Studien zur Sprach- und Volksforschung. 18

Sobolev, A. N. (2003–2005). *Little Dialect Atlas of Balkan Languages*. Munich: Biblion Verlag.

Stör, B. (2005). *Sprachregion München. (Bayerischer Sprachatlas, Regionalteil 6, SOB, Ergänzungsband.)*. Heidelberg: Winter.

Swiggers, P. (2010). Mapping the romance languages of Europe. In A. Lameli, R. Kehrein and S. Rabanus (Eds.), *Language and space: Volume 2: Language mapping* (pp. 269–300). Berlin/New York: De Gruyter Mouton.

ten Kate, L. (1723). *Aenleiding Tot de Kennisse van het Verhevene Deel der Nederduitsche Sprake*. Amsterdam: Wetstein.

Thun, H. (2000). Altes und Neues in der Sprachgeographie. In W. Dietrich and U. Hoinkes (Eds.), *Romanistica se movet... Festgabe für Horst Geckeler zu seinem 65. Geburtstag*. Münster: Nodus, 69–89.

Thun, H. (2010). Pluridimensional cartography. In A. Lameli, R. Kehrein and S. Rabanus (Eds.), *Language and space: Volume 2: language mapping* (pp. 506–524). Berlin/New York: De Gruyter Mouton.

Ushiyama, H. (1969). *Toozai Hoogen-no Kyookai [The Boundaries between Eastern and Western Dialects]*. Nagano: Shinkyoo Insatsu.

van der Merwe, I. J., & van der Merwe, J. H. (Eds.). (2006). *Linguistic atlas of South Africa: Language in space and time*. Stellenbosch: Sun Press.

Weinreich, U. (1954). Is a structural dialectology possible? *Word, 10*, 388–400.

Wenker, G. (1877). *Das rheinische Platt. Den Lehrern des Rheinlandes gewidmet*. Düsseldorf: Self-published.

Wenker, G. (1878). Sprach-Atlas der Rheinprovinz nördlich der Mosel sowie des Kreises Siegen. Nach systematisch aus ca. 1500 Orten gesammeltem Material zusammengestellt, entworfen und gezeichnet. Manuscript (hand-drafted onto printed base maps) held in the archives of the Forschungszentrum Deutscher Sprachatlas.

Wenker, G. (1881). *Sprach-Atlas von Nord- und Mitteldeutschland. Auf Grund von systematisch mit Hülfe der Volksschullehrer gesammeltem Material aus circa 30000 Orten. Text. Einleitung*. Strasbourg/London: Trübner.

Wenker, G. (1886). [Vortrag: Über das Sprachatlasunternehmen, mit Vorzeigung von Karten]. In Verhandlungen der achtunddreissigsten Versammlung deutscher Philologen und Schulmänner in Giessen vom 30. September bis 3. Oktober 1885. Leipzig: Teubner, 187–194.

Wenker, G. (1889–1923). Sprachatlas des Deutschen Reichs. Manuscript (hand-drafted onto printed base maps) held in the archives of the Forschungszentrum Deutscher Sprachatlas and in the Staatsbibliothek zu Berlin.

Wenker, G. (1895a). Über den Sprachatlas des Deutschen Reichs. *Verhandlungen der 43* (pp. 34–43). Versammlung deutscher Philologen und Schulmänner. Leipzig: Teubner.

Wenker, G. (1895b). Herrn Bremers Kritik des Sprachatlas. *Der Sprachatlas des Deutschen Reichs* (pp. 3–30). Elwert: Dichtung und Wahrheit. Marburg.

Wurm, S. A., Mühlhäusler, P., & Tryon, D. T. (Eds.). (1996). *Atlas of languages of intercultural communication in the Pacific, Asia, and the Americas*. Berlin/New York: Mouton de Gruyter.

Landscape: The *Thing* About Landscape's Nature: Is It a Creature/Monster of the Map?

Kenneth R. Olwig

Abstract Is the nature of landscape something that can be mapped, or is the landscape itself a thing or creature of the map? Or perhaps even a "monster" of the map? These are questions around which landscape studies have revolved in recent years. In this chapter I trace the two sides of the question and provide a capsule history of contemporary geographical scholarship, focusing on the contributions of Carl Sauer and European geographers. This landscape approach still dominates much of continental and especially German geography, but in Anglo-America it has declined and landscape has come to be seen not so much as some *thing* you can map, but rather as a thing of the map, that is, a creature born of cartography. I suggest a third alternative, which opens up new ways of thinking about things, nature, landscape and mapping. Maps are foundational pieces in the study of traditional and also postmodern and "non-modernist" landscape which in contemporary geography is concerned with the social bases for things governing and historically developing inter-relationships between society and nature—this is the thing about lansdscape.

Keywords Landscape · Mapping · Nature and culture · Landscape theory

1 Landscape as Mappable Thing

In the beginning of the 20th century the study of geography and related fields (such as anthropology, ethnology and archaeology) in Europe and North America was focused in great measure upon the interpretation of 'landscape.' Carl Sauer, when founding the geography department in UC Berkeley, California, summarized this landscape tradition in his classic 1925 text, *The Morphology of Landscape*. Here, he treated landscape basically as a thing, a material assemblage of phenomena on the

K.R. Olwig (✉)
Department of Landscape Architecture, Planning and Management,
Swedish University of Agricultural Sciences, Alnarp, Sweden
e-mail: Kenneth.Olwig@slu.se

S.D. Brunn and M. Dodge (eds.), *Mapping Across Academia*,
DOI 10.1007/978-94-024-1011-2_11

earth's surface, the *forms* or *shapes* of which (morphology) could be mapped. The suffix—*scape* in land*scape*, as Sauer noted, is basically a variant of *shape* (Sauer 1925: 25; see also, Merriam-Webster 1996: ship, shape), and since, for Sauer a defining characteristic of geographers was that they make maps (Sauer 1969b (orig. 1956): 391), then the task of the geographer was to map the shape and form of landscape. For Sauer (1925: 26) the facts of landscape were "place" facts and landscape geography was thus concerned with "chorography." Chorography was a branch of geography that had existed since the time of the ancient Greek thinkers, the root of which was *choros* (spelled "Chore" by Sauer and "chora" by many philosophers), the Greek word for place, country, area or *land*—the *land* in *land*-scape. For Sauer (1925: 46), thus, the "cultural landscape is the geographic area in the final meaning *(Chore)*." Sauer appreciated that ancient Greeks such as Herodotus had largely approached chorography through a narrative approach that merged with history and ethnography. Sauer nevertheless saw the emergence of scientific mapping techniques during the Renaissance age of exploration, or "Age of Surveys" as Sauer called it, as central to the cartographic chorography that he applied to the study of landscape. As he put it: "However much may have been added since in precision of measurement, in many ways we have retained the chorologic content as formulated in the maps of this period beginning the "Age of Surveys" (Sauer 1925: 23). Thus, according to Sauer: "One very great step in synthesis certainly took place at this time, that of the development of cartography into a real chorologic discipline" (Sauer 1925: 23). *Choro*graphic cartography focused largely on the development of *choro*pleth maps which essentially indicate the bounded areal extent of given phenomena, such as a pine forest or open plains of grass or heather, through the use of shading or markings, to indicate the size and shape of an area as outlined on the map.

In the landscape tradition about which Carl Sauer wrote cartography was used primarily to make thematic choropleth maps of both natural phenomena (such as geology, climate, flora and fauna) and cultural phenomena (such as animal husbandry, crops, field patterns, the built environment and population characteristics). Choropleth maps of differing phenomena can be layered so that one might come up with a spatial correlation between the natural phenomena and the cultural patterns that characterize a *particular* landscape type, such as a typical Northwestern European heath landscape of the kind that once dominated the western half of the Jutland peninsula, Denmark (Olwig 1984) (Fig. 1). An example of the relevant natural phenomena would be, for example: relatively flat glaciated topography; treeless sandy, leached and podzolized sandy soils; heather flora and a wet and windy Atlantic climate. The corresponding examples of cultural phenomena would be, for example: extensive sheep and cattle grazing; rye grain cultivation near stream meadows as well as intermittent cultivations scattered in places where the vegetation of a vast heathland commons has been burned; low population density; squat small houses; moonshine brewing and a knitting home industry. These elements together added up to form a landscape type of a particular areal extension and location and with a characteristic natural and cultural form and appearance (Olwig 1984).

Fig. 1 Jutland heathscape from Bærentzens Denmark, 1855

Once a correlation of natural and cultural landscape forms has been made, one can begin to ask questions about causality. Sauer was opposed to the "environmental determinism" that was popular at the time, arguing that the world's landscapes tended to be shaped in the image of the culture of the people that inhabited them. Those minded toward environmental determinism, however, argued that landscapes such as that of the Jutland heath were largely determined by nature. The sheep, the knitting, the rye, the low population density and the squat houses were thus an adaptation to the cool wind and rainswept, sandy, glaciated open heathland plains of the North Atlantic (Olwig 1984). It could, however, also be argued that the characteristic soil, open heather vegetation and even the climate of the landscape had been significantly "land-shaped" by the people living in the area. The development of extensive forms of pastoralism and agriculture thus might have resulted in the destruction of previous forest cover, inducing change in the local climate by exposing the land to the wind and rain, which in turn favored the growth of heather which then led to the leaching of the soil due to acidification of the rainwater as it percolates through the acidic peat soils generated by heather. But why, one might ask, did the population develop extensive forms of agriculture? Was this, for example, because of cultural perceptions of desirable livelihoods? Were the original settlers pastoralists whose culture was, like that of the Sami today, rooted in animal herding to the degree that they largely shunned agriculture and forestry and favored open grazing environments instead? Or, alternatively, is it possible that the extensive agriculture and its products (e.g. sheep grazing/wool products cultivation and rye/moonshine) was determined by spatial economic factors because the area is far from major urban centers and this required an adaptation to a peripheral location in space? Such a location would favor the production of goods that can be walked to distant markets (e.g. cattle and sheep) or easily carried (e.g. knitted goods and alcohol). Extensive land use would also be favored by the low price of land in economically peripheral areas. It is thus possible that distance from major urban markets explains the development of the extensive forms of agriculture that led to the spread of the distinctive heath landscape. Such questions can, in turn, prompt in-depth archaeological and anthropological studies based on ethnographic methods and natural scientific research into the history and development of such an area. One might, thus, study whether or not the earliest settlers were pastoralists, and whether or not the area had originally been forested. Once one understands how a given landscape came into being this knowledge can facilitate either the conservation or the development of the area and its landscape. In Jutland, for example, such knowledge facilitated a modernization movement to cultivate and afforest the heathlands that significantly transformed the local economy. This process, however, also left a few remaining heaths that are now conserved as nature reserves. This kind of landscape study, furthermore, need not be of only practical value, it can also prompt theoretical and philosophical speculation concerning the relation of humanity to nature, and to the nature of progress and modernity (Olwig 1984). Did science, modernity and urbanism bring about a better world or, as Sauer felt, did it sometimes lead to the impoverishment of natural and human environments?

The above approach to landscape still dominates in much of continental Europe in general, and Germany in particular, but in Anglo-America it has declined and, in many ways, been turned inside out. Here landscape has come to be seen by many not so much as some *thing* you can map, but rather as a *thing* of the map, a creature born of cartography.

2 Landscape as a Creature of the Map

Several factors worked to change the Anglo-American approach to landscape. To begin with, Richard Hartshorne, the dominant English language post-WWII geographical theorist, was extremely critical of the landscape paradigm in geography (Hartshorne 1939: 149–174, 250–284). Carl Sauer, as noted, identified landscape geography with chorography. Hartshorne, however, argued that the concept of landscape, as understood in English, was not chorographical. To understand his critique it is necessary to take a closer look at the meaning of the concept of landscape as used by Sauer and others in the 1920–1930s.

During the formative years of Sauer's life, before the cataclysm of WWII, German geography played a progressive role in geographical thinking worldwide, and it had done so since the days of Alexander v. Humboldt, whose explorations of the world's landscapes, and progressive politics, gave him heroic status in many places, including the United States. German geography had provided much of Sauer's inspiration and in Germany, along with much of Northern Europe, the term landscape (or *Landschaft* in German) had historically been used at least since the Middle Ages to designate a quasi-independent regional polity ruled largely according to locally characteristic customary law (Olwig 2002). The closest North American analogy to the historical landscape polity is a New England township, a form of polity brought to New England from old England by the first settlers. The spelling of landscape has changed through time and it also varies from place to place, but the *-ship* in town*ship* is basically just a spelling variant of the *-scape* in land*scape*, which is also, as Sauer noted, related to the word *shape* (Merriam-Webster 1996: ship). The suffix *-ship* can refer to "the body of persons participating in a specified activity [readership] [listenership]" (Merriam-Webster 1996: ship), which is a sense of *-ship* that we use when we say, for example, "the township voted to protect its watershed." The same usage could equally be applied to the historical landscape as a polity (which normally would comprise several towns), e.g. "the *landscape* voted to protect the sea dikes protecting its towns and their surrounding hinterlands." The suffix *-ship* can also suggest: "something showing, exhibiting, or embodying a quality or state [town*ship*] [fellow*ship*]" (Merriam-Webster 1996: ship). Simply put, one could thus say that a *township* is an area *shaped* by the polity of a town (as when the "township votes" to protect its watershed), and which thereby can be seen to exhibit the quality or state of being a town polity, and this is likewise the case for a landscape polity (as when it votes to develop its dikes) (for a more detailed etymological and historical elaboration see, Olwig 2002). By Sauer's day, however, most (but not all) of these landscape polities would have been absorbed into centralized modern nation states and would have often ceased to exist as quasi-independent political entities, as was the case with Jutland in Denmark. Nevertheless, the areas shaped by these polities would continue to exhibit their distinctive landscape identities in terms of culture and environment even if the effects of modern transportation, industrial development and agricultural practices were diminishing regional difference. This meant that it made sense for Sauer and others to continue to think of landscapes as areal, chorographic phenomena and as historically evolving places in which human culture played an active role in shaping the land. In this way the development of the historical landscape territory of Europe provided an analogy for understanding, in the abstract, the development of similar phenomena throughout the world. Hartshorne, however, disagreed.

Hartshorne argued that in English, as opposed to German, the word landscape did not mean an area or region but a visual scene of infinite unbounded depth, as perceived, for example, in a perspectival landscape painting (Hartshorne 1939: 149–174, 250–284). Hartshorne, therefore, felt that the chorographic approach should be shorn of landscape and made into a spatial science, *chorology,* which would make geography the comparative study of areas according to criteria chosen by a specialized geographer, be it a physical geographer or, for example, as in Hartshorne's case, an economic geographer. Anglo-American geography should thus abandon its interest in landscape as understood in Germany and thereby also abandon, with it, the concern with the historical evolution of society/nature relations (Hartshorne 1939: 263, 275, 1958, 1959: 48–64). This opened the way for physical and cultural geography to develop separately. Hartshorne sought to define geographical scholarship as a modern "spatial science" (Hartshorne 1958) and he successfully established this as geography's predominant paradigm. The discipline, however, eventually evolved beyond the bounded chorological regions favored by Hartshorne to a study of unbounded functional regions and relations between locational nodes in space. Attention thus shifted from the bounded areas one could draw on a map to the locational nodes one could pinpoint and plot on a map's grid, thereby enabling the "locational" and "spatial" analysis" of the relations between such nodes (Abler et al. 1971).

Though Hartshorne's early 20th century critique eventually led to the marginalization of landscape as a focus in mainstream social science oriented geography, his thinking nevertheless opened up for new humanities oriented ways of conceptualizing landscape as a spatial phenomenon. David Lowenthal, who had studied with both Sauer and Hartshorne, thus pointed out, together with Hugh Prince, that landscape, understood as visual scenery that can be perceived, for example, in landscape art, was not at all irrelevant to the study of geography (Lowenthal and Prince 1964, 1965). He, along with other humanities oriented geographers such as Yi-Fu Tuan, emphasized that spatial perceptions had a significant effect on human geographical behavior, and hence the shaping of the geographical environment as landscape scenery (Lowenthal 1961; Tuan 1974). Humanistic geographers thereby opened the door for new approaches to landscape within the realm of Anglophone geography. It was in this context that the geographer Denis Cosgrove was able to re-conceptualize landscape as a phenomenon which was essentially a thing or creature of the map rather than some *thing* to be mapped.

3 Cosgrove and the Cartography of the "Modern" Landscape

Drawing upon inspiration from studies within art history and the history of cartography, Cosgrove argued, using the historical example of the Veneto region inland from Venice, that the perception of landscape as scenery originally derived from cartography. There was a clear link, he contended, between the use of map-making in the enclosure and drainage of these areas, and the design of the parks and buildings which he termed "the Palladian landscape" after Andrea Palladio, the Venetian renaissance master architect who designed many of the characteristic estates in the area (Cosgrove 1984, 1993). The Renaissance was marked by the rediscovery of the cartographic techniques of the second century A. D. Alexandrian astronomer, astrologer, cosmographer and geographer Claudio Ptolemy. His cartographic technique involving the projection of locations upon the spatial coordinates of a grid like graticule was foundational to scientific cartography. Moreover, Cosgrove argued that by employing these cartographic methods as developed by surveyors and mapmakers Renaissance cartographers and artists were able to represent the world in terms of perspectival scenery. Put very simply, if you

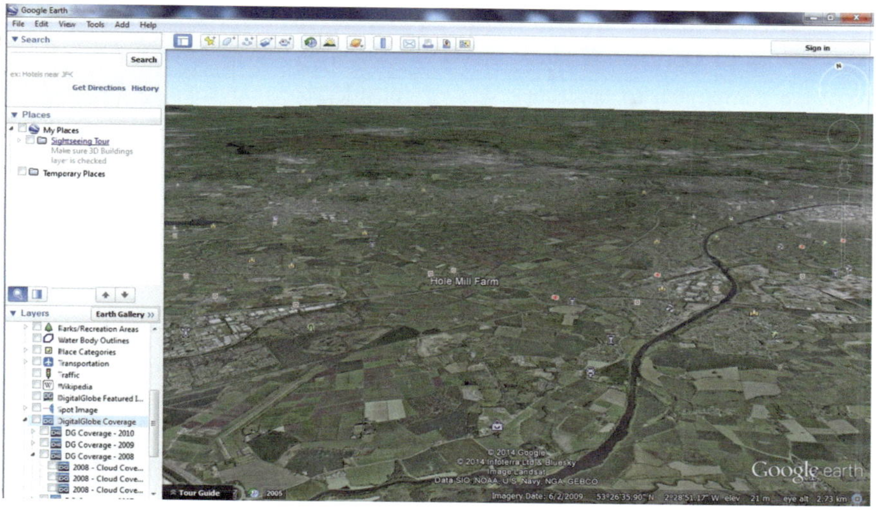

Fig. 2 Google Earth screenshot

alter the top-down projection of the map to a more horizontal projection you get a landscape scene, such as those often reproduced in the maps of the time. (This can be easily done on Google Earth for example; Fig. 2.) The Renaissance map makers and artists not only, however, made representations of existing spatial scenes, they also used these techniques to create designs for previously non-existent ideal scenes and to construct new environments following these ideals, thus giving birth to landscape architecture and landscape planning.

Cosgrove argued that the characteristic landscape of the Veneto did not evolve on the basis of ancient cultural tradition, but was largely the result of the colonization of the area by Venetian merchants who with the aid of surveyors and architects transformed the territory into a landscape shaped according to Palladian Renaissance ideals. By the same token, he showed that the famous English landscape garden ideal did not evolve on the basis of any ancient Anglo-Saxon cultural heritage, as many thought, but was transformed using the techniques for representing landscape in perspectival space that came together with the importation of Palladian architectural ideals from Italy. Cosgrove was a modernist who deeply admired both the retro-modernism of Britain's engineering monuments in the form of huge bridges and dams and the futurism of the Apollo space project (Cosgrove 2001). In his eyes the Palladian landscape was not only a thing of beauty that came about with the ability to visualize and modernize the environment in scenic terms, but also a model of agricultural improvement through drainage and rationalization that took place through advances in cartographic science (Cosgrove 1993). Landscape was thus, in his eyes, an often beautiful creature of the map. This however, as will be seen, was only a partial truth.

4 Is the Creature of the Map Really a Monster?

Cosgrove's approach to landscape was of considerable relevance to the fields of environmental design, planning and landscape architecture (Corner 1999), and it nestled well within the spatial orientation of late twentieth century geography. It also resonates with the critique of Sauer begun by Hartshorne (Olwig 1996) and continued by British spatially oriented geographers (Cosgrove and Jackson 1987; Olwig 2010). The problem with this approach to landscape was that it was an edifice partially built upon the idea that the English language concept of landscape which differed radically in its origin and development from the German because it referred to open spatial scenes, rather than chorographic areas. This idea, in turn, depended upon the argument that the idea of landscape grew out of the perspectival techniques of spatial representation that developed on the basis of Renaissance cartography. This, however, was not entirely true. The modern idea of landscape as scenery did largely grow out of Renaissance techniques of representation, though not exclusively those of Ptolemaic cartography (Olwig 2011a), but the English idea of landscape did not develop entirely from these Renaissance roots. English is fundamentally a Germanic language and it therefore has much in common with other Germanic languages such as Dutch, Danish, Swedish and German. Historically the word landscape in English, as in the other Germanic languages (though spelled in a variety of similar ways), had a shared meaning in which the *land* in *land*scape referred to a polity and its place, be it a *choros*, or a *country*. Thus, according to Samuel Johnson's classic dictionary, the first meaning of landscape is "(1) A region; the prospect of a country." It is only in the second meaning that landscape comes to be understood as a pictorial phenomenon in which, as on a map, information is plotted into a pre-given geometrical space. From this perspective landscape comes to mean": (2) A picture, representing an extent of space, with the various objects in it" (Johnson 1755 [1968]: landscape). Johnson had a serious interest in chorography (Mayhew 2004) and his definition displays insight, as will be seen, into the evolving meaning of landscape in relation to cartography.

If we return to the historical meaning of landscape as a polity, somewhat like a New England township, it becomes easier to elucidate the developing relationship between landscape and cartography. The historical landscape polity, as in a New England township, was governed and shaped by a type of meeting of the people with rights in the land of the landscape which was historically called a *thing* or *moot* (the root of *meeting*) (Merriam-Webster 1996: thing, meet). This meeting would often take place on, or adjacent to, a central commons to which the polity of the landscape had ready access. It was at such meetings that the customs of the landscape were formalized as common law and legal cases could be brought to court. The jurisdiction of such landscape polities was thus essentially shaped from the inside out, rather than defined by an external, administrative boundary on a map. Furthermore, in many places such landscape polities would be separated from other polities by extensive common wastelands, like those of the Jutland heaths, or inhospitable mountains. This meant that the extent of the jurisdiction and protection of the landscape polity's law would be blurred and there would be no clear-cut boundary (it was in such areas that the "outlaws" lived). With the rise of the centralized Renaissance state and its support of the development of cartography for administrative and military purposes, it became possible to circumscribe polities as regions defined according to lines on a map. It could even be argued that the definition of landscape polities as *regions* on a map went hand in hand with the subsumption of these polities under a central state and the decline of self-rule (Olwig 2002, 2008b). This is suggested by the very term *reg*ion in which the preface *reg*—comes from the Latin *regere* meaning "*to rule, direct,*" as in the words *reg*ent and *reg*ulate (Merriam-Webster 1996: region). As in Johnson's dictionary definition, however, the landscape as an extant polity and place (definition 1) preceded its representation within the space of the map and the perspectival drawing (definition 2).

Even though Sauer, as noted earlier, appreciated the ancient Greek narrative approach to chorography that combined historical, ethnographical and geographical description, he nevertheless uncritically favored, in *The Morophology of Landscape*, the use of cartography as a basis for the chorographic analysis of landscape. He thereby overlooked the significance of the fact that neither the historical landscape polity, nor the Greek polity of the *choros*, were originally defined from the outside in by boundaries drawn in the space of the map. Both were rather shaped from within by representative discursive bodies meeting, for example, by the town green or, in the Greek case, within the *agora* (Olwig 2008a). Sauer thus did not consider that the cartographic approach to chorography might give a misleading picture of both the ancient Greek *choros* and the later landscape polities of Europe (Olwig 2008a, b). By using the representational techniques of cartographic chorography one will have a tendency to favor those phenomena that lend themselves to being represented within encircling lines on a map, such as areas physically bounded by water or mountains or marked changes in geomorphology, as in the case of the glacial outwash plains of the Jutland heaths. The result thus is a tendency to not only emphasize physical environmental determinants, which Sauer otherwise opposed, but also to choose the same sort of barriers that were favored by the centralizing states that emerged in the Renaissance along with cartographic chorography. If, on the other hand, one focuses on the processes by which human environment interactions develop not from the outside in, but from the inside out, it becomes clearer that sharp natural and cultural boundaries generally tend to emerge primarily as a consequence of the political, property or planning lines drawn on maps (Latour 1999: 24–79; Leighly 1937; Olwig 2004). The effects of these cartographic lines, or striations, can thus also be seen in the development of the "Palladian landscape," where maps were used to survey and transform the common lands, to which the commoners had use rights, into the sharply delineated fields, forests and pastoral parks that became the private property of wealthy estate owners through enclosure (Olwig 2002). Perhaps, if Sauer had written *The Morphology of Landscape* several decades later, after he had made his path breaking landscape studies of the former Spanish colonies in the Americas in the more narrative historical chorographic mode of Herodotus, he might have been more circumspect concerning the scientific neutrality of a chorography and been more aware of its role in imperial conquest and the colonial appropriation of indigenous common lands (cf. Harley 1988). In this context the landscape creature of the map might rather be regarded as a monster of the map hiding, in the case of pastoral landscape parks of the estate owners, in sheep's clothing.

5 The 21st Century Non-modern Landscape

We have seen that, viewed from one perspective, landscape as a scenic creature of the map might be regarded as a harbinger of modern progress, bringing with it efficient productive agricultural lands, beautifully designed parklands and well planned cities. From another vantage point, however this thing/creature of the map might be regarded as a monster, a Leviathan swallowing once semi-autonomous polities into the bowels of a central state, or taking the commons of the commoners and privatizing them as the idealized idyllic and exclusive pastoral landscape gardens surrounding the mansions of the wealthy. In fact, the two seemingly opposed narratives might well be regarded as two sides of modernism's narrative coin, envisioned as a Faustian bargain in which material progress requires the destruction of the traditional and the small scale, or as Karl Marx (paraphrasing Shakespeare) put it, that everything solid must "melt into air" (Berman 1982; Olwig 2011a). There is, however, an emergent "non-modern" alternative to the Mephistophelean alternative to the modernist narrative's dualistic alternative between either a nostalgic traditionalism or the juggernaut of modernity.

A key figure in the questioning of modernity's narrative has been the anthropologist Bruno Latour, who has argued that "we have never been modern" (1993) and who has inspired progressive contemporary geographers to break down the society/nature, subject/object dualism characteristic of modernism that as we have seen, has bedeviled geographical scholarship since it became a modern spatial science (Whatmore 2002). Key to Latour's present thinking is the institution of the *thing* which, as has been seen, was the discursive representative body that governed the ancient landscape polity. Following the etymological analysis, if not the philosophy, of Martin Heidegger (1971), Latour draws a distinction between a *thing* in the modern sense of a material entity and *thing* in the original sense as an ancient form of parliament or meeting called a "thing" (or "moot") in English, a *Ding* in German, and a *ting* in the Scandinavian languages. A *thing* was a place where people met to *moot* and discuss *things*, or *public matters*, and agree upon the laws and customs that were to govern the *res publica*, or *commonwealth*, of the polity (Latour 2005). Of this, Latour writes:

> Now, is this not extraordinary that the banal term we use for designating what is out there, unquestionably, a thing, what lies out of any dispute, out of language, is also the oldest word we all have used to designate the oldest of the sites in which our ancestors did their dealing and tried to settle their disputes? (Latour 2004: 233)

Latour's point is that when discussing things in the physical material sense (e.g. the things that are the object of environmental science) it is important to understand that these things first gain meaning when mooted in social discourse and debate, not the least within the modern institutions (such as universities, research centers and disciplinary conferences) and media that, like the ancient judicial thing, shape the basis for law and policy through that which Latour calls *Dingpolitik* (Latour 2005).

When seen in the light of the *thing* as understood by Heidegger and Latour, and in relation to its historical importance to the constitution of the landscape polity, it becomes clear that the thing about landscape is not necessarily a thing to be mapped, nor a thing that is a creature/monster of the map, but a "hybrid" commonwealth of people and the things that matter as constituted by the "Dingpolitik" of a *res publica* (Krauss 2010). It also becomes apparent that a landscape need not correspond to a given territory inscribable on a map. Rather, given the enlarged sense of *thing* as understood by Latour, the political landscape of a contemporary polity can take place in many contexts mushrooming from local sites to the cyber-places of the Internet (Tsing et al. 2009) or to the deliberative processes undertaken by the countries that have ratified the *European Landscape Convention* (Olwig 2007).

6 Conclusion

The map has been foundational both to traditional landscape study as described by Carl Sauer in *The Morphology of Landscape*, and in the approach taken to landscape by the early work of Denis Cosgrove and his followers. The landscape in both cases is arguably a thing-like creature of the map, whether or not geographers have been aware of the degree to which mapping and its Siamese twin, perspectival drawing, have shaped their perception of landscape. Furthermore, it is hard to deny the usefulness of cartography and pictorial depiction both to the analysis of landscape as mapped and to the design of new landscapes (Corner 1999). The problem, however, is that the map structures space in a particular way by plotting the lived things of our world within an ideal, absolute, static space. The map, however, is not the territory, and the earth is not a globe (Olwig 2011b). When the landscape that we generate through our social and bodily practices (Ingold 2000) is reduced to the space of the map, and that of perspectival pictorial representation, it is impoverished, and the landscape becomes a scenic stage upon which we act out the roles we are given (Olwig 2011a). It is thus important to note that Sauer's *The Morphology of Landscape* was written early in his career as a preliminary overview of the many facets of landscape study anno 1925. Even at that time he was aware of alternatives to cartographic representation, such as the narrative chorography of ancient Greek geographers such as Herodotus, and in his later work he went on to develop approaches to landscape that were more in line with this earlier, non-cartographic, approach to chorography and landscape (Sauer 1969a). Denis Cosgrove, by the same token, as his thinking matured and developed, changed his mind about landscape being fundamentally a creature of the map and became interested in the lessons to be learned from the historic meaning of landscape as polity and place (Cosgrove 2004, 2006). Modernity's key representational forms, the map and the perspectival drawing, have created a representational hall of mirrors in which the object and its representation define each other in an endless pattern of circulating (self)reference (Latour 1999; Olwig 2004), but now there are also those who question the representational foundations of geography and have begun to explore "non-representational" approaches (Thrift 2007). The landscape, in a "non-representational" context, is not the postmodern product of its representation, a creature of the map. It is a human creation shaped in differing social and physical contexts, according to the customs, experiences and affections of a given polity, be it the political landscape of a place as expressed in town meetings, or the disciplinary landscape of a scholarly field like geography, as expressed at the meetings of the Association of American Geographers. The "non-modern" approaches to landscape discussed here question the need to bifurcate geography into warring divisions of social and physical geography, and provide argumentation for the resuscitation of the study of landscape as a discipline concerned with the social basis for the things that govern the inter-relationship between society and nature—this is the thing about landscape.

References

Abler, R., Gould, P., & Adams, J. S. (1971). *Spatial organization: The geographer's view of the world*. Englewood Cliffs, New Jersey: Prentice-Hall.

Berman, M. (1982). *All that is solid melts into air: The experience of modernity*. New York: Simon and Schuster.

Corner, J. (1999). *Recovering landscape: Essays in contemporary landscape architecture*. New York: Princeton Architectural Press.

Cosgrove, D. (1984). *Social formation and symbolic landscape*. London: Croom Helm.

Cosgrove, D. (1993). *The palladian landscape: Geographical change and its cultural representations in sixteenth-century Italy*. University Park, PA: The Pennsylvania State University Press.

Cosgrove, D. (2001). *Apollo's eye: A cartographic genealogy of the earth in the western imagination*. Baltimore: The Johns Hopkins University Press.

Cosgrove, D. (2004). Landscape AND Landschaft [lecture delivered at the Spatial Turn in History Symposium, German Historical Institute, February 19]. *German Historical Institute Bulletin, 35*, 57–71.

Cosgrove, D. (2006). Modernity, community and the landscape idea. *Journal of Material Culture, 11*, 49–66.

Cosgrove, D., & Jackson, P. (1987). New directions in cultural geography. *Area, 19*(2), 95–101.

Harley, J. B. (1988). Maps, knowledge, and power. In S. Daniels & D. Cosgrove (Eds.), *The iconography of landscape* (pp. 277–312). Cambridge: Cambridge University Press.

Hartshorne, R. (1939). *The nature of geography*. Lancaster, PA: Association of American Geographers.

Hartshorne, R. (1958). The concept of geography as a science of space, from Kant and Humboldt to Hettner. *Annals of the Association of American Geographers, 40*, 97–108.

Hartshorne, R. (1959). *Perspective on the nature of geography*. Chicago: Rand McNally.

Heidegger, M. (1971). The thing. In A. Hofstadter (Ed.), *Poetry, language, thought* (pp. 165–182). New York: Harper & Row.

Ingold, T. (2000). The temporality of landscape. In T. Ingold (Ed.), *The perception of the environment: Essays in livelihood, dwelling and skill* (pp. 189–218). London: Routledge.

Johnson, S. (1755 [1968]). *A dictionary of the English language*. London: W. Strahan.

Krauss, W. (2010). The '*Dingpolitik*' of wind energy in northern German landscapes: an ethnographic case study. *Landscape Research, 35*(2), 195–208.

Latour, B. (1993). *We have never been modern*. Cambridge, MA: Harvard University Press.

Latour, B. (1999). *Pandora's hope essays on the reality of science studies*. Cambridge, MA: Harvard University Press.

Latour, B. (2004). Why has critique run out of steam? From matters of fact to matters of concern. *Critical Inquiry, 30*(2), 225–248.

Latour, B. (2005). From realpolitik to dingpolitik or how to make things public. In P. Weibel & B. Latour (Eds.), *Making things public: Atmospheres of democracy* (pp. 4–32). Cambridge, MA: MIT Press.

Leighly, J. (1937). Some comments on contemporary geographic method. *Annals of the Association of American Geographers, 27*(3), 125–141.

Lowenthal, D. (1961). Geography, experience and imagination: Towards a geographical epistemology. *Annals of the Association of American Geographers, 51*(3), 241–260.

Lowenthal, D., & Prince, H. C. (1964). The English landscape. *Geographical Review, 54*(3), 309–346.

Lowenthal, D., & Prince, H. C. (1965). English landscape tastes. *Geographical Review, 55*, 186–222.

Mayhew, R. J. (2004). *Landscape, literature and English religious culture, 1660–1800: Samuel Johnson and languages of natural description*. Basingstoke, England: Palgrave Macmillan.

Merriam-Webster. (1996). *Collegiate dictionary*. Springfield, MA: Merriam-Webster.

Olwig, K. R. (1984). *Nature's ideological landscape: A literary and geographic perspective on its development and preservation on Denmark's Jutland heath*. London: George Allen & Unwin.

Olwig, K. R. (1996). Recovering the substantive nature of landscape. *Annals of the Association of American Geographers, 86*(4), 630–653.

Olwig, K. R. (2002). *Landscape, nature and the body politic*. Madison, WI: University of Wisconsin Press.

Olwig, K. R. (2004). This is not a landscape: Circulating reference and land shaping. In H. Palang (Ed.), *European rural landscapes: Persistence and change in a globalising environment* (pp. 41–66). Dordrecht: Kluwer.

Olwig, K. R. (2007). The practice of landscape 'conventions' and the just landscape: The case of the European landscape convention. *Landscape Research, 32*(5), 579–594.

Olwig, K. R. (2008a). Has "geography" always been modern?: *Choros*, (non)representation, performance, and the landscape. *Environment and Planning A, 40*, 1843–1861.

Olwig, K. R. (2008b). The Jutland Cipher—Unlocking the meaning and power of a contested landscape terrain. In M. Jones & K. R. Olwig (Eds.), *Nordic landscapes: Region and belonging on the northern edge of Europe* (pp. 12–49). Minneapolis: University of Minnesota Press.

Olwig, K. R. (2010). The 'British Invasion': The 'new' cultural geography and beyond. *Cultural Geographies, 17*(2), 175–179.

Olwig, K. R. (2011a). All that is landscape is melted into air: The "aerography" of ethereal space. *Environment and Planning D: Society and Space, 29*, 519–532.

Olwig, K. R. (2011b). The earth is not a globe: landscape versus the "globalist" agenda. *Landscape Research, 36*(4), 401–415.

Sauer, C. (1925). The morphology of landscape. *University of California Publications in Geography, 2*(2), 19–53.

Sauer, C. (1969a). The education of a geographer. In J. Leighly (Ed.), *Land and life: A selection from the writings of Carl Ortwin Sauer*. Berkely: University of California Press.

Sauer, C. (1969b [orig. 1956]). The education of a geographer. In J. Leighly (Ed), *Land and life: A selection from the writings of Carl Ortwin Sauer* (pp. 389–404). Berkeley, CA: University of California Press.

Thrift, N. (2007). *Non-representational theory: Space, politics, affect*. London: Routledge.

Tsing, A., Faire, L., Hathaway, M., Inoue, M., Satsuka, S., & Choy, T. (2009). A new form of collaboration in cultural anthropology: Matsutake worlds. *American Ethnologist, 36*(2), 380–403.

Tuan, Y.-F. (1974). *Topophilia: A study of environmental perception, attitudes, and values*. Englewood Cliffs, NJ: Prentice-Hall.

Whatmore, S. (2002). *Hybrid geographies: Natures, cultures, spaces*. London: Sage.

Mapping Power

Edoardo Boria

Abstract In opposition to the classical assumption that would see maps as neutral and objective products, deconstructionist critique has long explained their ideological and instrumental nature. This decisive intellectual reorientation of the conception of the map has clearly shown that maps conceal within a power of persuasion and that they have served discourses of power. But the deepening of the relationship between maps, authority and scholars via this approach has granted prominence to the first two elements, leaving the third in a distinctly subordinate position. Reassessing the figure of scholars and their cartographic practices, this chapter addresses the issue from a still largely unexplored angle, looking into graphical solutions chosen to depict power, in the belief that these solutions have helped to shape the interpretation of the spatiality of power, and influence the very exercising of it. Thus reversing the perspective, and that is investigating the maps of power rather than the power of maps, or the power over maps, it is reconstructed here by means of many empirical examples: those scholars and cartographic genres that have told the history of the spatial representation of power. In this way it also intends to favour an approach to the history of cartography that can organically place it within the more general framework of the history of visual culture and visual arts.

Keywords Mapping · Power · Cartographic practices · Borders · Stein rokkan · Geopolitical maps

1 Authority, Scholars and Maps

A 1671 engraving by Sébastien Le Clerc, portraying Louis XIV as he visits the Académie des Sciences (Fig. 1), is emblematic of the relationship between cartography and power. In a large hall overlooking the Louvre gardens, the King is

E. Boria (✉)
Department of Political Science, Sapienza University of Rome, Rome, Italy
e-mail: edoardo.boria@uniroma1.it

S.D. Brunn and M. Dodge (eds.), *Mapping Across Academia*,
DOI 10.1007/978-94-024-1011-2_12

Fig. 1 Engraving by Sébastien Le Clerc, Louis XIV visites l'Académie des Sciences (King Louis XIV visiting the Académie des Sciences) (*Source* Mémoires pour servir à l'Histoire Naturelle des Animaux 1671, frontispiece)

surrounded by scholars and scientific instruments, including two large maps and a spherical astrolabe. All three elements in the relationship between power and cartography are, thus, present: political authority, scholars and maps. This is the triad around which the present study unfolds.

Based on the premise that the map is an instrument of territorial control, the empiricist approach focused on the first two elements of the triad, i.e., political authority and scholars. The relations between these two agents have therefore been considered essential to every study on the subject (Konvitz 1987; Buisseret 1992; Godlewska and Smith 1994). The advent of radical perspectives in geography in the 1970s and, more so, in the late 1980s, with poststructuralism and deconstructionism, favoured a different approach, one that focused more attention on the third element—the map. This shift in the focus of analysis initially produced studies centred on the political and semiotic value of the map. Such studies aimed to expose the symbolic and rhetorical devices used to place the map at the service of partisan views (Lacoste 1976; Harley 1989; Monmonier 1991; Casti 1998). Attention was later extended to include not only the object but the entire process of its production, moving from the study of maps to that of mapping (Kitchin and Dodge 2007).

Seemingly, then, the evolution of the discipline did not overlook any of the three elements involved in the relationship between cartography and power. Once more, however, Le Clerc's engraving provides us with some food for thought regarding the role of scholars. The latter are portrayed standing near their technical instruments (a telescope, a spherical astrolabe, a burning glass) and the products of their work (some maps, a globe). Le Clerc thus alludes to their competence in the fields of geodesy and topography, as well as to their original task of measuring the Earth's surface and producing cartographic representations thereof. This same perspective characterized a long line of studies on cartography, in which scholars were essentially presented as mere technicians. This simplistic interpretation of the scholar's contribution to the history of cartography is attributable to a long-standing idea—that of progress. Those who subscribed to it considered the accuracy of maps as the yardstick for evaluating the quality of a scholar's contribution, and the evolution of cartography as a process of continuous, linear improvement owing to scientific and technological progress (Edney 1993).

This representation, however, obscures another facet of the scholar's figure, one more intellectual and creative in nature, one that coexists, or rather precedes the figure of the technician-scholar. Indeed, the moment cartographers translate spatial concepts into maps, they inevitably take an active part in the development and dissemination of these concepts. This occurs either directly, when a scholar introduces a new concept (e.g., Mackinder's map of the heartland 1904: 435) or indirectly, by virtue of the dialectical nature of the relationship between maps and reality, a relationship in which the map plays an active role in promoting specific perceptions of physical space and stimulating certain practices on the ground (Wood 1992; Edney 1997). In the context of the spatiality of power, for example, concepts such as linear borders or national identity were developed with the help of images representing, and thus also affecting the understanding of those same concepts.

Underscoring the centrality of the scholar in the development of political spatial concepts, the present study addresses the issue of the relationship between cartography and power, following a pragmatic approach mindful of the author's aims in drawing the map (Besse 2008). Such a perspective does not, however, neglect the map's "internal power," that is, its ability to transmit meaning that goes beyond the author's communicative aims, through universally accepted conventions, procedures and models able to evoke crucial implicit assumptions (Harley 2001: 112). Nor does the perspective adopted forget the pressure exerted on cartographers by political authority. Despite the undeniable importance of these two elements—the map itself and political authority—to the relationship between cartography and power, the figure of the scholar remains nevertheless central to the analysis of the social and historical significance of spatial representations of power. Numerous aspects deserve to be studied: questions like what, how and when. Information on the most frequently represented subjects, for example, will tell us which aspects of spatiality are perceived as most relevant. Knowledge of the ways in which spatial representations for political analysis are realised can tell us which cartographic models turned out to be the most useful and coherent with the theoretical frameworks adopted. Data on the periods and intellectual contexts in which the most appreciable results had been concentrated will inform us as to the scientific paradigms (Kuhn 1962) most favourable to the development of graphical forms for the representation of space.

The following is a systematic review of the major graphic solutions and cartographic genres used for political analysis, through an approach that aims to contextualize the history of cartography in the more general framework of the history of visual culture and visual arts.

We shall start by identifying the symbols used in maps to denote power, and attempting a preliminary classification of these symbols. One possible solution could be typological, consisting of the classification of symbols based on the kinds of elements they denote: symbols referring to political actors (e.g., the states with their power centres, their organisation, etc.); symbols representing what is at stake (e.g., natural resources); others denoting the instruments of political competition (e.g., electoral support); and then those pointing to strategies (e.g., alliances, hegemonies, or more generally, relations between political actors) and so forth, for all political symbols present in the maps.

Such a taxonomy, legitimate and comprehensive as it may be, would still be inadequate if not accompanied by a criterion that would allow an analysis of the long and fascinating evolution of the relationship between mapping and political studies. A relationship that stems from the parallel evolution of political science on the one hand, and of the use of maps in social sciences on the other. Chronology would thus appear to be more suitable as the main criterion, beginning with the observation that, while elements and topics relevant to political power have always been present in maps, their choice, the frequency in which they have been used, and the way in which they have been portrayed have varied over time. The method followed, therefore, consisted of reconstructing the history of the relationship between mapping and political science, discussing its strengths and weaknesses, and finally speculating on possible future scenarios for its development.

When the presence of political symbols in maps is studied with this approach in mind, the most striking phenomenon to emerge is that of progressive diversification of cartographic symbology. At first, the symbols are confined mostly to material elements, and then gradually extend to include more and more abstract information. As we shall see, this progression reflects the evolution of the sensibilities of scholars throughout the twentieth century.

Indeed, the evolution of the presence of political symbols in maps shares a number of characteristics with the history of cartography, in general, in the modern era. Initially, following the Cartesian paradigm, cartography favoured material elements, that is, elements that are visible and physically present on the ground, and are therefore easily transposed on a map that follows the principles of Euclidean geometry. These are maps in which place is essentially considered a portion of the Earth's surface, in the most physical, material sense.

Later on, the scientific revolution of the twentieth century surpassed the rationalist paradigm, shaking blind faith in the Cartesian cartographic model and paving the way for the appearance of abstract elements in maps. And finally, recent trends in the social sciences regarding representation required that spatial aspects of personal and collective perceptions, attitudes, values, identities and topophilia (i.e., sentiments, emotions and ideas concerning place) be taken into account, and—wherever possible—represented graphically. In the field of political science, such tendency translated into greater attention to the subject of political behaviour, and the ways in which power is perceived and reproduced.

The present chapter will, thus, follow the evolution of cartographic symbols referring to the political sphere of society. The first section examines symbols referring to material elements ("the objects of power"), the second deals with symbols referring to abstract elements ("the factors of power"), and the third and last section addresses those referring to elements that affect the way power is perceived, or evoke these perceptions symbolically ("perceptions of power"). Although such symbols maybe found on the same map, this classification is not only convenient for descriptive purposes, but can also help discern the increasing level of both abstraction and sophistication of cartographic display, paralleling a similar phenomenon in political science, with both fields increasingly intent on fully describing the relationship between physical place, and the emotions and values that it evokes.

2 Mapping the Objects of Power

Traditionally, the cartographic symbols most clearly associated with power have been the seats of institutional power. Maps have always been dotted with symbols that conjure up the norms of social tradition and the values of political power. The weight of authority—a central concept in political science—is thus reflected in maps also through an abundance of symbols that reify and qualify it. To quote Louis Marin (2001: 212–13):

> The map *decides* what is representable and what [is] not [to] be represented. There is thus a criterion for the representable [and the] non-representable. In fact, this decision is pragmatic: what is [there,] is what is judged worthy of representation, what is noteworthy. Representation thus comes full circle and rejoins the norm. (italics in original)

This is why the map of a city displays and often names churches and government buildings, but not private houses or dissident hideouts (Harley 1988a, b). Symbols of temporal authority have, thus, always been dominant in maps, reflecting the maps' full compliance with existing social norms, as well as a subordination of cartographers to the interests of their clients. What is more, the graphic representation of physical locations of power, such as government buildings or institutions, often violated basic principles of scale. Castles, for example—clear physical symbols of power—were systematically drawn larger than any nearby village, no matter how large, even where the castle was so small it should never have appeared on the map in the first place (Vayssière 1980: 257). This visual fallacy affected people's perceptions in two important ways: it increased the strength of institutional power in their eyes, and gave rise to spatial hierarchies, whereby a larger castle on the map (real or advertised) became associated with greater power.

Obviously, the cartographic symbols on a map reflect the period in which it was created. In the past, when ecclesiastical authority strongly influenced political and social life, the wealth of conventional symbols referring to religious institutions reflected its weight within society at the time. So, for example, Cassini's eighteenth century topographic map of France, the archetype of Euclidean cartographic representations, features many more religious symbols than what one might find on any modern map.

Admittedly, this was facilitated by the conviction that important things in society had to be imposing. Churches, government buildings, aristocratic mansions have, for centuries, been ostentatiously majestic, competing with one another in physical size and architectural grandeur. This correlation is no longer valid, however. Michel Foucault (1981–82; 2004: 346) was the first to show that, in contemporary society, the exercise of power has lost its visibility and tends to be concealed from those who are subject to it. Elements of reduced dimensions can wield considerable power (e.g., a financial stock exchange or the headquarters of a transnational corporation or even a digital telecommunications switch in an anonymous server building). Conversely, large elements may be nothing but the vestiges of past prominence (e.g., a royal palace or an aristocratic estate).

This rationale, favouring the physical locations of institutional power in mapping, dominated the cartographic scene for centuries, adapting to new circumstances as they emerged. With the rise of nation states, the central power, in order to strengthen its control over its territory, established an all-pervasive structure of local government. From the eighteenth century on, this process found its cartographic expression in the highlighting of these subsidiary institutions on maps. The map, through appropriate symbols, must clearly show their function, their jurisdiction and their level within the administrative hierarchy of the state.

The ultimate cartographic expression of the above phenomenon can be found in diagrams which, as often happens also in political parlance, use the capital to denote the entire state or its government. So, for example, one might say that "Berlin is displeased by the attitudes of London and Paris" to mean that "Germany is displeased by the attitudes of Great Britain and France." The cartographic equivalent would leave out the countries themselves with their messy geographical extents and meandering borders showing only the name and the symbol of the capital city. Some of the first such sketches in the scientific literature are perhaps those published in 1921 by Swedish political scientist Kjellén (1921; Fig. 2). Here, the

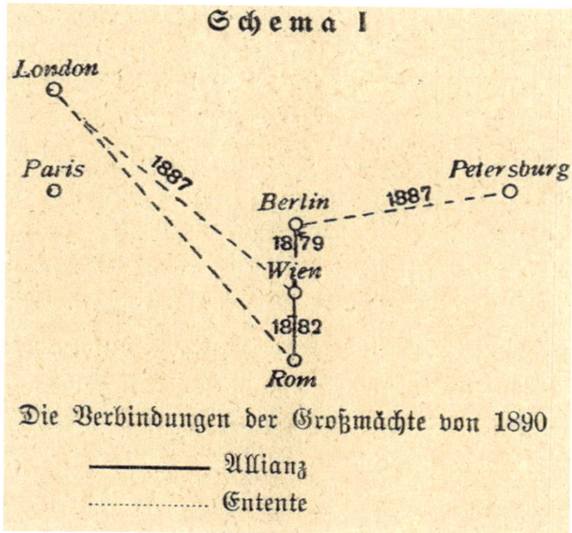

Fig. 2 Die Verbindungen der Grossmächte von 1890 (The Relations between the Great Powers in 1890) (*Source* Kjellén 1921: 51) Favouring the physical location of institutional power in mapping—an extreme example: the capital alone is used to denote the entire state

relations between the main European governments are designated "alliance" (such as between Rome and Vienna and Berlin), "entente" (London-Vienna) or none (all those concerning Paris).

In any event—be they traditional or "specialized," like the diagram by Kjellén (Fig. 2)—political maps commonly give prominence to the state's organisational structure. This is the first violation of a fundamental principle of Euclidean cartography—a principle that renders every place and geographical element equally eligible for representation, determining its dimension on the map through an across-the-board, strict mathematical operation of reduction in scale regardless of its political role.

The concentration on material elements and the centrality of the state made Euclidean cartography a perfect model for two cartographic genres: colonial cartography and military cartography. Both were based on the authority's intent—territorial control—and thus on its interest in the region's physical geography. Even today the tendency remains to identify power as exclusively referring to one specific socio-political organisation—the modern centralized state. Consequently, the political map automatically becomes the map of the state. In the 1970s Jean-Michel Brabant (1977: 14) wrote, "To map power is first to map the power of the State." Today, his words are echoed by Guntram Herb: "The territorial State and its subdivisions still seem to reign supreme" (Herb et al. 2009: 333). Classical cartography often fails to recognize political frameworks that lack a vast and pervasive deployment across the land similar to that of the modern nation state. Supranational organisations like the European Union or Mercosur, for example, are rarely represented in maps. Moreover, cartographic representations do not seem to have been adapted to reflect the increasing complexity of relations between central and local authorities, such as recent decentralizing tendencies that have induced many states (Italy and the UK, for example) to cede power to local authorities.

The importance afforded to the state in political maps is intimately related to modern Western political thought, in which the territorial principle forms the basis of the theoretical framework of the state. Indeed, the modern state itself is based on the following principle: the state holds the monopoly on power within a territory defined by recognized borders. Once the state has been given that monopoly, there remains no room for any rival political authority or institutional power. The territorial basis is, thus, the spatial container of the state, while the principle of sovereignty guarantees it exclusivity over political space in cartographic representation.

Accordingly, the convention governing the cartographic representation of power has been the recognition of a single power for each place. Even today—despite the growing importance of both supranational and intra-national powers rendering the territorial state a clearly inadequate cartographic model—attempts at representing the multiplicity and stratification of powers have been feeble. The dominant model remains that of the classical shaded area map, where the border defines the container within which the state is the only power at work. This approach is clearly state-centred, as indicated by the fact that the importance of mapped borders grew as modern states asserted themselves. In the sixteenth century, mapped borders were still very approximate, and it was only from the eighteenth century on that maps were attached to diplomatic agreements (Black 1997: 15–17). This is also due to the fact that, initially, borders were not a linear element. In sparsely populated areas and difficult terrain, the border was simply 'no man's land' that eluded political power altogether. Only later, once the concept of the fixed linear border had taken root, and each state was able to exercise its sovereignty throughout its territory, all the way up to the point where the sovereignty of the next state began, were borders consistently and carefully mapped and demarcated on the ground. This is the space that political maps have always portrayed.

The importance of the territorial basis of the state, therefore, made the border a central element in maps, and one of the most prominent. An intricate web of borders demarcates the world's territories. Border variations over time have been conveniently represented using isolines (Fig. 3) or the more original method of presenting a sequence of maps (Fig. 4).[1]

The political space defined by the linear border is perceived as homogeneous within each state (in terms of authority, jurisdiction, language, ethnicity). The border, marking the passage from one homogeneous state system to another, different, but equally homogeneous state system, thus plays a fundamental role in the description and very creation of political space. It follows that borders in maps are

---

[1]Normally referred to as the 'multiple static map' in English (Monmonier 1990: 30–45), this method dates back to the beginning of the twentieth century and was probably inspired by motion pictures, an innovation at that time. Numerous important studies in the field of graphic display have devoted attention to this method (Bertin 1967: 354–355; Tufte 1983: 34–36, 170–175) since its full introduction into information science (Neurath 1939).

Fig. 3 The territorial growth of Greece from the time of independence to the present day (*Source* Bowman 1928, 398)

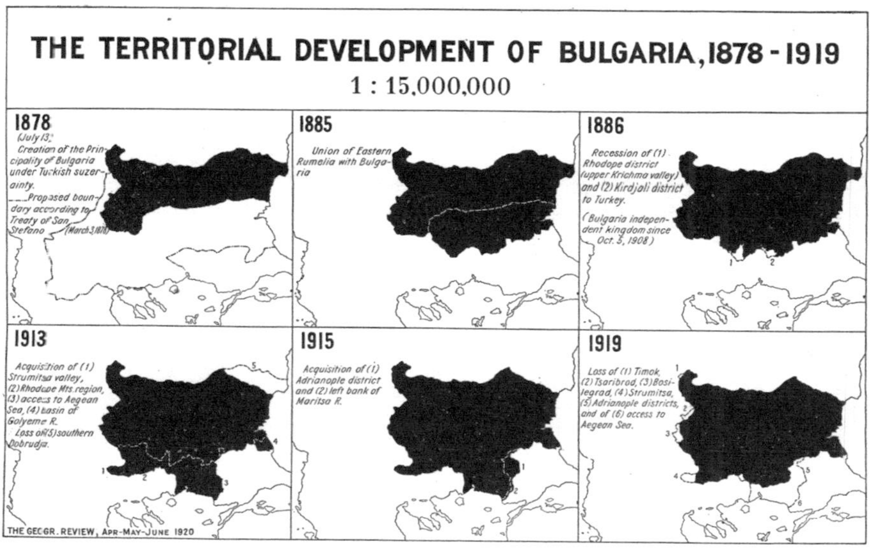

Fig. 4 The territorial development of Bulgaria (*Source* Bowman 1928: 382)

seen as repulsive, rather than attractive. The border is basically an element of defence, often highly militarized and closely guarded; not a place favouring political relations and economic exchanges potentially beneficial to both countries. It marks fracture rather than suture.

The above idea of the border persisted for a long time in political science, and consequently also in maps. Only in the second half of the nineteenth century did alternative views emerge, that gave rise to original cartographic experiments (Boria 2015). Anarchic theories, for instance, with their aversion to borders in the name of universal freedom, tended to prefer avoiding boundary lines altogether in mapping. A case in point is the maps in Élisée Reclus most well-known work, *Nouvelle Géographie Universelle* (1876–1894).

Another alternative interpretation of the border was proposed by Organic State Theory (Ratzel 1903: 3–45, 123–128, 537–620). Here, the border was conceived as a temporary balance point in a power relationship between two states. The resulting experiments attempted to convey the dynamic nature of political reality through maps. As we shall see below, the innovations of geopolitical cartography in the 1920s are an example of this trend. Nevertheless, the rigid perception of the border is still overwhelmingly dominant.

The meaning of borders and the way in which they are perceived have changed considerably over time, but these changes have not been accompanied by an adaptation of maps. The effects of the explosion of cyber traffic, for example, have been staggering. With internet and the new long-distance communication technologies (mobile phones, VOIP, video conferencing, social media, etc.) modern man is cast into a new spatial dimension of topological nature, detached from, and independent of the material reality of the land and the borders of the states (Crampton 2003; Brunn et al. 2004; Zook and Dodge 2009). The debate over the implications of this virtual dimension for our perception of space is still far from concluded, but it has already raised a number of stimulating questions for cartographers and lay mappers. Would it be acceptable, for example, to trace a straight line to link the cities of two parties to an online communication in order to represent their exchange, knowing full well that their communication follows non-linear trajectories even involving satellites in extra-atmospheric space? How should one represent online organisations on a map? And then there is the allure of the social networks, due mainly to their extraordinary potential, theoretically allowing everyone to communicate with everyone on anything. This medium develops an anarchic-like conception of social relations that goes beyond any political boundary.

Furthermore, the traditional portrayal of the border creates the impression that it encloses a compact territorial bloc, evenly controlled throughout by the central authority. This, however, is obviously not the case, especially in the numerous politically unstable countries (e.g., many African countries). Traditional maps, with their definite and static borders, are thus more of an idealization of the exercise of power than a realistic description of it.

Moreover, while in theory the concept of authority may appear obvious, the actual exercise of authority may clearly assume many different forms. For example, Antarctica and the wide expanses of the oceans are shared by all states, but this joint responsibility is rarely, if ever, mapped.

The difficulty that cartography deployed by political scientists encounters in expressing the increasingly complex meanings of borders today and, more generally, the difficulty in describing the political dimension of the world, arise essentially from the fact that the standard political map in current use originates from Euclidean topographic cartography, which—since the sixteenth century—has established itself as the only scientifically legitimate model. This is a model that strongly favours visible, material elements that are permanent and easily surveyed. Clearly, a descriptive method that focuses on tangible elements would tend to overlook the decisive role of abstract elements in political life. So, for example, political disputes do not always revolve around material concerns. When a material element, such as a natural resource, is at stake (e.g., mineral deposits, the control of a river basin), a topographic map is able to show them adequately. Sometimes, however, this is not the case. The control of a region may be important for some symbolic value related to the past, to founding myths of national identity and a sense of belonging to the community (e.g., Kashmir, Kosovo). In such cases, an adequate map should tell the reader about the underlying reasons for the conflict, rather than its physical setting.

Furthermore, this cartographic model imposes not only that represented elements should have a spatial nature, but that the latter correspond to that of the state, identifiable in a continuous territory within clear borders. This condition makes it difficult, for example, to map any form of non-institutional opposing power, be it a criminal organisation with cells scattered around the globe or a group of hackers operating in the virtual realm of cyberspace and whose targets are neither material nor territorial. Such actors do not conform to the territorial principle so dear to political science, and are thus difficult to interpret in our reading of political events using traditional representational tools (Crampton 2003; Lima 2011). Moreover, although traditional warfare is still practised, virtual conflict is likely to become increasingly frequent (so-called 'cyberwars', such as those between Estonia and Russia, between China and… Google, or between the United States and Wikileaks). It is, therefore, not surprising that the representation of wars is less and less frequently entrusted to maps.

Over the years, attempts have been made to adapt the original topographic map with new graphic solutions, so as to allow it to capture not only material but also abstract aspects of political phenomena. These abstract notions, and the new graphic solutions developed to represent them, will be discussed in the following section.

3 Mapping the Factors of Power

As outlined in the discussion above, maps tend to focus primarily on material elements, corresponding to the "restricted" vision adopted by political science, principally dedicated to institutions. Political science evolved, however, over the course of the twentieth century, widening its scope to include the values and ideals guiding political behaviour. The study of surface political structures alone no longer seemed adequate, and the discipline undertook the analysis of deeper processes, their explanation and representation.

The progressive inclusion of abstract elements in maps bears witness to an effort to adapt mapping to meet the emerging needs of the evolving field of political science. The mapping of the factors of power comprised:

1. The instruments of political struggle (e.g., popular support and military force);
2. Strategies and relations between political actors (forms of control, pressure, social conditioning, etc.);
3. Political actors' attributes (e.g., the form of government of states, their role on the international arena);
4. Resources and other elements at stake, not only material (e.g., mineral deposits) but also abstract (e.g., the consolidation of a specific ideological agenda or policy programme);
5. The spatial configuration of political systems (e.g., the unipolar, bipolar or multipolar nature of the international system).

By the end of the nineteenth century, intermediate bodies below the state (parties, unions, activist groups, NGOs, etc.) became central in the political organisation of Western societies. As a result, political science started taking this phenomenon into consideration, increasing the number of studies on the subject. Where necessary, the spatial dimension of these actors was also analysed. The best known example was that of electoral studies, and the long tradition of their cartographic counterparts, electoral maps. Garrigou (1990) documents the existence of such maps as early as 1830. Not surprisingly, these maps, detailing poll results by electoral district, emerged in France, during the same period that saw the advent of 'statistique graphique'. There, in 1826, the economist Charles Dupin devised the choropleth map, in which different areas (usually administrative regions within a state) are distinguished on the basis of a quantitative variable (Friendly and Denis 2001; Palsky 2008: 413–25; Crampton 2004: 41–53).

As noted, until the nineteenth century, only two methods were used to differentiate places following a political criterion: differentiation according to the weight of authority, was done by giving more visibility to the seats of power, and differentiation according to the territorial principle of sovereignty, which distinguished vast areas (states and regions) from one another using the border sign. Electoral maps introduced a new political criterion in mapping, one that expresses the principle of representation based on the choices of the electorate. Not surprisingly, such maps appeared with the establishment of this principle in political practice.

As often happens, a spatial visualization of a phenomenon induces one to reflect upon the characteristics of the phenomenon in question and its links to other phenomena. The first electoral maps, showing the geographical distribution of the vote, soon attracted the attention of researchers and scholars, who undertook scientific studies of electoral behaviour and the production of electoral maps. First and foremost among these scholars was André Siegfried, first president of the Fondation Nationale des Sciences Politiques (that now manages the famous Sciences Po of

Paris). In his *Tableau politique de la France de l'ouest sous la troisième république* (Siegfried 1913) the author employs dozens of maps in which he superimposes and compares electoral results to the distribution of different kinds of variables: geographic (geological distribution, types of terrain), social (patterns of settlement, levels of education) and economic (land tenure, productive structure). The finding that electoral behaviour in given areas remains constant over time prompted Siegfried to explore the reasons for this phenomenon, seeking underlying spatial patterns to voter preferences. These representations reflect, in some sense, the methodological determinism characteristic of that period, and a general simplification of the analysed phenomenon.

Siegfried's ideas exerted a considerable influence, and both the use of spatial patterns and of maps became popular in electoral studies (Key 1949). The "behavioural revolution" that swept political science in the 1960s gave further impetus to the field. A key spatially-related concept proposed during that period to help explain electoral behaviour was that of "neighbourhood effect"—the existence of a widespread practice of imitation in political choices (Cox 1969). This renewed interest in electoral studies benefited electoral cartography as well: electoral atlases and other publications richly illustrated with maps have proliferated since then, and despite recent criticism of the adoption of quantitative methods in the social sciences, electoral maps remain widespread, in scientific as well as lay contexts. Moreover, the advent of GIS and the availability of interactive, searchable maps with real-time updates have rendered electoral mapping all the more attractive.

Less academic, yet certainly pertinent to this topic, are electoral district maps. These maps have been at the centre of attention in cases of gerrymandering, that is, the manipulation of administrative boundaries to favour a particular candidate. Although in this context maps are seemingly passive, in as much as they merely reflect administrative decisions, in fact they become weapons of propaganda by virtue of their intrinsic persuasive power, here veiled by their official and institutional role (Monmonier 2001).

Another traditional object of study in political science is military force, or hard power, as it is called in its subfield international relations. This, most blatant expression of power, also gave rise to a number of areas of study within those disciplines such as war-studies and peace-studies.

Throughout the history of modern cartography, military mapping, an essential element in any war arsenal, has bred copious amounts of cartographic material, mostly the work of cartographic institutes within the armed forces. In this context, maps have been used essentially to visualize the forces involved in the conflict and the role of geographic and physical elements in the military campaigns. The genre benefited from the development of topographic and hydrographic survey methodologies. At the end of the nineteenth century, military cartography introduced a number of significant graphic innovations such as, the combination of the then new method of isochrones with arrows and curves to map troop movements (widely used in war colleges for the didactic analysis of strategy and troop movements); or the application of time-charts to representations of combat. Two original representations by J.F. Horrabin deserve mention here. These maps, entitled

"Time Chart of the Great War" (Wells 1921: 1052–53), essentially consist of a map superimposed on x and y axes, as if on a graph. Showing time on the y-axis and space on the x-axis, the maps describe the course of World War One pointing out salient events and dates. Most importantly, however, Horrabin's maps indicate the movement of the front in time—the advancement and retreat of the two parties to the conflict—along a vertical dotted line at the centre of the map. The geographic dimension is expressed in that the powers at war are positioned according to their geographic collocation.

It is not the intention here, however, to address the subject of cartography as an auxiliary tool for military purposes, or as an instrument of professional, didactic or popular information regarding the development of a conflict. The focus, rather, is on military might as perceived by political science, interested in force not so much as an instrument to be employed during a conflict but as an expression of political power.

Such an approach brings to mind the views of Thomas Hobbes, who saw the political arena as one of eternal conflict. The idea that only force and power are able to deter potential aggressors pervades Western political thought from Thomas Hobbes through Carl Schmitt and Hans Morgenthau to modern day Neoconservatives, all of whom see conflict as a natural and inevitable condition of political reality (Williams 2005; Evrigenis 2008).

These ideas, however, found no direct expression in maps until the end of World War One, a conflict that brought about an important innovation in the field of cartography itself: the methods previously used to present conflict were now applied to the political map, allowing it to reflect the state of permanent tension believed to characterize the system of international relations. In other words, the political strategy of each of the powers in the struggle was described in a cartographical way, using abstract elements such as penetration and expansion routes, strategic axes and force lines (Fig. 5). The methods previously used to present conflict were now applied to the political map, allowing it to reflect the state of permanent tension believed to characterize the system of international relations.

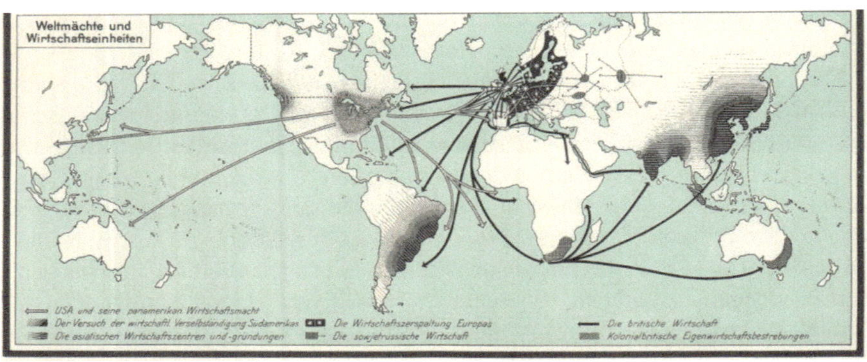

Fig. 5 Weltmächte und Wirtschaftslinien (World Powers and Economic Lines), by A. Hillen Ziegfeld (*Source* Braun and Ziegfeld 1930: 47)

This practice implied strong conceptual criticism of traditional political maps, concentrated on that which is present within the borders of the state (its territory, administrative units, and transport networks centred on the capital, etc.). Traditional maps were being accused of having forgotten that, to the extent that a national interest in foreign policy exists, there also exists a national space that breaks out of the borders of the state, and which is in some way cartographically representable.

Such requirements entailed, almost automatically, two other fundamental cartographic innovations. The first is the possibility of differentiating places and regions based on their geopolitical value. As we have seen, electoral maps were the first to classify geographic areas on the basis of a political criterion other than sovereignty and authority. Further progress was made in the 1920s, when a new political criterion was introduced—a qualitative, rather than a quantitative one, this time. Places and regions were evaluated on the basis of their supposed importance in world politics, and mapped accordingly. In such maps, for example, places such as the Strait of Malacca or the Suez Canal—control over which would confer power—would be highlighted, or even (a more radical choice) be the only places portrayed on an otherwise "silent" map, where the geographic background serves only to help the reader identify the region in question. The political criterion for the selection of objects to be mapped is a function of the basic objective of the map: to highlight the nerve centres of international politics able to contribute to the understanding of strategies and relations between states (Fig. 6). Geopolitical maps aim to highlight the nerve centres of international politics and the dynamics of interstate power relations.

The second fundamental novelty of this new cartography stems from the observation that political reality is dynamic and constantly redesigned on the basis of the moves and counter-moves of the opposing parties. In cartography, this

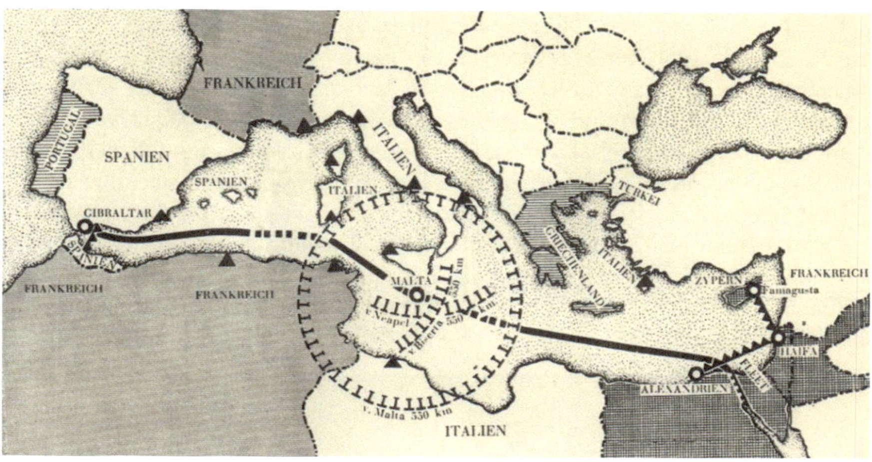

Fig. 6 Mittelpunkt des Mittelmeers (Malta. Heart of the Mediterranean) by Georg Stritt & Co., Malta (*Source* Jantzen, n.d.: 10)

challenge was initially met through the adoption of symbols already in use in historic atlases to express movement (e.g., arrows and curves to indicate the movement of peoples, armies) and, subsequently, by devising a novel symbology for abstract concepts made up of arrows, axes, circles and semicircles and an abundance of shadings and delimitations. Some of these symbols are present in Fig. 6. The effort made by Von Schumacher (1934, 1935) to standardize this syntax bears witness to the will to provide a theoretical basis to this novel cartographic language.

Until then, cartographic symbology had been strictly limited to the representation of fixed elements in the landscape. Now, the political-geographical elements (first and foremost, states) were made to move on the scene and interact. For the first time, the reader could perceive, on a map, their impetus to act, and the intrinsic dynamism of international politics.

This new kind of cartography depicted the political scene in its entirety, rather than single parts thereof. Indeed, while it is possible to read a portion of a topographic map without compromising its message, these maps constitute and must be read as, an organic whole. This is because the object to be mapped is not a geographical area or terrain but its political reality and its incessant dynamism. Space is no longer neutral and uniform, but active and heterogeneous.

The political criterion for the selection of elements to be mapped, and the transformation of the map from a static representation of elements in a given geographical area to a dynamic expression of power relations between political forces represented a net break with the past and paved the way for a new genre: geopolitical cartography or, as it was called in Germany, where it originated in the 1920s, *Geopolitische Kartographie*. Since then, this genre, imitated also in other countries (particularly in Italy), was destined to develop, diversify and enrich the landscape of cartographic communication, appearing in magazines, books and atlases.[2] In English-speaking countries today, the expression "geopolitical map" may sound somewhat odd, but it is commonly used in other countries, notably in France. Science, however, was often very critical of geopolitical cartography, as it did not appreciate many of its aspects: its persuasive vocation, its ideological compromises, its prescriptive tendencies, its methodological inadequacy, but most of all, the politicization of its content and objectives. German geopoliticians, led by Karl Haushofer, created a very dangerous convergence between scientific inquiry and state propaganda that undermined the credibility, and therefore the prospects, of this cartographic genre for future applications. It should be acknowledged, however, that their efforts yielded a very significant result. They were able to demonstrate that cartography could describe international political reality efficaciously and intelligibly, not only for experts but also for lay readers.

[2]For a historical account of the political content and graphic characteristics of geopolitical cartography see Herb (1997) and Boria (2008).

This objective was shared by a completely different cartographic model popularized in the United States—bird's-eye view maps—by Richard Edes Harrison, who jokingly called them "unorthodox maps" (Harrison 1944), and others such as Charles Owen (Schulten 1998; Cosgrove and Della Dora 2005). These are very special top-down perspective views, usually of extensive geographical areas (Fig. 7).

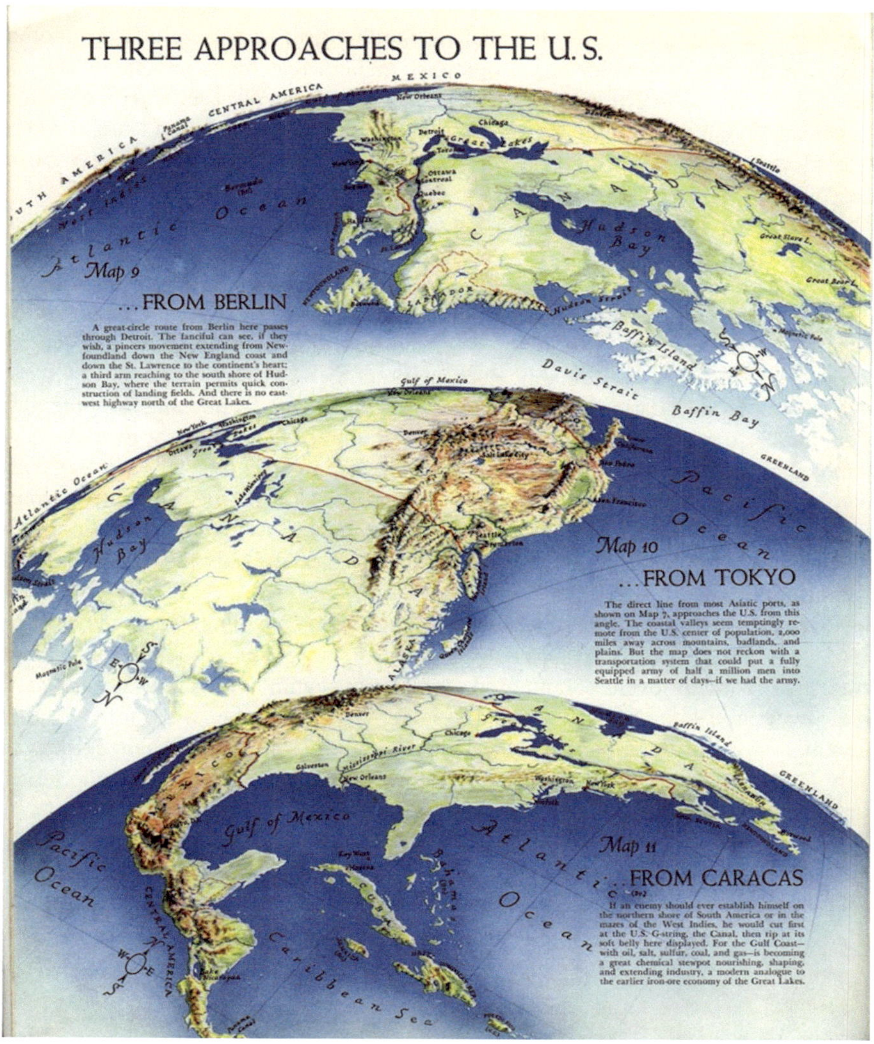

Fig. 7 Map, Three Approaches to the U.S., by R.E. Harrison (*Source* Harrison, Atlas for the U.S. Citizen, September 1940, republished in Harrison 1944: 22) Harrison drew his maps in perspective, as though seen from a single viewpoint. He also proposed novel directional orientations, violating the convention that maps should follow the direction of the Earth's rotation around the sun (the East-West axis)

Harrison's cartography was a full-fledged reflection of a new perception of the Earth's surface, a clear consequence of the conquest of the skies (Kern 2007: 307–313). His maps were more compelling than traditional Euclidean maps, in that they simulated a more natural view than the zenith perspective offered by the Euclidean model. While in Euclidean maps, each point on the Earth's surface is depicted as though seen from a point suspended straight above it, Harrison drew each map in perspective, as though seen from a single viewpoint. He also proposed novel directional orientations (e.g., diagonal or South-North views), violating another basic convention of Euclidean cartography, namely that maps should follow the direction of the Earth's rotation around the sun (the East-West axis). His objective was clear: "We have seen the world *not* globally…What we have not learned so far, is to adopt a constant flexibility in the process of forming a world view" (Harrison and Weigert 1944: 76–77 [italics in the original]).

Harrison's innovations, openly challenging the conventions of geometric cartography (Harrison 1944: 10–12), were picked up at all levels of cartographic production, from big reference atlases to school atlases. The proliferation of top-down perspective views called for a considerable "adaptive mental effort" (Siegfried 1950: Préface) in looking—for the first time in modern history—at the world deprived of its indisputably Eurocentric core.

Visualizing the Earth surface as a closed space and choosing to give less prominence to borders, these maps radically contrast the tendency to view geographical areas as separate and self-contained. Instead, they enhance a sense of their interdependence, echoing the analogous concept developed by liberal international relations theoreticians, whereby state interdependence, rather than division, is a major explanatory factor of international dynamics, and the only instrument capable of limiting conflict (Jackson and Sorensen 2006).

More than a few experts in the field of international politics were inspired by Harrison's maps to imagine alternative views of the globe. Robert Strausz-Hupé and Hans Weigert even co-authored publications with Harrison (Harrison and Strausz-Hupé 1945; Harrison and Weigert 1944). In the work of Nicholas Spykman, a major theoretician in the field at the time, one can recognize traces of both Harrison's bird's-eye-view cartography (Fig. 8) and German geopolitical cartography (Fig. 9).

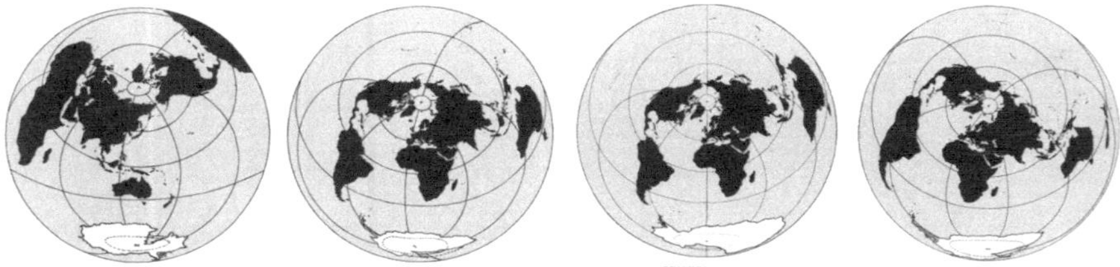

Fig. 8 The World centered on Tokyo, Berlin, London and Moscow, by J. McA. Smiley (*Source* Spykman 1944: 21)

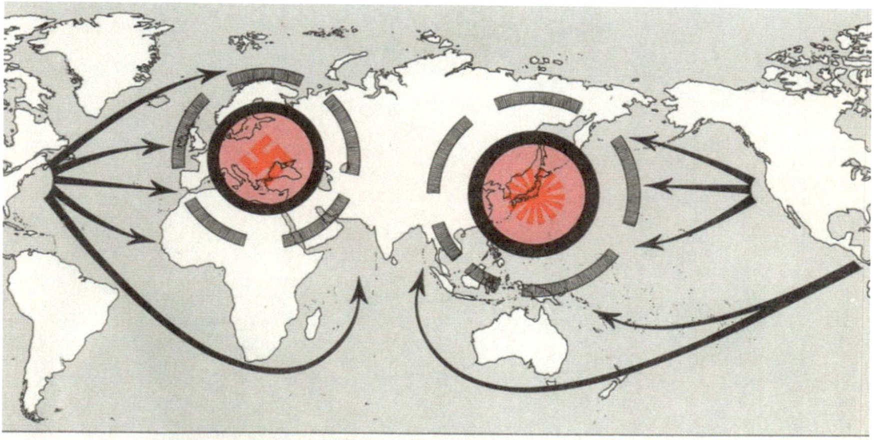

Fig. 9 Unity versus Plurality, by J. McA. Smiley (*Source* Spykman 1944: 47)

The proximity between different regions of the world seen in Harrison's maps can be interpreted in two ways (Schulten 1998: 186–187): on the one hand, the emphasis on the closeness of the other, in these maps, exacerbated the sense of fear from external aggression. On the other hand, the world seen as one and finite conveyed a sense of brotherhood between peoples, underscoring the need for international dialogue. Both these tendencies could be found in post-war political cartography. The first tendency found expression in the proliferation of North Pole maps on the gnomonic projection. These are maps that best display the global scale of the confrontation between the USA and the USSR. The two land masses—America and Eurasia—are distinctly juxtaposed in these maps, almost touching each other, rather than separated by the Atlantic Ocean, as they traditionally appear. The second tendency found expression in the experiments of anti-militaristic, politically committed scholars concerned with the dangers of a nuclear war (e.g., Bunge 1988) who launched a new genre of Atlases, entirely dedicated to the illustration of international political dynamics.

Among the most prolific pioneers in this field were two teams: the Anglophones Kidron and Segal (1981), Kidron and Smith (1983), Kidron and Segal (1987) on the one hand, and the Francophones Chaliand and Rageau (1983, 1984, 1988, 1989, 1991, 1993) on the other (Fig. 10). In later years, such publications evolved graphically and multiplied.

Similarly, in the field of peace studies, a context of increasing progressive idealistic sensitivities drew attention to North-South relations, widening the range of political issues explored, and producing a considerable amount of cartographic material covering poverty, underdevelopment, population growth, environment and resources. The desire to redefine North-South relations led a number of international organisations, such as UNESCO and UNICEF, to publish maps based on the

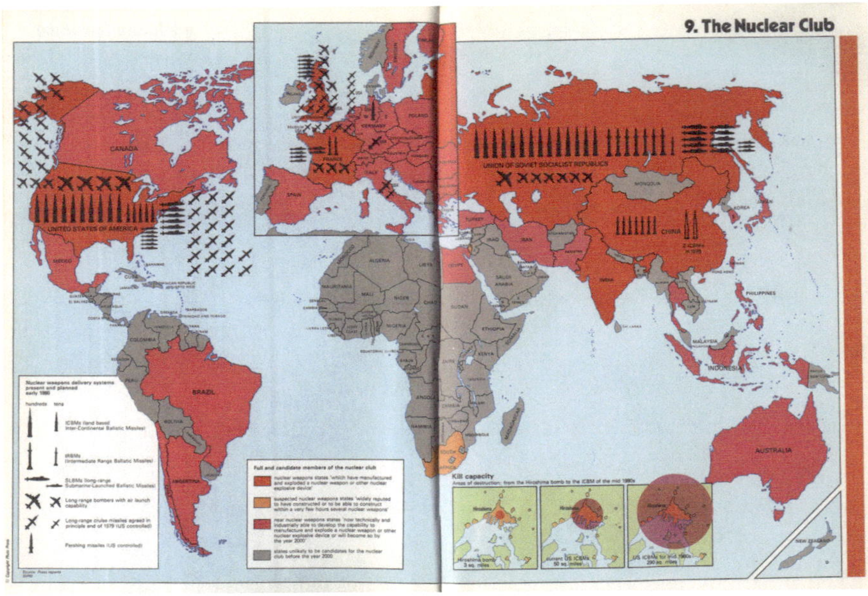

Fig. 10 The nuclear club (*Source* Kidron and Segal 1981: 9)

Peters Projection, which portrays each geographical area in proportion to real size, and is therefore considered more respectful of countries in the Southern Hemisphere (Loxton 1985; Vujakovic 1987, 1989; Crampton 1994; Monmonier 1995: 9–44).

The ideological tendencies of the time, which advocated acknowledging the value of non-Western cultures were, in some sense, inspired by an earlier school of thought in international relations—idealism, historically opposed to that of political realism (which focuses on factors of force). The idealist school of thought was ultimately based on the ethical and moral principles of Hugo Grotius and Immanuel Kant (1795), later found also in the work of eminent scholars such as Émile Durkheim. This approach asserts that international conflict can be dealt with by reproducing, on a supranational level, a situation analogous to that which exists within nations, namely, a state of general reconciliation where violence is banned and is an exception to the rule. This requires the enhancement of supranational bodies capable of limiting the authority and freedom of action of states—in turn called upon to submit to a recognized arbiter to settle disputes. For the collective will and international solidarity must prevail over national authority and egoism, and multilateralism over bilateralism.

This idealistic interpretation of international relations, which failed to find application before Wilson's Fourteen Points and the establishment of the League of Nations (1919), has less of a need to map, for idealism is, by definition, universalistic and thus not constrained by geography. The relevance of territoriality in political life is thus drastically reduced, unless one postulates the surpassing of the nation-state in favour of other political actors. This is, indeed, the aspiration expressed in the maps of thinkers broadly associated with this school of thought such as Giuseppe Mazzini,[3] Count Richard Coudenhove-Kalergi[4] and Leopold Kohr.[5] At any rate, this tradition has left us little cartographic material produced in scientific contexts.

The opposing conceptions held by these two schools of thought—political realism and international idealism—of the relationship between political actors brings us to the issue of the general geographical context in which political affairs take place. What is the spatial structure of the international political arena? In what way has the geographical context affected processes of political development in various countries?

One scholar who showed considerable sensitivity to geographical considerations in his studies on political systems was Stein Rokkan, who also served as President of the International Political Science Association (1970–1973). Rokkan sought to identify links between political behaviour and its spatial distribution. He was interested in the processes of political development, specifically state-formation and nation-building, and drew upon the four phases of the general theory on European political modernization elaborated by Almond and Coleman (1960). His work is a comparative diachronic study of the interdependencies and contiguity between the many regional political systems of Western Europe (Flora 1999).

Rokkan's studies, thus, naturally involved an innovative effort in the realm of graphic representation. His model is effectively illustrated with diagrams such as the conceptual map seen in Fig. 11. The map is so diagrammatic as to become almost tabular in form, yet it respects the relative positions of the elements, in latitude (North-South axis) as well as in longitude (East-West axis).

As Rokkan (1980) himself notes, the reduction of complex models involving numerous variables to simple diagrams risks appearing excessively abstract and mechanical. Such diagrams, however, are powerful descriptive tools, particularly well-suited to the comparative approach and long-term perspective of his analysis. They render the objective of the model instantly clear—to delineate parallel trends in the development of political systems in European countries by geographic region and pre-existing circumstances.

The map in tabular form, a graphic solution introduced by Rokkan into the political sciences, has been used in many fields of knowledge to represent the

---

[3]Inspired by the 'threshold principle'—the need for a sufficiently large territory postulated by the liberal interpretation of the concept of nation—Giuseppe Mazzini's 1857 map of Europe replaced nation states with a dozen large states or Multinational Federations, plus Italy (Hobsbawm 1990: 30–31).

[4]The map opening his most famous book, *Pan-Europa* (1923), laying the foundations of his movement for the unification of Europe, shows a united continent, from which only Great Britain is excluded.

[5]In *The Breakdown of Nations* (1957) by Leopold Kohr, a pacifist anarchist, a map is included in the appendix, where Europe is divided into many equally small states. This, he believed, was the only solution able to guarantee a more stable and reconciled world.

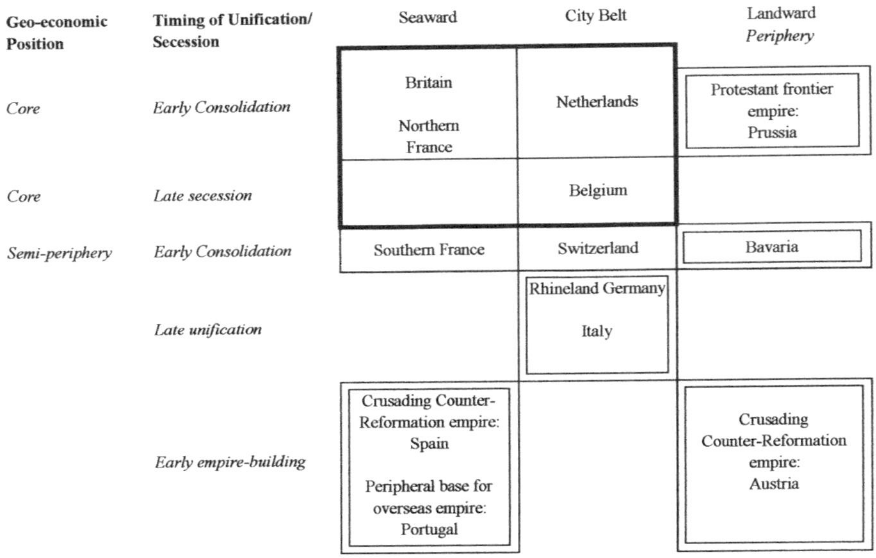

Fig. 11 Rokkan's Geo-economics, geopolitics, and territorial consolidation: a recasting of the 'conceptual map' for Western Europe (*Source* Flora 1999: 241) The map is so diagrammatic as to become almost tabular in form, yet it respects the relative positions of the elements, in latitude (North-South axis) as well as in longitude (East-West axis)

spatiality of phenomena. This method is particularly appropriate to show both the spatial and the temporal dimensions simultaneously (Boria 2012). In these cases, one axis of the Cartesian plane represents the geographical distribution of the phenomenon in question, while the other corresponds to its chronological evolution. Earlier archetypes have been applied to the fields of communication studies (Marey 1885: 20), religion (Lane-Poole 1894: XVII and XVIII), the history of civilizations (Wells 1921: 116, 142 and 1122–1126) and economics (Neurath 1939: 74 and 80).

Clearly, Rokkan goes beyond the mere forms of political organisation of the states, information that may appear in maps mainly for comparative purposes. Examples of such information may be: the form of government (representative democracy, presidential democracy, totalitarian regime, etc.); the ideological tendency of the politico-social system of each country, often presented in maps during the period of the Cold War to indicate the two rival blocs—the Communist and Capitalist binary, or the Iron Curtain dividing Europe; or characteristics, attributed to specific geographic areas based on subjective evaluations, such as political behaviour inclinations. One of the first examples of the latter is a map entitled "Tempérament politique de la région vendéenne" (Siegfried 1913: 35). Here, the cantons of the mapped region are marked on the basis of their voting history for the various political parties ("right," "left," "bonapartists" and "undecided").

Other examples of cartographic representations of political characteristics are those referring to the value and role of a country on the international political scene. Today, maps presenting qualitative evaluations of the political role of certain regions of the world can be easily found in the literature published on international relations. Mention can be made of two that are significant as sources for a historical reconstruction of political cartography. The first shows a number of European states, considered by the author to be "buffer states," interposed between two great powers (Van Valkenburg 1949: 101). The second is in fact a series of four maps (Cole 1959: 226–27) which expresses, by means of circles of varying dimensions, the relative value of world powers, obtained through a quantitative evaluation of their factors of force.

When it comes to the characteristics of political systems, the discipline of international relations is interested primarily in the international system as a whole, rather than in each single component of the system (a country, a region). They ask themselves, for example: does the system have a unipolar, bipolar or multipolar structure? What spatial configuration would the world assume—that is, into how many competing, internally homogeneous regions would it be useful to divide the world, for the sake of political analysis?

Interesting spatial representations have been proposed on these subjects, both regarding models that the system may theoretically follow, and with respect to its empirical reality in particular historical moments. In the diagram shown in Fig. 12, for example, Geoffrey Parker (1985: 176) tried to summarize possible spatial conceptualizations of the international political system proposed by different scholars from the binary model, representing a confrontation between two superpowers, to the idealist model, proposing a harmonious world order.

Among these models we also find the zonal model, in which a band in the Northern Hemisphere identifies the location of the hegemonic powers. The spatialization of this model derives from the concept of pan-region, a term in vogue in geopolitical studies, used to describe large, essentially economically autarchic and, ideally, culturally homogeneous blocs, politically unified under the leadership of a single power. This concept has been attributed to German geopoliticians between the two world wars (O'Loughlin and van der Wusten 1990), and was probably inspired by a map published in the English-language German propaganda weekly newsletter Facts in Review (10 April 1941). Reference to the concept of pan-region in German publications of that period, however, are never direct (Haushofer 1931, 1934: 234–241). At any rate, according to the interpretation largely accepted in geopolitics, pan-regions organise global political space into discrete geographical entities and have the potential of minimizing global tensions. Such positions link the concept of pan-region to the theory of balance of power, currently at the centre of Kenneth Waltz's neo-realist approach (1979), and indeed, the balance of power rests on the conviction that the stability of the international system and the prevention of wars can be achieved only through an economic and military balance between the powers. In this sense, the concept of pan-region can be seen as a particular instance of balance of power theory—particular inasmuch as it posits two further geographical conditions: the blocs should have a more-or-less equivalent

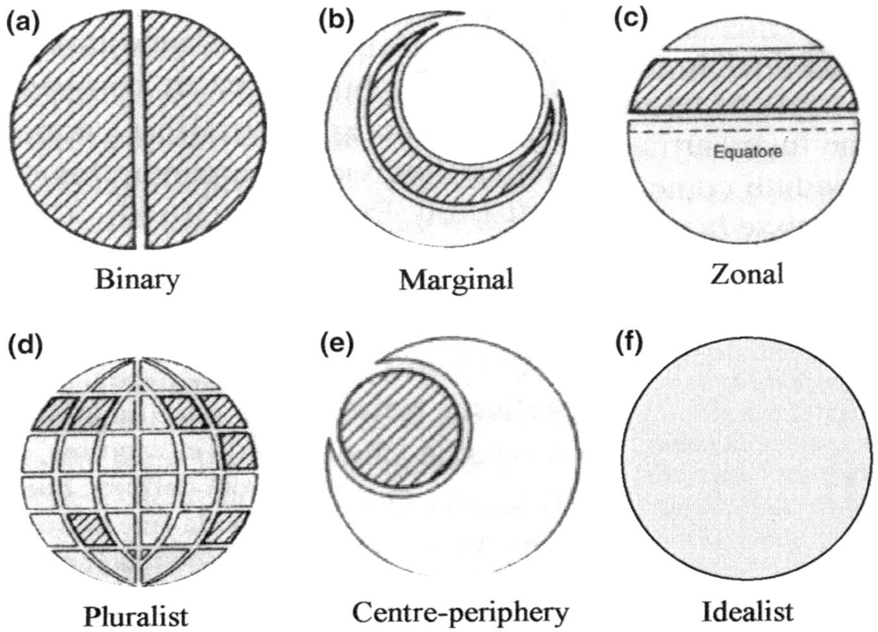

Fig. 12 Schools of Geopolitical thought (*Source* Parker 1985: 176) An example of a spatial conceptualizations of the international political system

territory in terms of surface area, and cover the entire globe. No area should be left without political value.

As for the actual structure that the system of international relations assumed at specific points in history, numerous interpretations have been proposed with respective cartographic appendices. Some, like Saul Bernard Cohen, have taken up and developed the concept of pan-region. Cohen conceptualized and proposed a cartographic representation of the planet divided into homogeneous geostrategic regions (Cohen 1964: 63), but the most famous theories in this respect were based on the premise that there is a region that is strategic, crucial to continental and global balance. The control of this region, therefore, becomes the element at stake in the conflict. According to one classical geopolitical theoretician, John Halford Mackinder, the world's strategic region was a vast Asian area he called 'Heartland', the Pivot Area (1904: 435, fig. 5). According to Nicholas Spykman it was 'Rimland', the coastal areas of the Asian continent that mattered most (1944: 52, map 46).

Clearly, then, from a theoretical point of view, both Mackinder and Spykman share the basic conviction that there is a region which is considered strategic, the control of which confers a considerable advantage in political competition. Yet, the decades separating their work brought about significant innovations in the cartographic representation of global geopolitics. Despite the extraordinarily innovative concept it attempted to convey, Mackinder's map was not particularly original or self-explanatory (Boria 2008: 304). Spykman's maps, on the other hand, many of which were drawn by Harrison, were decidedly more efficacious and appropriate. Ultimately, while recognizing Mackinder's considerable merit for having been the first to theorize the unity of global political space, it could be said that he may not have had much cartographic imagination. Rather than using an ordinary projection, which keeps the continents apart and the heartland in a marginal position, the idea of a compact and unified world could have been conveyed more clearly and compellingly on a polar projection, as done repeatedly by Spykman (1944: 16, 20, 56).

Following Mackinder and Spykman, many geostrategic models have been proposed in the literature, such as Zbigniew Brzezinski's (1997) *Arc of Crises* and Samuel Huntington's (1996) *World of Civilizations*. All of these theorists envisioned and mapped a division of the world into inexorably opposing blocs.

Despite the innovations introduced in the representation of international politics, however, many political scientists are still in the habit of using traditional looking maps, even in cases of material bound to have wide circulation and guaranteed success. For example, the cartography in Huntington's (1996) renowned *The Clash of Civilizations and the Remaking of World Order* is very basic in design terms, and suffers from an extremely rigid spatial perception of the world: a single characteristic corresponds to each area or place on the planet. We know full well, however, that political and social phenomena are rarely, if ever, so monolithic. What, then, is the point of rigidly attributing a religion or a culture to a country? An indication of the intensity of the phenomenon in each place would have been more correct. There seems to be a problem with the quality of cartographic representation.

Ultimately, regardless of one's opinion of the theory, Huntington's interpretation of political tensions in the world was innovative and caused quite a stir in the world of political science (among his most influential critics were Chomsky 2004; Kepel 2004; Said 2004). It was, however, accompanied by maps that were dishearteningly traditional.

The same can be said of the work of other prophets of the new world order (e.g., Thomas Barnett's "Non-Integrating Gap," 2004), so ambitiously original in their theories and conformist in their maps, revealing neither the pursuit of a "grammar of symbols," nor of an innovative graphic language. This was probably due to commonly held and deeply rooted mental models, which will be discussed in the following section.

4 Perceiving Power Through Maps

It is widely recognized that the relationship between power and space is governed by symbolic as well as material and discursive elements. In the *Oxford Handbook of Political Science* Göran Therborn (2009: 499) states: "Place is still crucial today to power and politics, in some respect arguably even more than before. Above all, it matters in three kinds of politics: democratic, military (war), and symbolic." The importance of place in the symbolic dimension of politics, is thus seen as being on a par with the other two, which, as we have seen, gave rise to two of the most important branches of cartographic practice in political science—electoral cartography and military mapping.

The symbolic dimension of politics refers to the vast array of symbols that evoke power and constitute a constant reminder of its presence. A topic that offers a good deal of food for thought in this context is the individual sense of belonging to a community, and the resulting strong bonds that the individual develops to their community and their native land. This is a subject of considerable interest in political science, for it is directly linked to two fundamental aspects of political modernity namely, the legitimation of state power from below, and its exclusive sovereignty over a territory, As Jean Gottmann (1952) and Benedict Anderson (1991) point out, such bonds are legitimized and perpetuated through symbols, icons and representations.

Maps are among those representations that bolster the sense of national identity. Unlike maps that portray a place as a physical, material object, these maps seek to express the association between the place and the emotions and values it evokes. It is not the intention here to indulge in a discussion of the genre of maps used to sustain nationalist, irredentist or secessionist political movements. Such maps, often deliberately manipulated, inevitably end up acquiring the characteristics of political propaganda, and have more to do with political rhetoric and the exploitation of cartography than with science and are thus outside the scope of the present study. Still, in the past few years, maps have assumed prominence within political studies, as a means to express political identity. The map as an expression of national sentiment is therefore a topic that merits some elaboration.

Due to the close links that exist between the land and the sense of national belonging, the map of the territory in which the members of a nation reside (and that radical nationalistic rhetoric considers assigned to the nation by natural right) plays a role in reinforcing identity, both collective and individual, acting on popular consciousness on the one hand, and that of individual citizens on the other. Similarly, maps that give prominence to places of considerable symbolic significance for the spirit of the nation contribute to the forging of a national memory.

A cartographic image that has aroused particular interest is that of the geographic profile of the state (Paasi 1986; Anderson 1991). It may be counted among the symbols that allow the national community to identify easily with its territory, and facilitate its instinctive acquisition in the collective consciousness, on a par with flags, crests, and landscape icons. The profile of the state may also become an expression of the spirit of the national community, a veritable logo of the nation, especially if characterized by an easily recognizable shape, such as the Italian "boot," the French "hexagon" or the Texas panhandle (Francaviglia 1995).

The logo of national territorial unity, frequently present in our daily lives, from weather maps on the nightly news to commercial advertisement, symbolizes and promotes cohesion. It is a typical case of visual allegory in which an abstract concept, that of nationhood, is represented by way of a visual symbol, the map, which, in turn, in semiotic terms, signifies the territory of the state. By presenting the territory of the state as a compact and discrete body, separated from the rest of the world by clearly-demarcated borders, the map is perfectly suited to the task of enhancing the sense of national unity and establishing its inviolability. In addition, it renders the very concept of nation natural, intrinsic to the state's territorial reality, to be taken for granted, despite the fact that the nation is, obviously, a pure social construct.

The same effect, that of rendering a political viewpoint natural, as it were, is achieved with the classical political world map present in every classroom: the world divided into numerous states, each separated from the others by clear borders and assigned its own, distinctive colour. Such maps present a very specific view of the international political situation, that of an orderly world, regular and uniform, where the political actors are states, and official borders are of great importance. An image of the world emerges, that is normalized, regulated, disciplined; an ideal world. Looking further back in time, this vision had been instrumental in the conception of the European powers' expansionist plans (O'Tuathail 1996).

The above considerations bring us to the dual nature of the relationship between the map and collective perceptions: the map is an expression of collective perceptions on the one hand, and an instrument shaping those same perceptions on the other. In this context, the political sciences have adopted the idea that representations contribute to preconceptions, and affect mindsets and mental habits which, in turn, influence people's choices. Post-positivist theories have contested rational choice theory, according to which, when provided with objective information, political actors will make choices that maximize benefits. Those who challenge the theory argue that an analysis of political actors' perceptions and preconceptions can provide useful clues to help explain the behaviours in question. For example, in a conflict between two countries, or even during international negotiations, the perceived force of the adversary strongly conditions the behaviour of both parties.

The map too, has been counted among the factors able to affect perceptions, highlighting its discursive and performative nature. A circular mechanism links spatial representations to behaviour: the first tend to affect spatial practices that, in turn, result in new cartographic representations. The application of this phenomenon to political studies regards the possible effect of mental maps on foreign policy decisions made by the leaders of a country. It all stems from Walter Lippman's (1965: 31) valid premise: "The world we have to deal with politically is out of reach, out of sight. It has to be explored, reported, and imagined… Gradually [one] makes for himself a trustworthy picture inside his head of the world beyond his reach." In the field of international relations, it follows, as pointed out by Nicholas Spykman (cited in Polelle 1999: 124) that: "Every Foreign Office, whatever may be the atlas it uses, operates mentally with a different map of the world." Whence the importance of investigating the questions,

> …are these maps accurate or distorted? Are they comprehensive and complete, or piecemeal and sketchy? Do they reveal emerging patterns, or do they merely show overt configurations? Do they indicate routes along which progress can be made, or do they point toward insurmountable obstacles and lead into dead ends? (Henrikson 1980: 496)

Studies on collective identities explore aspects of power that go beyond the classic research interest in the politico-institutional system—aspects that border on ethical and social norms, and that have become increasingly relevant in recent years (e.g., civil rights, social justice, sexual orientation, gender-based discrimination). Here, even basic social structures, such as the ethnic group and the family, become centres of power. Ultimately, an exhaustive representation of power would have to consider public as well as private space (Foucault 2004).

Despite the significant inroads these post-structuralist and post-modernist trends have made in the social sciences, cartography has continued to deal with politics primarily in the restricted sense of the term, favouring public institutions, and above all, the state. As we have seen, maps have primarily described the state's organisational structures, the instruments of political struggle and its strategies. Facing a rhizomatic conception of power, however, maps are in crisis, for this intellectual approach demands that power and politics be present in all manifestations of social life, in all forms of social interaction, on all scales from the local to the global. In short, maps are in crisis having to face the need to examine the informal dimension of politics.

To date, attempts to portray the informal sphere of politics cartographically are still in their embryo stages (Bhagat and Mogel 2008). Hopefully, these will develop further in the future; otherwise, the map risks becoming increasingly marginal as an instrument in political science publications.

5 An Agenda for the Future

The most important change that has occurred in recent years as far as political space is concerned, and which maps must take into account, is that it has become increasingly reticular in nature. Thus, authority, power and dominion are spread out over the territory, and the centres from which they emanate seem strongly interconnected. Power is no longer localized, but rather stems from such relationships. In modern political structures, power depends more on networks than on territory and is thus difficult to represent cartographically, since, as we have seen, political cartography is still strongly attached to the practice of depicting political entities, following the model of the modern territorial state. To rectify this situation, cartography would have to—in the words of Manuel Castells (1996)—stop looking at the "space of places" and concentrate on the "space of flows."

What would an effort to map the reticular nature of power involve? Many things. The realisation that political elites conceive of power increasingly in terms of the control of flows rather than that of territory; the realisation that today, more than ever before, political competition is often motivated by the desire to control intangible resources (computer and telecommunication networks, exchanges of knowledge and ability, etc.); the identification of the organisational models of such networks; the functional identification of network nodes and the assessment of connectivity levels; the awareness of the existence of reticular organisations of power (e.g., criminal and terrorist networks).

An approach that takes into account the reticular nature of political space has the potential to improve our understanding of power and political conflicts, as it allows one to grasp more effectively what is at stake and what the balance of power is in a conflict. For example, how can one analyse the ongoing Palestinian-Israeli conflict, a conflict which has already been subject to contradictory cartographic practices (Wood 2010: 231–255) without considering the tight Israeli control over both material (air, sea and land transportation) and intangible (electricity, telephone, internet,) networks in Palestinian space? Figure 12.13 shows how what was once an "island"—the West Bank—has really become a highly fragmented "archipelago" (white) due to Israeli control of the region's arteries and strategic nodes (grey).

It should be noted that the map appeared in a non-academic publication, as are the maps published in the French monthly Le Monde Diplomatique. Such maps circulate among specialists as well as the general public. They are forms of popular geopolitics (O'Tuathail 1999) that are introducing ground-breaking innovations into political cartography, on a par with scholars deeply committed to the development of novel cartographic practices, such as Jacques Levy (2008) and Daniel Dorling (e.g., Dorling et al. 2008). There are promising signs in the methodological arena as well, including the objections raised in the framework of activist cartography and resistance mapping (Crampton 2009; Cobarrubias and Pickles 2009; Wood 2010: 111–119).

These examples, as well as the long historical process described in this work, bear witness to the fact that over the years cartography sought to break free of its initial tendency to portray the most visible and institutional aspects of power. It did so with partial success, but that is not enough. The globalized world presents phenomena that are spatially very complex in terms of patterns, relations and structure. For the purpose of adapting maps to better represent the reticular nature of political space, one could draw on the extraordinary recent development of network science and information visualization (Lima 2011), and seek inspiration in other disciplines that have already begun to address this subject. For example, in archaeology, the idea that "ancient states are better understood through network models rather than bounded-territory models" (Smith 2005: 832), and that "new maps based on networks" (845) are required, is gaining ground. In-depth considerations regarding the need to increase and enhance the use of maps in research, adapting them to new scientific trends, can also be found among cultural anthropologists (Aldenderfer 1996) and historians (Baker 2003). Obviously, the mapping of networks has also been addressed in the field of urban studies (e.g., the work of the Globalization and World Cities Research Network), as well as in diffusion studies, from Torsten Hägerstrand (1967) on.

But perhaps more than attempting new ways to represent political space, it would be important to complete the recent epistemological turn that revolutionized the way in which we understand maps (Dodge et al. 2009; Pickles 2004). This is essential in order to overcome the scars of diffidence that deconstruction left on the map.

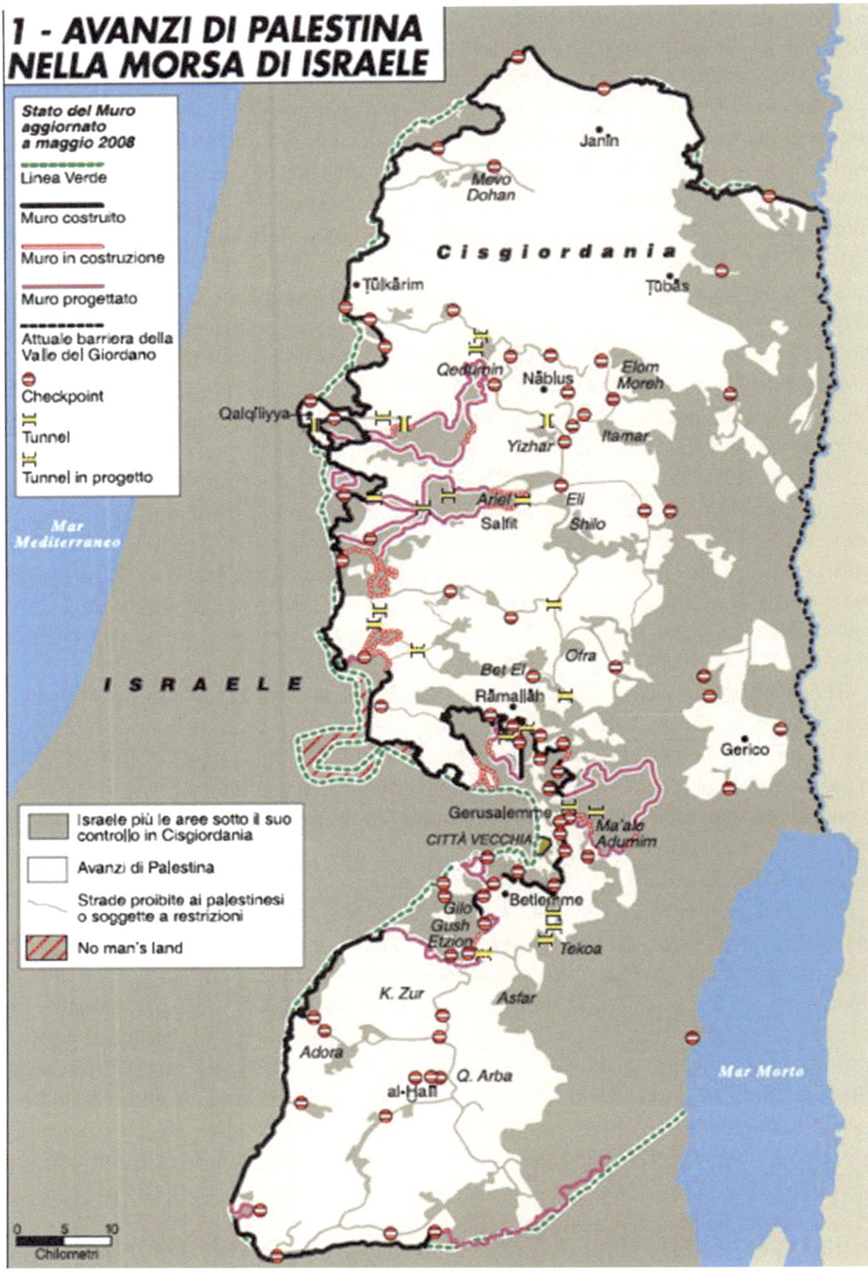

Fig. 13 Map by L. Canali, Avanzi di Palestina nella morsa di Israele The tight Israeli control over networks in Palestinian space (Scraps of Palestine in the Grip of Israel. *Source* Limes 2009)

Indeed, the impression remains that the reluctance of political scholars to use maps may be caused not only by the intrinsic limitations of traditional cartography, but also, and perhaps mostly, by the presence of critical deconstructionist voices (Herb et al. 2009: 334). Such voices are currently very strong in the social sciences and do not limit their disapproval to conventional mapping, but attack all forms of representation. They decry the way in which the map has long been presented—that is, as a neutral, objective tool—a mirror of reality, as it were: "The post-structuralist approach takes issue with the very concept of representation and the modern structure of knowledge of which it is an expression, emphasizing the fact that the 'issue of representation', raises not only epistemological but also moral issues" (Minca and Białasiewicz 2004: 56). This is the primary obstacle to be overcome by the geographical map today.

References

Aldenderfer, M. (1996). Introduction. In M. Aldenderfer & H. D. G. Maschner (Eds.), *Anthropology, space, geographic information systems* (pp. 3–18). New York: Oxford University Press.

Almond, G. A., & Coleman, J. S. (1960). *The politics of developing areas*. Princeton, NJ: Princeton University Press.

Anderson, B. (1991). *Imagined communities: Reflections on the origin and spread of nationalism*. London: Verso.

Baker, A. R. H. (2003). *Geography and history. Bridging the divide*. Cambridge: Cambridge University Press.

Barnett, T. P. M. (2004). *The Pentagon's new map. War and peace in the twenty-first century*. New York: Putnam.

Bertin, J. (1967). Semiologie graphique. Paris—Le Haye: Mouton Gauthier Villars.

Besse, J. M. (2008). Cartographie et pensée visuelle. Réflexions sur la schématisation graphique. In I. Laboulais (Ed.), *Les usages des cartes* (pp. 19–32). Strasbourg: Presses Universitaires de Strasbourg.

Bhagat, L., & Mogel, L. (Eds.). (2008). *An atlas of radical cartography*. Los Angeles: Journal of Aesthetics and Protest Press.

Black, J. (1997). *Maps and history. Constructing images of the past*. New Haven & London: Yale University Press.

Boria, E. (2008). Geopolitical maps: A sketch history of a neglected trend in cartography. *Geopolitics, 13*(2), 278–308.

Boria, E. (2012). Cartografare il tempo. In E. Casti & J. Lévy (Eds.), *Le sfide cartografiche* (pp. 95–115). Il Lavoro Editoriale: Ancona. Mapping time, In *EspaceTemps.net* http://www.espacestemps.net/articles/mapping-time/ Original version (2010) Jan, 2012.

Boria, E. (2015). Representing the politics of borders: Unorthodox maps in reclus, mackinder and others. *Geopolitics*, *20*(1), 142–170.

Bowman, I. (1928). *The new world. Problems in political geography* (4th ed.). NY and Chicago: World Book Company, Yonkers on Hudson.

Brabant, J. M. (1977). Des résponses aux questions de Michel Foucault. *Hèrodote* 6: 12–14. (English translation: "Response: Jean-Michel Brabant.") In J. W. Crampton, & S. Elden (Eds.), *2007. Space, knowledge and power. Foucault and geography* (pp. 25–28). Aldershot: Ashgate.

Braun, F., & Hillen Ziegfeld, A. (1930). *Geopolitischer Geschichtsatlas*. Dresden: Ehlermann.

Brunn, S. D., Cutter, S. L., & Harrington, J. W. (Eds.). (2004). *Geography and technology*. Dordrecth: Kluwer.
Brzezinski, Z. (1997). *The grand chessboard: American primacy and its geostrategic imperatives*. New York: Basic Books.
Buisseret, D. (Ed.). (1992). *Monarchs, ministers and maps. The emergence of cartography as a tool of government in early modern Europe*. Chicago: The University of Chicago Press.
Bunge, W. (1988). *The nuclear war atlas*. New York: Blackwell.
Castells, M. (1996). *The information age: Economy, society and culture. The rise of the network society* (Vol. I). Oxford: Blackwell.
Casti, E. (1998). L'ordine del mondo e la sua rappresentazione. Semiosi cartografica e autoreferenza. Milano: Unicopli. (English translation: Reality as representation: The semiotics of cartography and the generation of meaning. Bergamo: Bergamo University Press).
Chaliand, G., & Rageau, J. P. (1983). *Atlas stratégique*. Paris: Fayard.
Chaliand, G., & Rageau, J. P. (1984). *Atlas de la decouverte du monde*. Paris: Fayard.
Chaliand, G., & Rageau, J. P. (1988). *Atlas politique du XXe siècle*. Paris: Le Seuil.
Chaliand, G., & Rageau, J. P. (1989). *Atlas des Européens*. Paris: Fayard.
Chaliand, G., & Rageau, J. P. (1991). *Atlas des diasporas*. Paris: Odile Jacob.
Chaliand, G., & Rageau, J. P. (1993). *Atlas des empires*. Paris: Payot.
Chomsky, N. (2004). *Hegemony or survival. America's quest for global dominance*. London: Penguin Books.
Cobarrubias, S., & Pickles, J. (2009). Spacing movements. The turn to cartographies and mapping practices in contemporary social movements. In B. Warf, & S. Arias (Eds.), *The spatial turn. Interdisciplinary perspectives* (pp. 36–58). Abingdon: Routledge.
Cohen, S. B. (1964). *Geography and politics in a divided world*. London: Methuen.
Cole, J. P. (1959). *Geography of world affairs*. Harmondsworth: Penguin.
Cosgrove, D., & Della Dora, V. (2005). Mapping global war: Los Angeles, the Pacific, and Charles Owens's pictorial cartography. *Annals of the Association of American Geographers, 95*(2), 373–390.
Coudenhove-Kalergi, R. (1923). *Pan-Europa*. Vienna: Pan-Europa Verlag.
Cox, K. R. (1969). The voting decision in a spatial context. In C. Board, R. J. Chorley, & P. Haggett (Eds.), *Progress in geography* (pp. 81–117). London: Arnold.
Crampton, J. W. (1994). Cartography's defining moment: The Peters projection controversy, 1974–1990. *Cartographica, 31*, 16–32.
Crampton, J. W. (2003). *The political mapping of cyberspace*. Chicago: The University of Chicago Press.
Crampton, J. W. (2004). GIS and geographic governance. Reconstructing the choropleth map. *Cartographica, 39*(1), 41–53.
Crampton, J. W. (2009). Cartography: Performative, participatory, political. *Progress in Human Geography, 33*(6), 840–848.
Dodge, M., Kitchin, R., & Perkins, C. (Eds.). (2009). *Rethinking maps. New frontiers in cartographic theory*. London: Routledge.
Dorling, D., Newman, N., & Barford, A. (2008). *The atlas of the real world. Mapping the way we live*. London: Thames & Hudson.
Edney, M. H. (1993). Cartography without 'progress': Reinterpreting the nature and historical development of mapmaking. *Cartographica, 30*(2 & 3), 54–68.
Edney, M. H. (1997). *Mapping an empire. The geographical construction of British India* (pp. 1765–1843). Chicago: The University of Chicago Press.
Evrigenis, I. D. (2008). *Fear of enemies and collective actions*. New York: Cambridge University Press.
Flora, P. (Ed.). (1999). *State formation, nation-building, and mass politics in Europe: The theory of Stein Rokkan based on his collected works*. Oxford: Oxford University Press.
Foucault, M. (1981). As malhas do poder, *Barbàrie* 4: 23–27 and 5: 34–42 (English translation: J. W. Crampton, & S. Elden (Eds.), *2007. "The meshes of power." Space, knowledge and power. foucault and geography*) (pp. 153–162). Aldershot: Ashgate.

Foucault, M. (2004). *Naissance de la biopolitique. Cours au Collège de France, 1978–1979.* Paris: Gallimard-Seuil.

Francaviglia, R. V. (1995). *The shape of Texas: Maps as metaphors.* College Station, TX: University Press.

Friendly, M., & Denis, D. J. (2001). *Milestones in the history of thematic cartography, statistical graphics, and data visualization.* http://www.datavis.ca/milestones/. Accessed November 2, 2011.

Garrigou, A. (1990). Invention et usages de la carte électorale. *Politix, 3*(10–11), 33–44.

Godlewska, A., & Smith, N. (Eds.). (1994). *Geography and empire.* Oxford: Blackwell.

Gottmann, J. (1952). *La Politique des États et leur Géographie.* Paris: Colin.

Hägerstrand, T. (1967). *Innovation diffusion as a spatial process.* Chicago: The University of Chicago Press.

Harley, J. B. (1988a). Silences and secrecy: The hidden agenda of cartography in early modern Europe. *Imago Mundi, 40,* 57–76.

Harley, J. B. (1988b). Maps, knowledge and power. In D. Cosgrove & S. Daniels (Eds.), *The iconography of landscape* (pp. 277–312). Cambridge: Cambridge University Press.

Harley, J. B. (1989). Deconstructing the map. *Cartographica, 26*(2), 1–20.

Harley, J. B. (2001). *The new nature of maps. Essays in the history of cartography.* Baltimore: The Johns Hopkins University Press.

Harrison, R. E. (1944). *Look at the world. The fortune atlas for world strategy.* New York: Knopf.

Harrison, R. E., & Strausz-Hupé, R. (1945). Maps, strategy and world politics. In H. Sprout & M. Sprout (Eds.), *Foundations of national power* (pp. 90–93). Princeton: Princeton University Press.

Harrison, R. E., & Weigert, H. W. (1944). World view and strategy. In H. W. Weigert & V. Stefansson (Eds.), *Compass of the world: A symposium on political geography* (pp. 74–88). New York: Macmillan.

Haushofer, K. (1931). *Geopolitik der Pan-Ideen.* Berlin: Zentral Verlag.

Haushofer, K. (1934). *Weltpolitik von Heute.* Berlin: Zeitgeschichte.

Henrikson, A. K. (1980). The geographical 'mental maps' of American foreign policy makers. *International Political Science Review, 1*(4), 495–530.

Herb, G. H. (1997). *Under the map of Germany. Nationalism and propaganda 1918–1945.* London: Routledge.

Herb, G. H., Häkli, J., Corson, M. W., Mellow, N., Cobarrubias, S., & Casas-Cortes, M. (Eds.). (2009). Intervention: Mapping is critical! *Political Geography, 28,* 332–342.

Hobsbawm, E. J. (1990). *Nations and nationalism since 1780. Programme, myth, reality.* Cambridge: Cambridge University Press.

Huntington, S. P. (1996). *The clash of civilizations and the remaking of world order.* New York: Simon and Schuster.

Jackson, R., & Sorensen, G. (2006). *Introduction to international relations: Theories and approaches.* Oxford: Oxford University Press.

Jantzen, W. (n.d.). Geopolitik im Kartenbild. Heidelberg: Mittelmeer, Vowinckel.

Kant, I. (1795). Zum ewigen Frieden. Ein philosophischer Entwurf, Friedrich Nicolovius, Königsberg.

Kepel, G. (2004). *The war for Muslim minds. Islam and the west.* London: Belknap Press.

Kern, S. (2007). *Il tempo e lo spazio, Il Mulino, Bologna, 2007 (original edition: The culture of time and space 1880–1918)* (p. 1983). Cambridge: Harvard University Press.

Key, V. O. (1949). *Southern politics in state and nation.* New York: Knopf.

Kidron, M., & Segal, R. (1981). *The state of the world atlas.* New York: Simon & Schuster.

Kidron, M., & Segal, R. (1987). *The new state of the world atlas.* New York: Simon & Schuster.

Kidron, M., & Smith, D. (1983). *The war atlas.* New York: Simon & Schuster.

Kitchin, R., & Dodge, M. (2007). Rethinking maps. *Progress in Human Geography, 31*(3), 331–344.

Kjellén, R. (1921). *Dreibund und Dreiverband. Die diplomatische Vorgeschichte des Weltkriegs.* Munich: Duncker & Humblot.

Kohr, L. (1957). *The breakdown of nations*. London: Routledge & Kegan Paul.
Konvitz, J. (1987). *Cartography in France 1660–1848. Science, engineering, and statecraft*. Chicago: The University of Chicago Press.
Kuhn, T. S. (1962). *The structure of scientific revolutions*. Chicago: The University of Chicago Press.
Lacoste, Y. (1976). *La gèographie, ça sert, d'abord, à faire la guerre*. Paris: Maspero.
Lane-Poole, S. (1894). *The tables of Mohammedan dynasties. Twenty years' work upon the Arabic coins in the British Museum*. Westminster: Constable.
Levy, J. (2008). *L'invention du Monde: une Géographie de la Mondialisation*. Paris: Presses de Sciences Po.
Lima, M. (2011). *Visual complexity. Mapping patterns of information*. New York: Princeton Architectural Press.
Limes. (2009). *Rivista Italiana di Geopolitica* 1.
Lippmann, W. (1965). *Public opinion*. New York: Free Press.
Loxton, J. (1985). The Peters' phenomenon. *The Cartographical Journal, 22*(2), 106–108.
Mackinder, J. H. (1904). The geographical pivot of history. *The Geographical Journal, XXIII*(4), 421–445.
Marey, E. J. (1885). *La Méthode Graphique*. Paris: Masson.
Marin, L. (2001). *On representation*. Stanford, CA: University Press.
Mémoires pour servir à l'Histoire Naturelle des Animaux. (1671). Paris: Imprimerie Royale.
Minca, C., & Białasiewicz, L. (2004). *Spazio e politica*. Padova: CEDAM.
Monmonier, M. (1990). Strategies for the visualization of geographic time-series data. *Cartographica, 27*(1), 30–45.
Monmonier, M. (1991). *How to lie with maps*. Chicago: The University of Chicago Press.
Monmonier, M. (1995). *Drawing the line: Tales of maps and cartocontroversy*. New York: Holt.
Monmonier, M. (2001). *Bushmanders & Bullwinkles. How politicians manipulate electronic maps and census data to win elections*. Chicago: University of Chicago Press.
Neurath, O. (1939). *Modern man in the making*. New York: Knopf.
O'Loughlin, J., & Van der Wusten, H. (1990). Political geography of panregions. *Geographical Review, 80*, 1–20.
O'Tuathail, G. (1996). *Critical geopolitics: The politics of writing global space*. Minneapolis, MN: The University of Minnesota Press.
O'Tuathail, G. (1999). Understanding critical geopolitics: Geopolitics and risk society. *Journal of Strategic Studies, 22*(2/3), 107–124.
Paasi, A. (1986). The institutionalisation of regions: A theoretical framework for understanding the emergence of regions and the constitution of regional identity. *Fennia, 164*, 105–146.
Palsky, G. (2008). Connections and exchanges in European thematic cartography. The case of the 19th century choropleth maps. *Belgeo, 3–4*, 413–425.
Parker, G. (1985). *Western geopolitical thought in the twentieth century*. London: Croom Helm.
Pickles, J. (2004). *A history of spaces: Cartographic reason, mapping and the geo-coded world*. London: Routledge.
Polelle, M. (1999). *Raising cartographic consciousness: The social and foreign policy vision of geopolitics in the twentieth century*. Lanham, MD: Lexington Books.
Ratzel, F. (1903). *Politische Geographie oder die Geographie der Staaten, des Verkehres und des Krieges* (2nd ed.). Munich and Berlin: Oldenbourg.
Rokkan, S. (1980). Eine Familie von Modellen für die vergleichende Geschichte Europas. *Zeitschrift für Soziologie, 9*(2), 118–128.
Said, E. W. (2004). *From Oslo to Iraq and the roadmap*. London: Bloomsbury.
Schulten, S. (1998). Richard Edes Harrison and the challenge to American cartography. *Imago Mundi, 50*, 174–188.
Siegfried, A. (1913). *Tableau politique de la France de l'ouest sous la troisième république*. Paris: Colin.
Siegfried, A. (1950). *Atlas international*. Paris: Larousse.

Smith, M. L. (2005). Networks, territories, and the cartography of ancient states. *Annals of the Association of American Geographers, 95*(4), 832–849.
Spykman, N. J. (1944). *The geography of the peace*. New York: Harcourt, Brace and Company.
Therborn, G. (2009). Why and how place matters. In R. E. Goodin (Ed.), *The Oxford handbook of political science*. Oxford: University Press.
Tufte, E. R. (1983). *The visual display of quantitative information*. Cheshire, CT: Graphics Press.
Van Valkenburg, S. (1949). *Elements of political geography*. New York: Prentice-Hall.
Vayssière, B. H. (1980). 'La Carte de France', in Cartes et figures de la Terre, Centre Georges Pompidou, (pp. 252–265).
Von Schumacher, R. (1934). Zur Theorie der Raumdarstellung. *Zeitschrift für Geopolitik, 11*(10), 635–652.
Von Schumacher, R. (1935). Zur Theorie der geopolitischen Signatur. *Zeitschrift für Geopolitik, 12*(4), 247–265.
Vujakovic, P. (1987). The extent of adoption of the Peters projection by "Third World" organizations in the UK. *Society of University Cartographers Bulletin, 21*(1), 11–15.
Vujakovic, P. (1989). Peters atlas: A new era of cartography or publishers con-trick? *Geography, 74*(3), 247–254.
Waltz, K. N. (1979). *Theory of international politics*. New York: Random House.
Wells, H. G. (1921). *The outline of history being a plain history of life and mankind*. New York: MacMillan.
Williams, M. C. (2005). *The realist tradition and the limits of international relations*. Cambridge: Cambridge University Press.
Wood, D. (1992). *The power of maps*. New York: Guilford.
Wood, D. (2010). *Rethinking the power of maps*. New York: Guilford.
Zook, M. A., & Dodge, M. (2009). Mapping cyberspace. In N. Thrift, & R. Kitchin (Eds.), *The international encyclopedia of human geography* (pp. 356–367).

"Now – Well, Look at the Chart": Mapping, Maps and Literature

Andrew Frayn

Abstract This chapter examines the resistance in literary criticism to making maps. Literary analysis is deeply invested in the construction of space and associated theories, but these have rarely been cartographical. Recent work that discusses the development of maps and literature in the early modern period is surveyed briefly, before focusing on the early 20th century as a key moment. The relationship between the map and the physical world is discussed in the context of realist fiction, highlighting the impossibility of realist fiction being real just as the map is not the territory. Maps, however, are used in literature to enhance the reality of the fictional world, from science fiction to novels prophesying war. (Post)colonial literatures are seen as a vital site of literary mapping, using Conrad's *Heart of Darkness* as a case study. The chapter concludes by examining theoretical standpoints, engaging with the plausibility of Fredric Jameson's concept of cognitive mapping and his spatialising of literature through the semiotic rectangle. Franco Moretti's work on mapping is invoked as one potential way forward for literary cartography, albeit one which needs further critical refinement. Literary critics are encouraged to shed their inhibitions about using and making maps, examining the work of Barbara Piatti's cohort on *The Literary Atlas of Europe* as another avenue for exploration. Interdisciplinary collaboration will be invaluable in the development of literary cartrographies.

Keywords Literary cartography · Modernism · Realism · Cognitive mapping · Franco moretti

Literary studies is a polymorphous, even amorphous discipline. Literary criticism has focused recently on methods which aim to move away from the traditional anglo-centric, male-dominated canon, or which reassess these texts. Poststructural and deconstructive criticism moved away from neat structures and teleological understandings towards a more diverse, inclusive and consequently complex

A. Frayn (✉)
Department of English and Acting, Edinburgh Napier University, Edinburgh, UK
e-mail: A.Frayn@napier.ac.uk

S.D. Brunn and M. Dodge (eds.), *Mapping Across Academia*,
DOI 10.1007/978-94-024-1011-2_13

understanding of literary texts and their production. Gender and sexuality studies, and particularly queer studies, draw on Marxist frameworks for their understanding of society and continue to question interpretations of culture rooted in hegemony; postcolonial studies similarly seeks to recover the work of previously ignored, glossed over, or elided racial others and to reconsider works written from Imperial centres. Literary studies is an inclusive discipline, and with the growth of inter-disciplinary and transdisciplinary studies it now reaches out tendrils into the most unlikely of areas. There is an enduring traditionalist resistance to the empirical and quantitative studies which underpin the digital humanities, an emerging area that attempts to answer one of the key questions for twenty-first century literary studies: how is the discipline to deal with the proliferation of information, the availability and accessibility of which has increased exponentially over the last fifteen years? What opportunities does this offer us, and how can we deal with the problems it causes? These methods are developing continually, and will continue to do so in the coming years; one opportunity offered, which is yet to be exploited fully, is to use this data to map and to represent literary texts visually.

Literature is intimately connected with space, and literary criticism often looks to mapping as an explanatory metaphor; there has been fruitful cross-pollination between literature and theories of space from a variety of disciplines. A Marxist conception of the productive qualities of space (Lefebvre 1991 [1974]; Harvey 1989) has become increasingly influential in literary studies which use geographical methods (e.g. McCracken 2007; Thacker 2003, 2005). There remains a resistance in literary studies to the creation of maps, diagrams and charts, and more generally to visual representations as analytical tools, although this is weakening as digital humanists start to make effective use of textual analysis software, following the pioneering work of Franco Moretti. This chapter probes that resistance, and also looks at the work of those academic literary critics who engage with ideas about space such as Fredric Jameson, Graham Huggan, and Andrew Thacker. So invested is literary studies in ideas of space that Jameson (2003: 696) has noted that "statistics on the volume of books on space are as alarming as the birthrate of your hereditary enemy." He observes in the corresponding footnote that 5000 such volumes had been published in the previous three years, although the terms used to generate that number remain a mystery.

To discuss these ideas I use examples from early twentieth-century European literature, which map both verbally and graphically. There is a significant body of cartographic criticism about the early modern period (see Conley 1997; Klein 2001; Matei-Chesnoiu 2012, 2015; Peters 2004; Sanders 2011; Sanford 2002). A recent edited collection examines medieval geographies from 300 to 1600 (Lilley 2014), while Moretti (1998, 2005) has based his analyses on the nineteenth century. Less attention has been paid in these terms to the period usually known as modernism—although recent studies tend to discuss modernisms, looking outside the high modernist canon of T.S. Eliot, James Joyce, Ezra Pound, Virginia Woolf, and others (e.g. Mao and Walkowitz 2006)—but it is bound up in ideas about space. Even the designation of "high" modernism, against "low" or popular culture, sees the relationship between culture and society spatially (see DiBattista and

McDiarmid 1996; Huyssen 1988). The understanding of space was radically revised in the early twentieth century as a result of scientific, social and technological developments and their corresponding displacement into narrative. Furthermore, theories of postmodern space such as Jameson's idea of cognitive mapping, which I discuss below, either implicitly or explicitly define themselves against the modern and modernism. In spite of the work being done by the distinguished scholars mentioned already there is still more we can do with maps, and I conclude by examining literary cartographies and geographies while arguing, after Moretti and Barbara Piatti's Zürich-based *Literary Atlas of Europe* project, for the creation and use of cartographical representations in literary criticism.

1 Why not Map?

So, why the lingering resistance of literary critics to the use of maps and charts? Let's jump over the disciplinary fence into geography and see what somebody there has to say. Mark Monmonier (1993: 204) ponders the reasons:

> Why biographers and literary analysts rarely use maps puzzles me. Perhaps, as "word people," the idea of providing a geographic summary of an important person's life never occurs to them. Perhaps they merely lack the tools, skills, and self-confidence to attempt a map themselves, or the funds to hire a cartographic illustrator. Or, perhaps by tradition, publishers, critical reviewers, and readers don't expect maps in these genres.

Perhaps Monmonier is right: words are our stock in trade and we do not want to devalue them by including diagrams. With words as the primary focus of the discipline, it is certainly difficult to do anything as witty or innovative as Denis Wood and John Krygier's graphic essay "Ce N'est Pas La Monde," which they call "a proposition about maps as propositions and about comic books as academic discourse in the form of a comic book of propositional maps" (2009: 212), but we can do more than at present. Pictures are increasingly permitted and even desired in works both of fiction and criticism, so why not maps? Even in a world which is now so resolutely interdisciplinary—most English Literature departments have become departments of English Studies or Literary Studies, and theories and works from Philosophy, Sociology, Art History, Theatre Studies and Religious Studies are regularly invoked—the quantitative and positional analysis of literature is still largely taboo, although the swift and continuing emergence of digital humanities poses a challenge to the entrenchment of traditional methods (see Hand 2011; Torget and Christensen 2012). There is still a separation, it seems, between literary criticism which writes about geography and critical geography which writes about literature. Is the resistance to cartography a hangover from the days of connoisseurship? Does the heady whiff of (social) scientific method make us feel nauseated within that hangover? Muehrcke and Muehrcke (1974: 331) argue that "a map is by nature interdisciplinary, and all imaginable fields of learning may be brought to both making it and reading it." The same is surely true of literature. The logistical issues which Monmonier outlines are a factor in the resistance to maps, but the onus is on the critic to push for the inclusion of such diagrams to evolve the discipline, and to begin to engage productively and dynamically with cartography; opportunities to do so are continuing to increase as publishing technologies develop and e-books become more common.

Maps are resisted for their perceived reality, for the sense that a map exists only to show. Monmonier's (1993) description of the map as "providing a geographic summary" serves only to reinforce that viewpoint. Eric Bulson (2006: 21) asserts that "early literary maps were a way to advertise the novel's realism," but we are yet to move significantly past this state. Angharad Saunders (2010: 440–1) observes that literature "is not searching for public endorsement of accuracy, but rather self-awareness of life's uncertainties and possibilities. [...] Literature gets us to think anew, it knows about the 'other' and motivates us to contemplate different spatial and social orders, which would otherwise remain concealed or suppressed." While Geography in the academy is taking the "cultural turn" (see Barnett 1998; Harley 2009; Philo 1991; Wood and Fels 2009), the impression of maps in the rest of the academy has yet to catch up. Literary studies, and the humanities more generally, is taking the spatial turn (see Gilbert 2009, and the other studies in the same volume; Dear et al. (2011), particularly Part II: Spatial Literacies), but it is unsure whether it had the map the correct way up when it did so. Maps are still too often associated with the tourist box-ticking of literary guidebooks such as the *Blue Guide to Literary Britain and Ireland*, and similar works like Grevel Lindop's (1993) rather more elegant *Literary Guide to the Lake District*, which genre Bulson (2006) describes and critiques (see also Thacker 2005: 58–59). These volumes tend to conflate fictional representations such as Hardy's Wessex or Dickens' London with the physical world. Similarly *Lit Long*, a developing project at the University of Edinburgh, textmines literary references to Edinburgh places on a large and automated scale. Ian Ousby (1990: 50) writes in the *Blue Guide* that:

> Hardy began the creation of "Wessex" with a mild inaccuracy, borrowing its name from the ancient Saxon kingdom whose centre lay east of the region he had in mind. In all else, however, he kept as near the facts as possible. Indeed, he possessed the rarest sort of artistic tact: an instinctive knowledge that imagination is sometimes superfluous and mere literal accuracy is eloquent enough.

Behind the praise of Hardy's mastery lies a paradox: how can a work of fiction be literally accurate? It recalls the impossibly large map in Borges' "On Exactitude in Science" (1975) or the 1:1 scale map of the land won by the Allied troops in the BBC TV series *Blackadder Goes Forth*. Hardy's Wessex was not and could not be literally accurate. In fact, he acknowledged his own "inaccuracies" late in his career by revising his novels to make their geography more consistent (see Draper and Fowles 1984: 11–12, 25; Hawkins 1983: 194–195). Geoff King (1996: 3–4) notes perceptively in his excellent *Mapping Reality* the commercial imperative which underlies maps aimed at the literary traveller: they often attempt to encourage literary tourism and to market places as literary brands. Literary maps have been championed enthusiastically for many years (e.g. Fuson 1970), but only recently has there been a more concerted drive towards literary criticism which uses and reads maps.

2 Maps and Literary Criticism

Literary criticism is apparently suffused with maps, at least from scanning the titles of books and articles. Drawing on the often false certainties offered by the map as understood above, the term mapping is often used as a synonym for "explaining fully and clearly"; geographers Neil Smith and Cindi Katz point out the problems of spatial metaphors (1993). Works such as Simon Gikandi's *Maps of Englishness* (1996), Christopher GoGwilt's *The Invention of the West: Joseph Conrad and the double-mapping of Europe and empire* (1995), Michael Kane's *Modern Men: mapping masculinity in English and German literature, 1880–1930* (1999), Ricardo Quinones' *Mapping Literary Modernism* (1985) and Andrew Radford's *Mapping the Wessex Novel* (2010), to go back to Hardy, map verbally and metaphorically; these non-graphical mappings heavily outweigh the number of works which engage directly with cartography such as cultural geographer Richard Phillips' *Mapping Men and Empire* (1997). Even King's *Mapping Reality: an exploration of cultural cartographies* has no figures. Geographer Chris Perkins (2003: 341) has pointed out pithily that "it has become increasingly fashionable for researchers informed by concerns with critical social theory to use the 'm' word, but to have little appreciation of how maps work as tools." For example, Gikandi and Quinones use the term almost wholly in the abstract. GoGwilt's is a more rigorously theorised approach to the experience of space, as are Andrew Thacker's *Moving Through Modernity* (2003) and Scott McCracken's *Masculinities, Modernism and the Urban Public Sphere* (2007). These works draw extensively on the spatial theories of Michel de Certeau, Walter Benjamin, Maurice Blanchot, Gaston Bachelard and, particularly, Henri Lefebvre to situate literary works in a sophisticated conception of urban space. While these are valuable analyses of the spaces of modernism, only McCracken dares to cross the disciplinary line by mapping the teashops which he argues are intimately connected with the development of modernist literature (Fig. 1), and which were the preserve not only of shoppers but of workers (2007: 101). However, these are the only two maps in the monograph, and he notes (2007: xv) that he was persuaded to make them by geographer Miles Ogborn.

When attempts have been made to use cartography in conjunction with scholarly literature, too often the map has been an afterthought. *The Atlas of Literature* (Bradbury 1996) maps authors, locations, and journeys in a wide variety of periods and spaces. Thacker and Brooker (2005: 1) are charitably inclined towards the project, commenting that "much can be gleaned from these maps about the intimate connections between cultural texts and geographical location." Much can certainly be learned from thinking about such connections, but the reader of the volume is left to make those links. The maps are accompanied by short essays written by

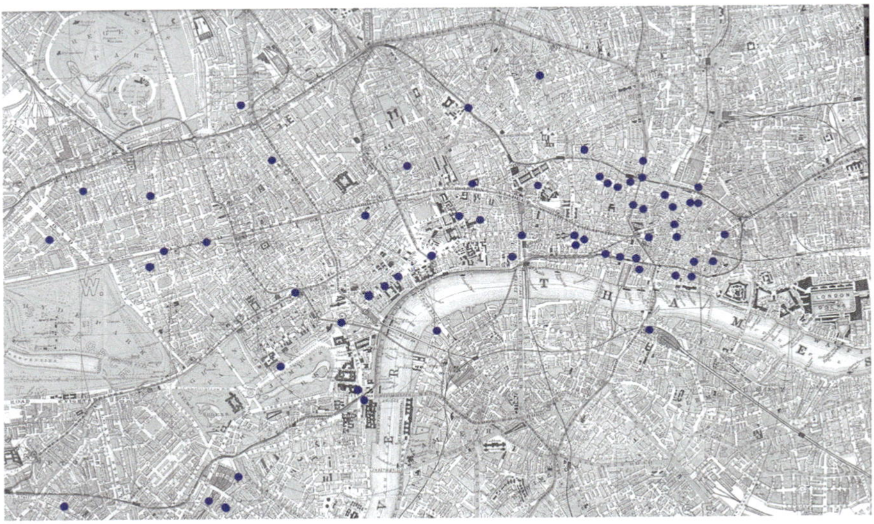

ABC teashops in central London, 1890

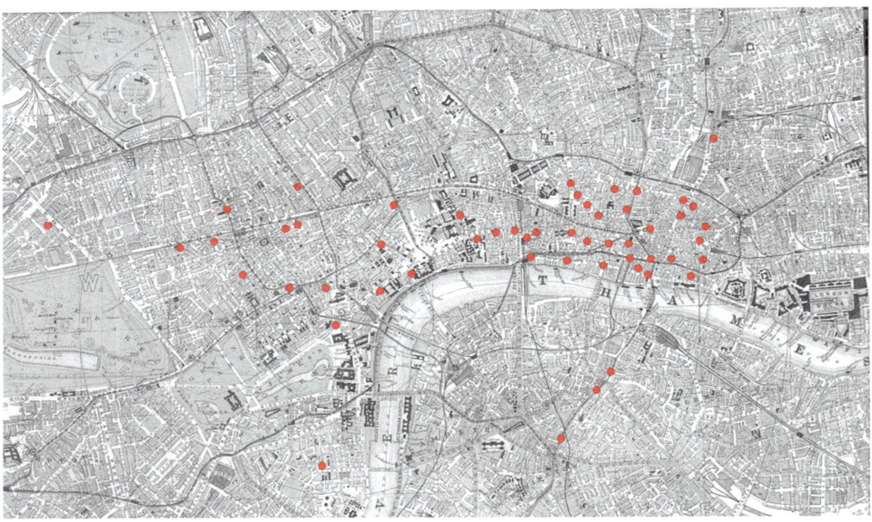

Lyons teashops in central London, 1890

Fig. 1 Maps of London's teashops in Scott McCracken, *Masculinities, Modernist Fiction and the Urban Public Sphere* (Manchester: Manchester University Press 2007: 102)

academics or media authors about the personalities who lived there and the events which took place there, but the text fails to speak directly to the maps. Consequently, I side with Moretti's (1998: 7) assessment of this and similar works: "maps play in them a wholly peripheral role. Decorative. There are quite a few of them, by all means, especially in the more recent books: but they are colorful appendixes, that don't intervene in the interpretive process: at times, they even show up at the end of the text—when the discourse is over, done with." In Bradbury's volume there is no attempt to use the maps critically: they are simply for plotting places and people, and there is no apparent rationale to explain what, where and who is included. To take the example of the map and chapter "Paris in the Twenties" (Fig. 2) there is no consistency in the labels on the map, and many have unclear referents. The map is overpowered by labels, particularly in the lower right quadrant, indicating the primacy of text over image, and places indicated on the map become obscure suggestions of, well, something. As it is not mentioned in the essay it takes either some local knowledge or an educated guess to realize that "Falstaff" does not refer to any conflation of literature and history relating to Shakespeare's character, but to the Parisian restaurant of the same name. Similarly, does "Coco Chanel" refer to a boutique or a residence of the *modiste* herself? The key tells us that "dates refer to the period of time a person or institution was there, or the date of a visit. A date is not given if something or someone was in a place sporadically, or if the information is not known" (Bradbury 1996: 9). This inconsistency fails to elucidate further than a general sense of activity; and undated labels

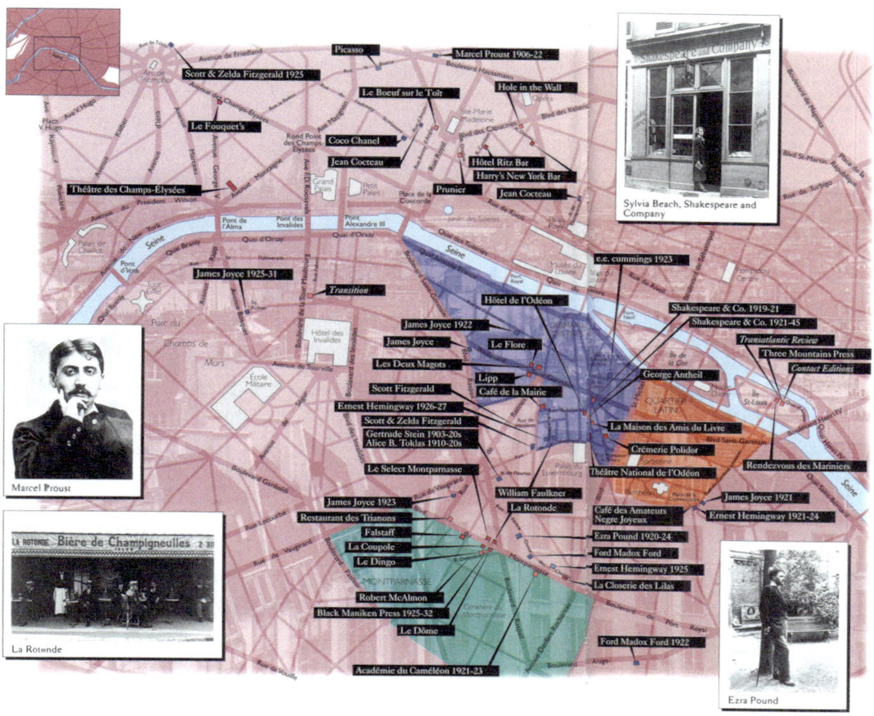

Fig. 2 'Paris in the Twenties' in *The Atlas of Literature*, Malcolm Bradbury (ed.) (London: de Agostini 1996: 174–75)

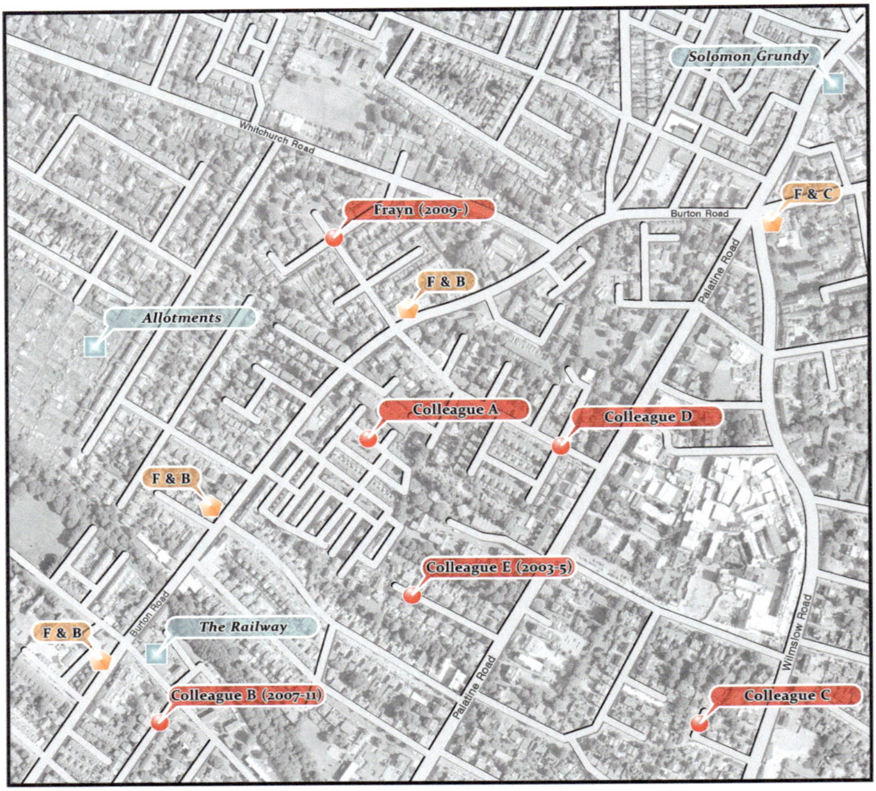

Fig. 3 Frayn and colleagues in South Manchester. *Red dots* place of residence; *blue* denotes building or defined space for meeting, *orange* denotes passing in street

lead, as in the example of Chanel, to further confusion. Such maps are at best frustrating, and at worst misleading. For example, I mapped my residences alongside five colleagues living in a former neighbourhood, marking street locations and buildings where we met in the two years I lived there. The map (Fig. 3) suggests a vibrant sense of academic community, where in reality these meetings and intersections are the exceptions which prove the general rule that we rarely see each other outside of the department corridor.

Maps are texts which demand to be read, mediated representations of a wider reality, and their use—or, indeed, their disuse—as uncritical indicators of place needs to be rejected. Academic cartographers and map scholars know, having read Foucault, Barthes and Derrida, that mapping is bound up in power relations and that, as Wood and Fels (2009: 223) put it, "*there is nothing natural about a map*" (their italics), although most commercial map-makers and the GIS industry are yet to engage with these theories. Maps are collaborative endeavours, from the funding of the survey, to the collection of its data, to its rationalisation into legible design, to printing and even where, for and by whom they are sold. Brian Harley (2009 [1989]: 285) argues that "the steps in making a map—selection, omission, simplification, classification, the creation of hierarchies and 'symbolization'—are all inherently rhetorical," and that rhetoric "may be all the more potent for not generally being recognized" (King 1996: 21). There is as much subtext in a map as in a literary text. Maps are not neutral, and they are not mirrors reflecting the world, as Harley (2009 [1989]: 281) points out: "By accepting their textuality we are able to embrace a number of different interpretative possibilities. Instead of just the transparency of clarity we can discover the pregnancy of the opaque. To fact we can add myth, and instead of innocence we may expect duplicity." As Piatti and Hurni (2011: 222) put it, "the mapping of fiction allows a better, deeper understanding of how fiction works—the mapping process supports the interpretation; it opens new horizons for literary scholarship because some maps make aspects visible which have been invisible before." I would go even further, and say that the mapping process not only supports the interpretation, but should be considered a part of the interpretation. So let's put to rest the myth about maps and start to work with them.

3 Maps in Literature

Despite having argued that maps are not and cannot be perfectly mimetic, they are often used in fiction to enhance a perception of reality. Philip and Juliana Muehrcke (1974: 319) wrote over thirty-five years ago that the map "is used frequently, not only in literature but in real life, to convey the idea of clarification," and their analysis remains pertinent. Maps commonly appear in genres where accurate representation is particularly prized, such as historical fiction and histories, and also in genres where maps heighten the reality of unlikely, even far-fetched events and places. Annika Richterich (2011: 244) asserts that "readers perceive the maps themselves as realistic objects, since they are familiar tools of orientation and overview in daily life. Moreover, such maps localize the story in a factual geography and act as a topographic authentication of the literary space." Authors of science fiction have used maps to great effect to reify fantastical worlds from Robert Louis Stevenson's *Treasure Island* to Jules Verne's *20,000 Leagues Under the Sea*, through Tolkein's *The Hobbit* and *The Lord of the Rings*, to contemporary authors such as Kim Stanley Robinson, China Miéville, Jim Grimsley and Ursula Le Guin. Graham Huggan (1994: 22) posits that "writers of fantasy are attracted by the pictorial extravagance of the map, which invents even as it attempts to 'document.'" The maps in these works tend to be positioned at the front of the volume, allowing the reader literally (or, perhaps, literarily) to orientate him or herself in the fictional space of the novel. The imagined worlds of these authors have started to attract attention from literary critics (Harpold 2005; Padrón 2007; Habermann and Kulm 2011), and the mapping of fantasy worlds is becoming as much an object of academic interest as reading the realist against the real.

Works of prediction and prophesy such as Erskine Childers' scaremongering pre-First World War novel *The Riddle of the Sands* (2007 [1903]) invoke maps to similar effect. Maps emphasise a detailed knowledge of the antagonistic power, heightening the reality of the threat which had yet to develop, and which now seems prescient ahead of the First World War. Two maps and two charts precede the novel alongside the preface by the purported editor, the end of which claims that the charts are "based on British and German Admiralty charts, with irrelevant information omitted" (Childers 2007 [1903]: 13). The protagonist Davies resorts to maps to make sense of the dangerous situation:

> Now – well, look at the chart. No, better still, look first at this map of Germany. It's on a small scale, and you can see the whole thing." He snatched down a pocket-map from the shelf and unfolded it [see Map A]. "Here's this huge empire, stretching half over central Europe – an empire growing like wildfire, I believe, in people, wealth, and everything. They've licked the French, and the Austrians, and are the greatest military power in Europe. I wish I knew more about all that, but what I'm concerned with is their sea-power. [...] They've got no colonies to speak of, and *must* have them, like us. They can't get them and keep them, and they can't protect their huge commerce without naval strength. The command of the sea is the thing nowadays, isn't it? (Childers 2007 [1903]: 80)

We must assume that the map (Fig. 4) is not an exact reproduction of what Davies and the other main character Carruthers are looking at, as it gives no real sense of empires and does not show central Europe. However, it reinforces the importance of naval power: over half the map's area is taken up by the North Sea. The blankness by which it is represented suggests an area waiting to be filled, in this case by military force: the lines showing the limits of the German seaboard trail off threateningly into the blank space. England is always present in the novel in the form of the protagonists, but is a location only at the beginning and the end, and is

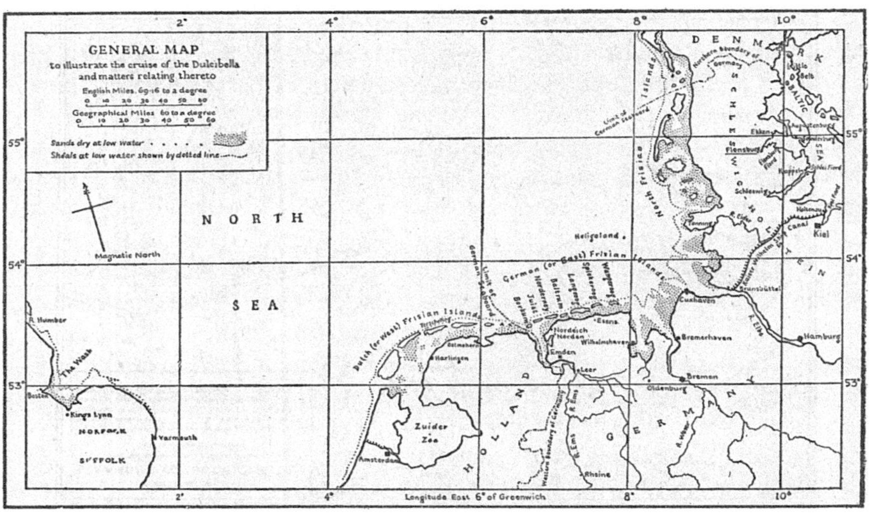

Fig. 4 Map A in Erskine Childers, *The Riddle of the Sands* (1903; London: Penguin 2007: 7)

therefore marginal on the map. This marginality echoes the small-scale agency of Davies and Carruthers, working against the full might of the German military to stave off the threat of invasion by maintaining the blank barrier of the North Sea. Maps are intimately linked with the military and with power struggles, this connection enshrined in the name of the UK's Ordnance Survey, founded to map Scotland in an attempt by the British Empire to control vulnerable areas and restless colonies (Hewitt 2010).

Postcolonial studies has frequently employed spatial metaphors as a fundamental part of literary analysis (Huggan 1994; Muecke 1992; Berland 2005). The focus on issues of boundary-making, conquest, delimitation, demarcation and naming means that geography and cartography are key components; however, Andrew Thacker (2005: 58) rightly observes that "though the rise of postcolonial criticism has emphasised the significance of geography as a paradigm for understanding culture and power, it is important not to identify a critical literary geography with a version of postcolonial studies." Jon Hegglund (2012) situates a reading of Conrad alongside King Njoya's map of Bamum, a kingdom in modern-day Cameroon, and in terms of the influential Scottish geographer Halford Mackinder. This comparative work in expanding the modernist archive suggests an important way forward for constructions and interpretations of literary space.

Early representations of actions taken in the name of imperialism demonstrate the interconnectedness of political and military power and cartography, such as Gillray's famous cartoon of Pitt and Napoleon carving up the globe (Fig. 5). Brian

Fig. 5 James Gillray, The Plumb-pudding in Danger: or State Epicures taking un Petit Souper, 1805

Friel's play *Translations* (1981) deals with the impact of English colonialism in Ireland in the form of the Ordnance Survey's mapping of the country, a dual translation of three-dimensional space onto the two-dimensional map, and of the Gaelic language into English (see Bulson 2006: 77–79). Conrad's *Heart of Darkness* (2007 [1899]) is regularly deployed as an example of the critique of colonialism; however, Con Coroneos (2002: 108) describes it as "colonialism's articulate enemy but also one of its most insidious accomplices." The protagonist Marlow is inspired by a fascination with maps:

> I would look for hours at South America, or Africa, or Australia and lose myself in all the glories of exploration. At that time there were many blank spaces on the earth, and when I saw one that looked particularly inviting on a map (but they all look that) I would put my finger on it and say, When I grow up I will go there. [...] I have been in some of them, and... well, we won't talk about that. But there was one yet—the biggest, the most blank, so to speak—that I had a hankering after.
>
> True, by this time it was not a blank space any more. It had got filled since my boyhood with rivers and lakes and names. It had ceased to be a blank space of delightful mystery—a white patch for a boy to dream gloriously over. It had become a place of darkness. (Conrad 2007 [1899]: 8–9)

The blank spaces were never blank, merely unexplored by and unknown to Imperial forces; Marlow's experience therein can be rendered only as an unverbalised ellipsis. Conrad (1924) later describes filling in the blank spaces on a similar map (Fig. 6), and a subsequent passage in *Heart of Darkness* describes the colouring of the map to indicate the division of Africa at the 1884–1885 Berlin conference. The white patch on the map becomes dark as the gleaming space is filled in; moral darkness descends as European empires take control of large swathes of Africa, exploiting the indigenous population and making them more 'white'. Karen Piper (2002: 16) suggests that a "fascination with the primitive (and its dangers) is popularized in the character of Kurtz, who travels into the 'darkness' of Africa and is transformed by it. It is this transformation, on the margins of the map, that is repeatedly described by explorers as both seductive and threatening." On the Mercator projection, the centre of Africa is anything but on the margins of the map, although as Arno Peters and others have notoriously observed, its size is diminished by the distortion of the projection (see Monmonier 2004: 15–16). Exploration is essential to cartography, particularly in the name of colonization, but can be hazardous. Jameson (1988: 12) suggests with specific reference to *Heart of Darkness* that "cartography is not the solution, but rather the problem, at least in its ideal epistemological form as social cognitive mapping on the global scale."

4 Fredric Jameson and Cognitive Mapping

Fredric Jameson has consistently seen literature in spatial terms, and argues for the importance of space in literary and cultural studies: his "postmodern readings tend to emphasize the chaotic and anomic nature" of space (Gilbert 2009: 106). Jameson's proposed project is "cognitive mapping," which perhaps has contributed

Fig. 6 Extract from Carte de L'Afrique dressee et dessinee sous la Direction de Mr. J. G. Barbie du Bocage. Prepared as plate no. 26 in Maison Basset's 1852 edition of the *Atlas Illustre*

to the proliferation of criticism which maps without using cartography. He draws on Kevin Lynch's *The Image of the City* (1960), particularly the notion of "imageability," but expands the terms under consideration:

> The conception of cognitive mapping proposed here […] involves an extrapolation of Lynch's spatial analysis to the realm of social structure, that is to say, in our historical moment, to the totality of class relations on a global (or should I say multinational) scale.

> […] A secondary premise must also be argued—namely, that the incapacity to map socially is as crippling to political experience as the analogous capacity to map spatially is for urban experience. (Jameson 1991: 416)

Even for Jameson the idea of cognitive mapping is intensely problematic, a possibility yet to be realised. Lynch argues that it is impossible for residents of a modern city to locate themselves even in their most local environment when there are no clear landmarks for orientation, and the whole city is an unimaginable totality. While maps allow people to orientate themselves, the map cannot fully depict the space in which the reader/traveller finds himself. Unquestioning reliance on cartographic technologies only adds to the disorientation felt in the postmodern city and, indeed, the postmodern world. The development of GPS and satnav systems attests to this separation between their users and physical space, and news outlets regularly feature stories about drivers trapped down dangerous or unpassable single-track lanes, hanging over cliff edges like extras from *The Italian Job*, or getting stuck on tidal causeways. Jameson (1988: 12) asserts that "the map, if there is to be one, must somehow emerge from the demands and constraints of the spatial perceptions of the individual," and this insistence on individuality is part of the problem. While the world seems smaller and smaller in terms of the ease of communication, the amount of information available is increasing exponentially. Cognitive mapping is Jameson's response to postmodernity's decentred subject, seeking a totality which is now generally deemed impossible or inaccessible, the unseen political unconscious: "what I have called cognitive mapping may be identified as a more modernist strategy, which retains an impossible concept of totality whose representational failure seemed for the moment as useful and productive as its (inconceivable) success" (Jameson 1991: 409). Jameson's historical divisions are too-neatly spatialised into blocks of realism, modernism, and postmodernism, but he is conscious of that problem; the shift from a focus on time in modernism to a focus on space in postmodernism is crucial for him.

Can the cognitive mapping Jameson desires ever happen? The information available to draw the map is inevitably incomplete, but the project's value is more widely seen in its attempt to link the personal with the social, and the local with the global. The power of mapping is invoked and at the same time is the totemic, monolithic concept which prevents the project's success:

> cognitive mapping, which was meant to have a kind of oxymoronic value and to transcend the limits of mapping altogether, is, as a concept, drawn back by the force of gravity of the black hole of the map itself (one of the most powerful of all human conceptual instruments) and therein cancels out its own impossible originality. (Jameson 1991: 416)

Cognitive mapping may become possible given suitable political conditions, but it is for now hypothetical. Maps are powerful instruments, but that should be as much a reason to use them, with appropriate caution, as not to use them. As critics, if we are clear about the principles on which selections are made and the map is plotted, then they can be valuable in assisting the work and play of interpretation. This takes us back to Borges' (1975) life-sized map: in any kind of mapping the finest of detail is and must be lost, and subtleties are eroded in the drive to achieve a coherent representation at a scale reduced from life-sized (see also King 1996: 13). That does not mean that complexity, paradox, and difficulty are lost altogether in literary interpretation. Maps do not replace analysis, but offer an alternative and supplementary viewpoint, and visual representations are a powerful tool for communication, both to a popular audience and in pedagogy (see Latour and Woolgar 1986: 45–53).

Jameson also sees literature in spatial and diagrammatic terms in other ways, and he regularly uses the semiotic rectangle (Greimas and Rastier 1968) to describe oppositions, problems, and paradoxes (Fig. 7); Moretti cites this as a precursor of his own work (1998: 9). As literary studies was developing, influential theorists such as I. A. Richards and John Crowe Ransom sometimes used diagrams, but the practice did not become widespread: notable exceptions include Deleuze and

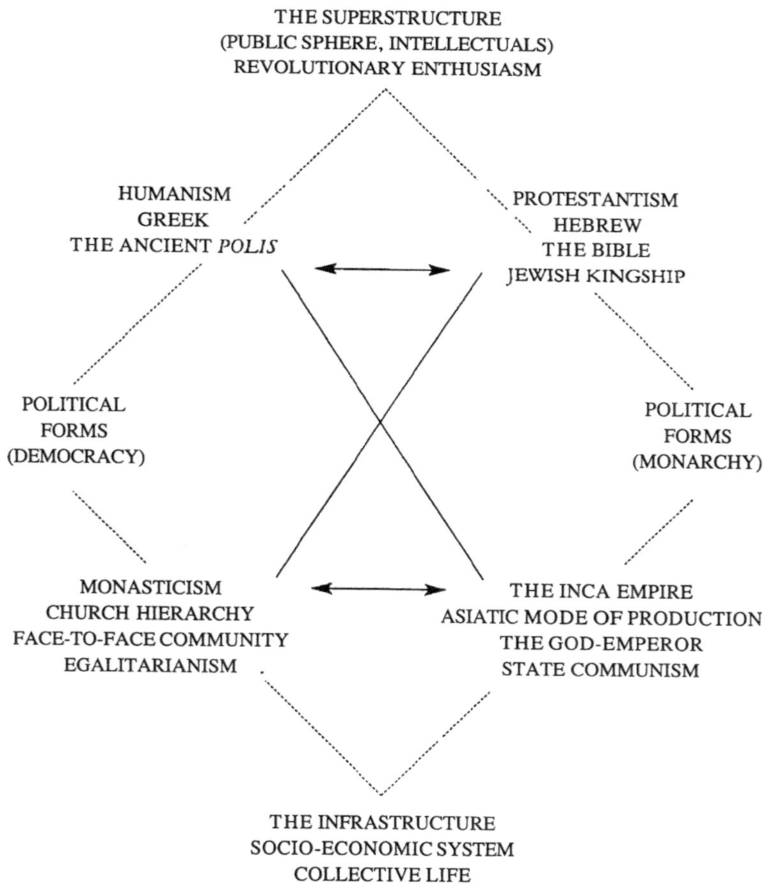

Fig. 7 Semiotic rectangle depicting the constituent parts and contradictions of Utopias in Fredric Jameson, *Archaeologies of the Future* (London: Verso 2005: 30)

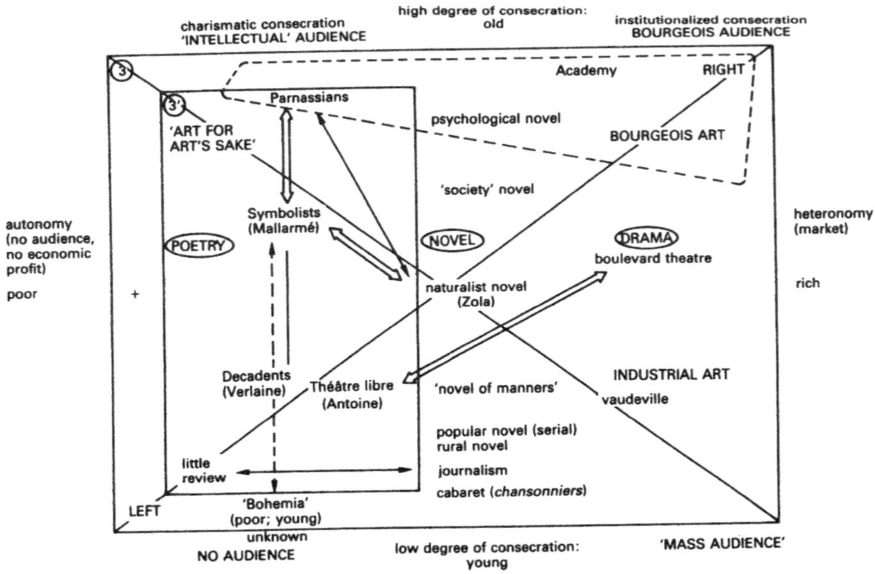

Fig. 8 French literary field in the second half of the 19th century (+ positive pole, implying a dominant position;− negative pole, implying a dominated position) in Pierre Bourdieu, *The Field of Cultural Production* (Cambridge: Polity 1993: 49)

Guattari (1988) and Pierre Bourdieu (1984 [1979], 1993). None of these critics was by training a literature scholar. A reading group to discuss Bourdieu's *Field of Cultural Production* with some of my colleagues and peers within the discipline at the University of Manchester proved an interesting informal case study for the value of, problems with, and resistance to the use of diagrams in literary studies. The diagram, based on the semiotic rectangle, grabbed the attention of the group and became a fetish object, controlling the discussion. Even when moving into talking about the text, conversation kept reverting back to the diagram (Fig. 8). To me it is clearly labelled, but that was not an opinion shared by all. It shows extremes of audience at the corners, defining the literary field in terms of its market, and Bourdieu develops the diagram further by placing the forms and genres commonly produced and read in late nineteenth-century France within that field. He shows the socio-political alignments of these forms, with the effect of making some surprising tacit claims for overlapping readerships and modes of production: I remain unsure, for example, whether I would include journalism and cabaret under what appears to be a "novel" column, but on the axes presented here they seem correctly placed. This attempt at non-geographical mapping, based on both qualitative and quantitative analysis, shows another way forward into developments of literary mapping in different directions and by different critical and representational methods. For example, a current project on *Mapping Metaphor* at the University of Glasgow is an ambitious attempt to create links between different linguistic and representational categories.

5 A Literary Cartography?

Franco Moretti (1998, 2005) has made the most concerted effort in literary criticism to use maps and diagrams, and he continues to push at reading's boundaries (2013). He maintains that maps can play a valuable role in charting the development and the social role of literature locally, nationally and internationally, and also in analysing the way authors represent and use space. Moretti (2005: 2) acknowledges his methodological links with Bourdieu, and asserts his desire to "place the literary field literally in front of our eyes." He chooses to draw his own maps rather than merely read existing ones, albeit acknowledging his reliance on cartographer Serge Bonin to do so (Moretti 1998: 8), and agitates for the use of maps "not as metaphors, and even less as ornaments of discourse, but as analytical tools: that dissect the text in an unusual way, bringing to light relations that would otherwise remain hidden" (Moretti 1998: 3). In his development of the project, Moretti (2005: 1–2) states that his aim is to move "from texts to models, then; and models drawn from three disciplines with which literary studies have had little or no interaction: graphs from quantitative history, maps from geography, and trees from evolutionary theory." Not only does he want to map, but to use other graphical representations; maps and graphs are naturalised, easily legible forms from school teaching. Trees require more work to interpret, but offer the opportunity to create "a theory—of diversity" (Moretti 2005: 30). Indeed, the only diagram in Susan Stanford Friedman's *Mappings: Feminism and the Cultural Geographies of Encounter* (1998: 85) is a mapping of hybridity theory. Moretti's desire to reach out and make new connections is laudable. The potential value can be seen in Fig. 9, which

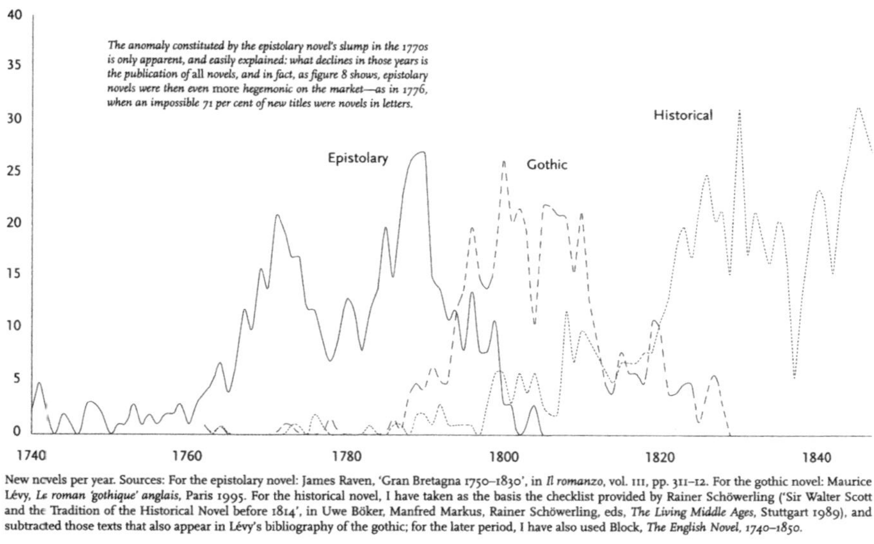

Fig. 9 British hegemonic forms, 1760–1850 in Franco Moretti, *Graphs, Maps, Trees: abstract models for a literary history* (London: Verso 2005: 15)

shows dramatically the shifts in genre which take place. The graph poses questions further to Moretti's thesis about the cyclical nature of genre hegemony. To take a different historical viewpoint, what happens to create these hegemonies? Why is the epistolary novel so dominant in the late 1780s? What happens at the points of intersection on the graph? Is there a crossing-over period of forms when novels contain aspects of both? Part of the problem with such sweeping surveys is that they raise as many questions as they answer, and the data still needs knowledgeable human interpretation. However, there is clearly a value in itself to charting those developments and being able to raise such questions.

Moretti's project has attracted attention from across the academy, and has been criticised in a variety of quarters, from conservative literary critics, to those who sympathise with Moretti's approach but do not agree with the specifics, to those working in the social sciences who Moretti is so keen to embrace (Goodwin and Holbo 2011). Thacker (2005–6: 60) comes in the second category, and particularly criticises Moretti's "faith in the objectivity of maps," while recognising the value of his *Atlas* as "a key work in the development of a new literary geography" (Thacker 2005–6: 61). David Harvey (2001: 86) dismisses Moretti in a footnote as banal and reductive. However, Moretti's methodology steps away from individual academic work. For the quantitative assessment of taxonomic categories he relies on scholarship from different eras, stretching from the early part of the twentieth century, up to the literary criticism of almost a hundred years later, and consequently with different standards and levels of access to information which might skew the sample; the source material, of novels from a range of European countries and in a variety of languages and genres, is totalised in the presentation. The more generalised claims which Moretti values (though not exclusively) require an extreme breadth of knowledge, which would necessitate a major shift in the way of working in literary studies. As he notes, "quantitative work is *truly* cooperation" (Moretti 2005: 5). To do large-scale analysis of publication and historical data would require either a markedly reduced output in the effort to gather information, a change in the funding and progression model for the humanities in order that undergraduates and postgraduates would serve an apprenticeship, becoming subsidiary names on research papers as in the sciences, or a willingness to work together in ways which literary studies does all too rarely. There are co-edited collections, of course, but these mostly divide labour in order to create the end product. There is also an aesthetic issue: as readers of literature, we are used to reading flowing text, but when maps and/or images are included, the reader has to choose when to stop reading the text and interpret the diagram. This is perhaps a failing of Moretti's own text: there are no clearly-marked breaks at which to consider the provocative diagrams he includes. Figure 9 is referenced in the middle of the sentence, and the reference also alludes to a further figure overleaf. When diagrams are orientated differently to the text it adds a further level of rupture between the text and the diagram which is unproductive. This is an issue rooted in publishing considerations, where in-line diagrams are more costly to set and process, as are colour plates. This volume itself has two different versions: the print version has inline black and white images with a separate section for a small number of colour plates; the online version includes all images in colour.

6 Challenges for Literary Criticism

Literary critics need to shed some inhibitions about the use of maps, as I have argued throughout. This does not mean being careless, but means using them while being aware of the issues. The notion of a literary geography (Thacker 2005; Hones 2008, 2014; Saunders 2010) is a useful starting point for a literary cartography. Only Thacker directly addresses the importance of the map, and then with reservations which I outlined above. In positing some tenets of a "*critical* literary geography" he states his desire "to stress the distance from an effortless mapping of represented landscapes in literary texts, and to raise more complex questions about space and power, and how space and geography affect literary forms and styles" (Thacker 2005: 60; his italics; see also Bushell 2012). The description of "effortless mapping" is perhaps a misnomer. It would be better to think about unreasoned mapping, cartography which is not clear about its underlying processes and methodologies, or simplistic representation, which plots geographic location without relational analysis. Graham Huggan (1994: 31) was an early critic to test the boundaries between cartography and literary criticism, and he notes astutely that "the function of maps and mapping strategies in literary texts is [...] frequently ambivalent: maps may be simultaneously perceived as useful tools and dangerous weapons." He focuses on the study of maps *in* literature, as discussed above in the example from *The Riddle of the Sands*, but his careful study of this trope also offers us a foundation for making maps *of* literature and analysing them. Huggan (1994: 32–33) posits the following "First Principles for a Literary Cartography":

- Basic concepts of the map; definition of theoretical positions, with particular reference to theories of spatial representation.
- Discussion of the historical and political significance of the map; of mapping, with reference to theories of territoriality and marginality.
- Delineation of the map's function within the literary text; exemplification of these functions and discussion of their syntagmatic (internal) and paradigmatic (external) operations.
- Analysis of the map topos in different literatures, with particular reference to literatures or literary forms considered 'marginal'; evolution and contemporary manifestations of the map topos in these literatures.

The necessity for a clear understanding of cartographic theory and cultural specificity is highlighted, as is the question of function; that question must be extended, for literary mapmaking, to how (and whether) the fictional can be plotted usefully for the purpose of analysis—as some scholars are starting to do.

Table 1 Spatial elements of a fictional text

Category	Explication/definition
Setting	Where the action takes place (i.e. a house, a village)
Zone of action	Several settings combined (i.e. a whole city, a region)
Projected space	Characters are not present there, but are dreaming of, remembering, longing for a specified place
Marker	A place which is mentioned, but not part of the categories above; markers indicate the geographical range and horizon of a fictional space
Route	Along which characters are moving by foot, by train, or on horseback, etc.

Source Barbara Piatti et al., Spatial elements of a fictional text from Mapping Literature: Towards a Geography of Fiction in William Cartwright et al. (eds.), *Cartography and Art* (Berlin: Springer 2009: 183)

The work of the *Literary Atlas of Europe* project, led by Barbara Piatti, has developed the work of literary cartography, taking as model regions Prague, Northern Frisia and Lake Lucerne. Moretti is acknowledged as an inspiration (Piatti et al. 2009: 180). Piatti, a geographer working across disciplines, acutely breaks down textual places into five key spatial entities (Table 1): setting, projected space, zone of action, marker and route (Piatti et al. 2009: 183). These categories differentiate effectively between types of reference in the text, and provide the basis for a rigorous literary mapping. Where Moretti uses maps in a variety of ways to suit the purpose of criticism, Piatti aims to form an overarching theory of literary geography and cartography; the work of analysis comes 'in the end, as a last step' (2009: 191). I argue, after Moretti, that the work of analysis and the creation of maps should interlink dynamically, remaining attentive to complexity and nuance whilst guarding against the reductiveness about which Thacker is concerned. For Piatti, 'the core of the project [...] is a database that stores information about fictional spaces' (2009: 189), but there are methodological questions to be answered about the efficacy of literary databases in some of the maps she creates (Fig. 10). The cultural and temporal specificity of the text needs to be acknowledged, in spite of the interest in mapping over centuries. On what do we map, if doing so over long periods? Can fourteenth-century texts be mapped alongside twenty-first century texts? Can they be understood as part of the same geography, or do we need to find a way to indicate changes in space over that length of time? Even, on a more fundamental level, does such a focus on location restrict the work of interpretation? Geocritics who are not cartographers *per se* such as Westphal (2011) would not see this as a problem, but I suggest that when making maps these are issues of methodology that deserve further attention. Another issue is the troubling echo of the intentional fallacy (Wimsatt and Beardsley 1954) in the drive to include data about the author (Piatti and Hurni 2011: 220; Reuschel and Hurni 2011: 295; see also Brosseau 1994: 336–7). Does it matter whether a distortion was intentional, and can we ever tell? The intention of the author can never be recovered, and the theorist Walter Benjamin (1999: 875) cautioned: "Never trust what writers say

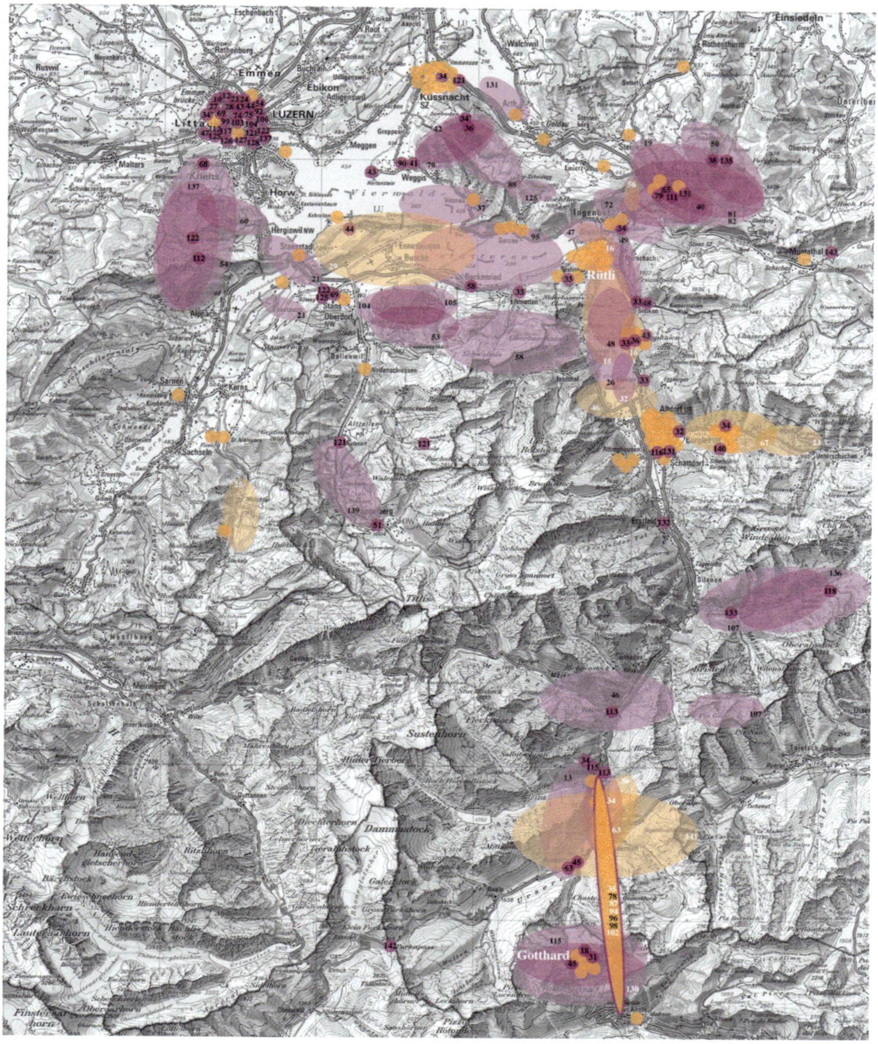

Fig. 10 Barbara Piatti, Historically unbound sites in *Die Geographie der Literatur* (Berlin: Wallstein 2008)

about their own writings." Once dissonance and distortion is identified, it can be analysed in terms of spatial politics, or compared with other texts; there is no need to resort to quasi-factual or counterfactual constructions of what the author might have meant.

Despite these reservations, Piatti's is much the best effort to map literature to date; in every theory there are such issues, and her work is invaluable in starting to bring literary cartography towards theoretical maturity. There are the beginnings of solutions to these problems in her work already. She and other researchers on the project set out clearly the principles for an up-to-date method of mapping: Reuschel and Hurni (2011: 294) argue that "in order to explore the multi-dimensional geography of literature, it is essential to exploit the capabilities of database enhanced, digital, interactive cartography." The backing of the database allows the mapping not only of the single text, but also of literary regions, both fictional and geospatial. The dynamic nature of interactive technologies can also show change over time clearly, adding rendering and movement to the two-dimensional map. This poses a problem for the enduring methods of dissemination of scholarly criticism, the printed book and journal. E-books are still widely viewed as a lesser cousin of print publishing, but a change in valuation seems likely to come soon. It is also worth considering how the development of three-dimensional televisual displays will impact on mapping. Maps might also show density, "hot-spots" (Reuschel and Hurni 2011: 298), or frequency through distortion, as Piatti and Hurni suggest (2011: 221), although I am uncomfortable with the qualitative judgment being derived wholly from quantitative analysis as they seem to imply in talking about "settings [which] bear *most meaning*" (their italics). Piatti and Hurni (2011: 221) are conscious, at any rate, that "for the prototype of the literary atlas it will be crucial to combine qualitative and quantitative evidence and to switch effortlessly back and forth between them." I retain my suspicion of any "effortless" critical work, but a combination of these methods seems absolutely vital. A closer link between cartography and the theories of geocriticism led by Westphal and Robert T. Tally, who also translated Westphal's work, seems likely to be fruitful. Tally's *Literary Cartographies* (2014: 1) collects a wide-ranging and convincingly-theorised series of essays that take as their starting point that "maps presuppose narratives, which in turn may function as maps." But again, this is a cartography without visualization. To continue an earlier metaphor, this seems an interdisciplinary fence that needs to be kicked down.

As we have seen above, some of the next steps in both literary geography and cartography are already being taken. Phillips and McCracken (2005) offer an excellent summary of a wide range of work on space in the introduction their special issue of *new formations*, and special issues of *The Cartographic Journal* (2011) on "Cartographies of Fictional Worlds" (2011) and "Cartography and Narratives" (2014), edited by Piatti and Hurni and Sebastian Caquard and William Cartwright respectively, show the range of recent research being conducted in the area. The first issue of a new journal of *Literary Geographies* appeared in 2015, and an associated blog maintains a thorough bibliography of recent work. The discussion of literary space is being visited and revisited increasingly fruitfully as the field develops, often underpinned by the theories of Lefebvre and work which follows him, led by David Harvey (1989) and Edward Soja (1989) and followed closely by Jameson (1991), although it is ironic that these key theorists do not engage with the pragmatics of visual representation. Lefebvre (1991 [1974]) sees space not as an empty container in which activity takes place, or even a constantly written palimpsest, but argues that "*(Social) space is a (social) product*" (1991 [1974]: 26). He sees space in a tripartite structure: "spatial practice," the experience of the social and physical world which ensures continuity and the cohesion of systems; "representations of space," effectively official and institutional spaces, broadly conceived; and the "representational spaces" of unofficial, artistic and cultural texts and interpretations (1991 [1974]: 33). Drawing on this, Thacker (2005: 63) develops Huggan's points by stating a need to "reconnect the representational spaces in literary texts not only to the material spaces they depict, but also reverse the movement, and understand how social spaces dialogically help fashion the literary forms of texts," summarising pithily: "literary texts represent social spaces, but social space shapes literary forms." (see also Sorum 2009) Where texts present fictionalised versions of real cities, or are set in named cities, it is also productive to ask if and how the text differs from the real, how spaces are re-shaped or how routes are altered, how the claims made in the text question or reaffirm structures of power. Literary spaces can mean not only those of the text, but also of writer and reader (Hones 2008; Saunders 2010), the coming together across distance of figures inside and outside the text itself.

There is a clear need for fully interdisciplinary work. It can be difficult enough to speak across intradisciplinary boundaries, as Bracken and Oughton (2006) point out in relation to human and physical geography. As Phillips and McCracken (2005) note, it is difficult to speak effectively across borders, however mutable they might be, and it is something at which we still need to work harder (see also Hones 2008: 1307). Words must necessarily remain the lifeblood of literary criticism and literary studies, but surely as cartographers become more prepared to engage with literary texts, and contemporary authors such as Paul Auster, Alasdair Gray, W.G. Sebald and others become more prepared to use typography and intertextual images, it is only right that literary critics should be more prepared to use cartography—its history, metaphorical language, and the images themselves—to enhance the work of interpretation for the twenty-first century.

Further Reading

The Cartographic Journal, 48(4) (2011). Special edition edited by Barbara Piatti and Lorenz Hurni. Consists primarily of contributions from Piatti and her cohort, whose work is difficult to find in English; her book *Die Geographie der Literatur* (2009) offers the most convincing and extensively theorised take on literary cartography.

Huggan, G. (1994). *Territorial Disputes: Maps and mapping strategies in contemporary Canadian and Australian fiction*. Toronto: University of Toronto Press. One of the early works in postcolonial studies to explicitly address the trope of mapping, and Huggan posits his "First Principles for a Literary Cartography" in his introduction.

Jameson, F. (1991). *Postmodernism: Or, the cultural logic of late capitalism*. London: Verso. Jameson's theory of postmodernism builds on the work done by Harvey (1989) and Soja (1989) and insists on the importance of space and materialism in interpretation. The final parts of the long conclusion deal with cognitive mapping.

Moretti, F. (2005). *Graphs, Maps, Trees: Abstract models for a literary history*. London: Verso. The most extensive attempt so far to use visual representations of place and data in literary criticism. Its idiosyncracies have provoked some criticism, and it is certainly a text which stimulates debate.

New Formations, 57 (2005). Special edition edited by Richard Phillips and Scott McCracken. Contains a range of criticism at the intersection of literature and geography written by academics from both disciplines, and includes Thacker's valuable "The Idea of a Critical Literary Geography".

References

Barnett, C. (1998). The cultural turn: Fashion or progress in human geography? *Antipode, 30*, 379–394.

Benjamin, W. (1999). *The arcades project*. Cambridge, MA: Harvard University Press.

Berland, J. (2005). After the fact: Spatial narratives in the Canadian imaginary. *New Formations, 57*, 39–55.

Borges, J. L. (1975). On exactitude in science. In *A universal history of infamy* (p. 131). Harmondsworth: Penguin.

Bourdieu, P. (1993). *The field of cultural production*. Cambridge: Polity.

Bourdieu, P. (1984 [1979]). *Distinction: A social critique of the judgement of taste*. London: Routledge & Kegan Paul.

Bracken, L., & Oughton, E. (2006). 'What do you mean?' The importance of language in developing interdisciplinary research. *Transactions of the Institute of British Geographers, 31*, 371–382.

Bradbury, M. (Ed.). (1996). *The atlas of literature*. New York: de Agostini.

Brosseau, M. (1994). Geography's literature. *Progress in Human Geography, 18*, 333–353.

Bulson, E. (2006). *Novels, maps, modernity: The spatial imagination, 1850–2000*. London: Routledge.

Bushell, S. (2012). The slipperiness of literary maps: Critical cartography and literary cartography. *Cartographica, 47*(3), 149–160.

Childers, E. (2007 [1903]). *The riddle of the sands: A record of secret service*. London: Penguin.

Conley, T. (1997). *The self-made map: Cartographic writing in early modern France*. Minneapolis: University of Minnesota Press.

Conrad, J. (2007 [1899]). *Heart of darkness*. London: Penguin.

Conrad, J. (1924). Geography and some explorers. *National Geographic Magazine, 45*, 241–274.

Coroneos, C. (2002). *Space, Conrad, and modernity*. Oxford: Oxford University Press.

Dear, M., Ketchum, J., Luria, S., & Richardson, D. (Eds.). (2011). *GeoHumanities: Art, history and text at the edge of place*. London: Routledge.

Deleuze, G., & Guattari, F. (1988). *A thousand plateaus: Capitalism and schizophrenia*. London: Athlone.

DiBattista, M., & McDiarmid, L. (Eds.). (1996). *High and low moderns: Literature and culture 1889–1939*. Oxford: Oxford University Press.

Draper, J., & Fowles, J. (1984). *Thomas Hardy's England*. London: Jonathan Cape.

Friedman, S. S. (1998). *Mappings: Feminism and the cultural geographies of encounter*. Princeton, NJ: Princeton University Press.

Friel, B. (1981). *Translations*. London: Faber and Faber.

Fuson, B. (1970). At halfway point: State literary maps. *The English Journal, 59*(1), 87–98.

Gikandi, S. (1996). *Maps of Englishness: Writing identity in the culture of colonialism*. New York: Columbia University Press.

Gilbert, P. (2009). Sex and the modern city: English studies and the spatial turn. In B. Warf & S. Arias (Eds.), *The spatial turn: Interdisciplinary perspectives* (pp. 102–121). Abingdon: Routledge.

GoGwilt, C. (1995). *The invention of the west: Joseph Conrad and the double-mapping of Europe and empire*. Stanford, CA: Stanford University Press.

Goodwin, J., & Holbo, J. (Eds.). (2011). *Reading graphs, maps, trees: Critical responses to Franco Moretti*. Anderson, SC: Parlor Press.
Greimas, A. J., & Rastier, F. (1968). The interaction of semiotic constraints. *Yale French Studies, 41*, 86–105.
Habermann, I., & Kulm, N. (2011). Sustainable fictions—Geographical, literary and cultural intersections in J.R.R. Tolkein's *Lord of the Rings*. *The Cartographic Journal, 48*(4), 263–273.
Hand, E. (2011). Culturomics: Word play. *Nature, 474*, 436–440.
Harley, J. B. (2009 [1989]). Deconstructing the map. In M. Dodge (Ed.), *Classics in cartography: Reflections on influential articles from Cartographica* (pp. 273–294). Hoboken, NJ: Wiley-Blackwell.
Harpold, T. (2005). Verne's cartographies. *Science Fiction Studies, 32*(1), 18–42.
Harvey, D. (1989). *The condition of postmodernity: An enquiry into the origins of cultural change*. Oxford: Blackwell.
Harvey, D. (2001). The cartographic imagination: Balzac in Paris. In V. Dharwadker (Ed.), *Cosmopolitan geographies: New locations in literature and culture* (pp. 63–87). New York and London: Routledge.
Hawkins, D. (1983). *Hardy's Wessex*. London: Macmillan.
Hegglund, J. (2012). *World views: Metageographies of modernist fiction*. Oxford: Oxford University Press.
Hewitt, R. (2010). *Map of a nation: A biography of the ordnance survey*. London: Granta.
Hones, S. (2008). Text as it happens: Literary geography. *Geography Compass, 2*(5), 1301–1317.
Hones, S. (2014). *Literary geographies: Narrative space in let the great world spin*. Basingstoke: Palgrave Macmillan.
Huggan, G. (1994). *Territorial disputes: Maps and mapping strategies in contemporary Canadian and Australian fiction*. Toronto: University of Toronto Press.
Huyssen, A. (1988). *After the great divide: Modernism, mass culture, postmodernism*. Basingstoke: Macmillan.
Jameson, F. (1988). *Nationalism, colonialism and literature: Modernism and imperialism*. Derry: Field Day.
Jameson, F. (1991). *Postmodernism: Or, the cultural logic of late capitalism*. London: Verso.
Jameson, F. (2003). The end of temporality. *Critical Inquiry, 29*(4), 695–718.
Jameson, F. (2005). *Archaeologies of the future: The desire called utopia and other science fictions*. London: Verso.
Kane, M. (1999). *Modern men: Mapping masculinity in English and German literature, 1880–1930*. London: Cassell.
King, G. (1996). *Mapping reality: An exploration of cultural cartographies*. Basingstoke: Macmillan.
Klein, B. (2001). *Maps and the writing of space in early modern England*. Basingstoke: Palgrave.
Latour, B., & Woolgar, S. (1986). *Laboratory life: The construction of scientific facts* (2nd ed.). Princeton, NJ: Princeton University Press.
Lefebvre, H. (1991 [1974]). *The production of space*. (Trans.) Donald Nicholson-Smith. Oxford: Blackwell.
Lilley, K. (Ed.). (2014). *Mapping medieval geographies: Geographical encounters in the latin west and beyond, 300–1600*. Cambridge: Cambridge University Press.
Lindop, G. (1993). *A literary guide to the Lake District*. London: Chatto & Windus.
Lynch, K. (1960). *The image of the city*. Cambridge, MA: MIT Press.
Mao, D., & Walkowitz, R. (2006). *Bad modernisms*. Durham, NC: Duke University Press.
Matei-Chesnoiu, M. (2012). *Re-imagining western European geography in English renaissance drama*. Basingstoke: Palgrave Macmillan.
Matei-Chesnoiu, M. (2015). *Geoparsing early modern English drama*. Basingstoke: Palgrave Macmillan.
McCracken, S. (2007). *Masculinities, modernist fiction and the urban public sphere*. Manchester: Manchester University Press.

Monmonier, M. (1993). *Mapping it out: Expository cartography for the humanities and social sciences*. Chicago, IL: University of Chicago Press.
Monmonier, M. (2004). *Rhumb lines and map wars: A social history of the Mercator projection*. Chicago, IL: University of Chicago Press.
Moretti, F. (1998). *Atlas of the European novel 1800–1900*. London: Verso.
Moretti, F. (2005). *Graphs, maps, trees: Abstract models for a literary history*. London: Verso.
Moretti, F. (2013). *Distant reading*. London: Verso.
Muecke, S. (1992). *Textual spaces: Aboriginality and cultural studies*. Kensington: New South Wales University Press.
Muehrcke, P., & Muehrcke, J. (1974). Maps in literature. *Geographical Review, 64*(3), 317–337.
Ousby, I. (1990). *Blue guide to literary Britain and Ireland*. London: Black.
Padrón, R. (2007). Mapping imaginary worlds. In J. Akerman & R. Karrow Jr. (Eds.), *Maps: Finding our place in the world* (pp. 255–287). Chicago, IL: University of Chicago Press.
Perkins, C. (2003). Cartography: Mapping theory. *Progress in Human Geography, 27*(3), 341–351.
Peters, J. (2004). *Mapping discord: Allegorical cartography in early modern French writing*. Newark: University of Delaware Press.
Phillips, R. (1997). *Mapping men and empire: A geography of adventure*. London: Routledge.
Phillips, R., & McCracken, S. (2005). Editorial. *New Formations, 57*, 7–9.
Philo, C. (1991). *New words, new worlds: Reconceptualising social and cultural geography*. Lampeter: St. David's University College.
Piatti, B., Bär, H. R., Reuschel, A.-K., Hurni, L., & Cartwright, W. (2009). Mapping literature: Towards a geography of fiction. In W. Cartwright, G. Gartner, & A. Lehn (Eds.), *Cartography and art* (pp. 177–192). Berlin: Springer.
Piatti, B., & Hurni, L. (2011). Cartographies of fictional worlds. *The Cartographic Journal, 48*(4), 218–223.
Piper, K. (2002). *Cartographic fictions: Maps, race, and identity*. New Brunswick, NJ: Rutgers University Press.
Quinones, R. (1985). *Mapping literary modernism: Time and development*. Princeton, NJ: Princeton University Press.
Radford, A. (2010). *Mapping the Wessex novel: Landscape, history and the parochial in British literature, 1870–1940*. London: Continuum.
Reuschel, A.-K., & Hurni, L. (2011). Mapping literature: Visualisation of spatial uncertainty in fiction. *The Cartographic Journal, 48*(4), 293–308.
Richterich, A. (2011). Cartographies of digital fiction: Amateurs mapping a new literary realism. *The Cartographic Journal, 48*(4), 237–249.
Sanders, J. (2011). *The cultural geography of early modern drama, 1620–1650*. Cambridge: Cambridge University Press.
Sanford, R. (2002). *Maps and memory in early modern England: A sense of place*. New York: Palgrave.
Saunders, A. (2010). Literary geography: Reforging the connections. *Progress in Human Geography, 34*, 436–452.
Smith, N., & Katz, C. (1993). Grounding metaphor: Towards a spatialized politics. In M. Keith & S. Pile (Eds.), *Place and the politics of identity* (pp. 67–83). London: Routledge.
Soja, E. (1989). *Postmodern geographies: The reassertion of space in critical social theory*. London: Verso.
Sorum, E. (2009). "The place on the map": Geography and meter in Hardy's elegies. *Modernism/Modernity, 16*(3), 553–574.
Tally, Robert T., Jr. (Ed.). (2014). *Literary cartographies: Spatiality, representation, and narrative*. Basingstoke: Palgrave Macmillan.
Thacker, A. (2003). *Moving through modernity: Space and geography in modernism*. Manchester: Manchester University Press.
Thacker, A. (2005–6). The idea of a critical literary geography. *New Formations, 57*, 56–73.

Thacker, A., & Brooker, P. (2005). Introduction. In A. Thacker & P. Brooker (Eds.), *Geographies of modernism: Literatures, cultures, spaces* (pp. 1–5). London: Routledge.
Torget, Andrew J., & Christensen, Jon. (2012). Mapping texts: Visualizing American historical newspapers. *Journal of Digital Humanities, 1*(3), 1–3.
Westphal, B. (2011 [2007]). Geocriticism: Real and fictional spaces. Translated by Robert T. Tally Jr. Basingstoke: Palgrave Macmillan.
Wimsatt Jr., W. K., & Beardsley, M. C. (1954). The intentional fallacy. In *The verbal icon: Studies in the meaning of poetry* (pp. 3–18). Lexington: University Press of Kentucky.
Wood, D., & Fels, J. (2009 [1986]). Designs on signs/myth and meaning in maps. In M. Dodge (Ed.), *Classics in cartography: Reflections on influential articles from Cartographica* (pp. 209–260). Hoboken, NJ: Wiley-Blackwell.
Wood, D., & Krygier, J. (2009). Ce n'est pas la monde. In M. Dodge, R. Kitchin, & C. Perkins (Eds.), *Rethinking maps: New frontiers in cartographic theory* (pp. 189–219). London: Routledge.

Music, Maps and the Global Jukebox: Culture Areas and Alan Lomax's Cantometrics Projects Revisited

John R. Gold, George Revill and Daniel Grimley

Abstract The activity of mapping culture areas has acquired notoriety in academic circles on account of the deterministic and nationalistic agendas with which that activity has often been associated. Yet such cartographic exercises need not only serve the narrow agendas of specific groups; they can also serve as creative and imaginative instruments for enhancing cultural understanding and advancing notions of equity. To illustrate this contention, this chapter considers the theory of 'cantometrics' devised by the American folk collector and musicologist Alan Lomax (1915–2002). 'Cantometrics,' which literally meant 'measurement of song,' was first suggested in 1959 and developed in the 1960s as an approach that effectively sought to define and categorise world music through mapping folksong styles and culture. In this chapter, which contains five main sections, we examine the nature and imputed meaning of the cartographic output from Cantometric analysis, especially the mapping of the world into 56 culture areas, as well exploring the products of the subsequent, computerised "Global Jukebox," with its proto-GIS cultural system. In outline, the first two sections supply contextual background about the theorising of folk music in relation to culture areas and about Lomax's work up to the mid-1950s. The third section analyses the nature, characteristics and flaws of cantometric inquiry, with the ensuing part commenting on the parallels between cantometrics and cognate areas of geographical inquiry. The

J.R. Gold (✉)
Department of Social Sciences, Oxford Brookes University, Gipsy Lane, Oxford OX3 0BP, UK
e-mail: jrgold@brookes.ac.uk

G. Revill
Department of Geography, Open University, Milton Keynes MK7 6AA, UK
e-mail: george.revill@open.ac.uk

D. Grimley
Faculty of Music, University of Oxford, St Aldates, Oxford OX1 1DB, UK
e-mail: daniel.grimley@music.ox.ac.uk

S.D. Brunn and M. Dodge (eds.), *Mapping Across Academia*,
DOI 10.1007/978-94-024-1011-2_14

conclusion briefly reflects on the significance and implications of Lomax's theory for the comparative study of world music.

Keywords Alan Lomax · Folk music · World music · Cantometrics · Culture areas · Equity

1 Introduction

The history of 'world music' can be told in two, very different ways. Defined *strictu senso*, its origins can be precisely traced to a meeting held at the Empress of India public house in North London on 29 June 1987. The meeting, which was attended by representatives of the music industry and other 'interested parties,' chose the term 'World Music' over contenders such as 'Tropical Music,' 'World Beat' and 'Hot Music' as a convenient marketing label for a rapidly expanding range of folk and fusion styles (Taylor 1997; Connell and Gibson 2003: 144–149; fRoots 2011). Yet as a broad genre, 'world music' *sensu lato* can claim far older antecedents. Bohlman (2002), for example, traces its beginnings back to eighteenth century theories of folk culture and thereafter to a diverse assortment of musical practices that extend from phonograph collections of so-called 'exotic' and 'oriental' music to the sonic extravaganzas of the Eurovision Song Contest and the development of ethnomusicology as an academic discipline.

Understandably, any prehistory of world music has its roll call of movers and shakers (Nidel 2004; Bohlman 2012), but few have contributed more than Alan Lomax (1915–2002). As a collector, field researcher and ethnomusicologist, he strongly influenced American folk and blues music in the 1940s and was subsequently prominent in promoting the folk revival of the 1950s and 1960s. As a gifted writer, he offered accessible accounts of the history and development of folk music, illuminated by the acuity of an insider's view. As a broadcaster, concert organiser, performer and political activist, he acted as an advocate and populariser, bringing a wide variety of roots music styles together in single events, venues and programmes. Finally, Lomax had an enduring academic interest in creating taxonomies of world music, most notably through the cantometrics project that he developed in collaboration with the musicologist Victor Grauer, the sociologist Conrad Arensberg and others in the 1960s.

In this chapter, we focus on cantometrics as an approach that effectively sought to define and categorise world music through mapping folksong styles and culture. At the outset, it must be stressed that this work largely bypassed the attention of geographers at the time despite, first, having sympathetic resonances with the quantitative-theoretical and behavioural approaches that then characterised human geography and, secondly, sharing North American cultural geographers' interests in the cultural hearths, cores, peripheries, cultural diffusion and migrant identities associated with music (e.g. see Carney 1990, 1998, 2003). Nevertheless, its development still bears scrutiny by geographers in terms of its innovatory approaches to the quantitative study of culture and the mapping of culture areas and for its potential in offering insight into the intellectual climate of a period of paradigmatic change in the social sciences.

The chapter contains five main sections. The two opening parts supply contextual background about the theorising of folk music in relation to culture areas and about Lomax's work up to the mid-1950s. The third section analyses the nature, characteristics and flaws of cantometric inquiry, with the ensuing part commenting on the parallels between cantometrics and cognate areas of geographical inquiry. The conclusion briefly reflects on the significance and implications of Lomax's theory for the comparative study of world music.

2 Culture Areas and Theories of 'the Folk'

Commentators have considered folk music to be intimately related to the localities in which it is practised (Nettl 1983), but in the present context the conceptual framework that exerted greatest historic influence was derived from the anthropological theory of *Kulturkreislehre* or 'cultural circles' (Bohlman 1988: 55; Kluckhohn 1936). Imbued with the pervasive fascination with cultural history that characterised Germany or, strictly speaking, German-speaking lands, since interest in the idea of the Volk pre-dated unification in 1871, *Kulturkreislehre* theorists in the nineteenth century sought to reconstruct cultural history in cases where written records were absent by postulating a spatialised conception of cultural development over time (Andriolo 1979: 133–144). Folk music, which occupied a prominent place in the analysis, was thereby seen as grounded in the specificities of place.

The key source for such ideas was von Herder's (1778, 1779; later 1880) two-volume folk song collection *Stimmen der Völker in Liedern*. Herder's work contains an important duality. It is commonly considered to provide the foundations for the collection of folk music in the service of nation-building (e.g. Skultans 1998; Francmanis 2002). Yet it is also seen as pointing the way to the first examples of world, rather than simply nationalistic, music (Bohlman 2002: 40). This apparent contradiction can be understood by disentangling Herder's ideas about 'the folk' (Bohlman 1988, 2002). When using that expression, Herder meant something akin to the term 'ordinary people'; an Enlightenment ideal of universal humanity and a conception that transcended national or other boundaries. Music as a universal attribute of the human condition provided a means of bringing together 'enlightened human beings' (Bohlman 1988: 38). Yet, there was 'ambivalence' in Herder's work, which helps explain 'how he manages to appear both liberal and proto-fascist' (Young 1995: 42). The way that Herder grounded human value in the specificity of language, for instance, makes it easy to interpret the idea of 'folk' in a more exclusionary and hierarchical manner. Certainly, many writers, musicians, politicians who read his work emphasised notions of locality, nation, and homogeneity of race and culture, and extracted anti-Enlightenment ideas about relativism, difference and the superiority of German culture (Bluestein 1972: 11).

Nevertheless, Bohlman (1988: 38) argues that universality and nationalism were not antithetic for Herder but aspects of an essentially democratic vision forged in a German political context reflecting the active unification of individual states and principalities. He therefore suggests that Herder was arguing for individual regions, or cultures, to each have a specific and valued 'voice' within an ecumenical sense of national or, indeed, global cultural difference.

Since the mid-1980s, scholars have pointed to the failure of theories such as *Kulturkreislehre* as being reductive even when explaining the development of folk music in all but the most isolated of societies (e.g. Pickering and Green 1987a; Bohlman 1988; Clayton 2003: 64–67; Aubert 2007). They challenge the idea of place-based folk music grounded in oral tradition and community participation at both theoretical and empirical levels. Here, one can point to important debates within the humanities and social sciences concerning both the 'invention of tradition' and the politics of classificatory systems. If conventional theories of folk music production rely on bounded notions of community and static and/or monumentalising conceptions of place, then there are clearly conceptual problems to be addressed when accounting for the cultural practices of modernising, urban and industrial populations for whom large-scale migration, industrialisation and commercial popular culture are part of everyday life. In such circumstances, the development of world music and the related proliferation of roots-based fusion styles confront the very senses of authenticity, purity and stability which make such styles seem so exotic and appealing to many audiences.

With these points in mind, Pickering and Green (1987b) developed the concept of 'vernacular culture' to account for the heterogeneity of everyday folk culture practices (cf. Richards 1992; Revill 2005). More recent theorists have drawn on the concept of hybridity, derived from post-colonial theory (Bhabha 1994), to shift attention away from theoretical and idealised constructs of cultural purity and instead embrace instead the messy and eclectic nature of lived experience (Gilroy 1993; Born and Hesmondhalgh 2000; Radano and Bohlman 2000; Biddle and Knights 2007). Here, spaces are characterised by flow and connection, with boundaries shaping culture and identity through contestation and transgression rather than exclusivity and stasis. At the same time, heterogeneous cultural practices may be interpreted as sometimes conscious and sometimes strategic sets of borrowings and adaptations rather than as evidence for cultural degeneration (Revill 2005).

Cantometrics, when seen against this background, occupy an ambiguous position. Its typical approach—isolating and classifying characteristics and then producing a typology of similarities and differences—harks back to culture area theories that conceived folk music as being produced within bounded places and static communities. At the same time, its attention to listening and music's performative characteristics looked forward to more fluid, improvisatory conceptions of cultural production.

3 The Politics of the Melting Pot

There is no doubting Alan Lomax's pivotal position in the history of folk music collection or, indeed, in the development of popular music during the mid-twentieth century (Szwed 2010). Born in Austin (Texas) in 1915, he was strongly influenced by working with his father John A. Lomax in developing the *Archive of Folksong* for the Library of Congress. Between 1933 and 1942, the Lomaxes undertook extensive trips across rural America, especially in Appalachia and the Deep South to assemble oral history interviews with blues, folk and gospel singers as well as making field recordings of their music. Alan Lomax undertook further collecting tours in these areas during the period 1959–1965 (Collins 2004), but added extensively to his range from the 1940s onwards, with trips to Italy and Spain, the United Kingdom, Ireland, the USSR, Romania, Morocco, the Bahamas, Haiti and the Eastern Caribbean. These trips eventually yielded an inventory now made available on over one hundred CDs (cf. http://www.culturalequity.org/alanlomax/ce_alanlomax_discography.php). In recognition of these efforts and other aspects of his work, Lomax was awarded the National Medal of the Arts by the United States Congress in 1986, and a Grammy by the Recording Academy in 2002 for his life-long contributions to music.

Though Lomax did not train as a professional musicologist, his activities as a folk collector and his later work on cantometrics were heavily influenced by his university education. He held a BA in Philosophy from the University of Texas at Austin (1936) and he undertook graduate study in Anthropology at Columbia University. While there, he attended and was inspired by the lectures of Curt Sachs, New York University's German-born professor of music (Cohen 2003: 98; Szwed 2010: 140–142). Sachs, along with the Austrian-born ethnomusicologist Erich Moritz von Hornbostel, had pioneered the creation of recorded archives embracing world musical traditions in the early twentieth century and developed wide-ranging theories of the history and diffusion of world musical styles. Lomax's radio broadcasts between 1946 and 1949, made for the Mutual Broadcasting System, typically showed the eclectic influence of Sachs and von Hornbostel. One programme entitled 'Dancing Around the World,' for example, juxtaposed recordings of Chicago blues, Appalachian and Cajun fiddling, Afro-Cuban music, West African drumming, Django Reinhardt's 'Djangology,' dances from Ireland, Finland, Israel, Armenia, a Russian chorus, a Balinese xylophone orchestra, klezmer wedding music and a Dixieland revival recording featuring Sidney Bechet and 'Wild Bill' Davison. Lomax's commentary included observations about instrumental and vocal styles, calling attention to the Middle Eastern influences apparent in a klezmer clarinet solo and the precise diction of the Finnish bandleader (Cohen 2003: 98–99; Szwed 2010: 233–234).

In many respects, the USA provided fertile ground for building the notion of world music. Waves of immigration, particularly from Europe and Africa, provided the raw material of a complex patchwork of folk music styles drawing from diverse cultural practices and traditions. Not only did folk music researchers believe it possible to find clear evidence of a multiplicity of migrant national folk music styles within North America, they also regarded the distinctive forms of music associated, say, with the Deep South, Appalachia or the Mid-West as active and creative fusions that incorporated music from countries of migrant origin into something uniquely new and American. In this way, it was believed that African-American Gospel music from the Deep South fused African practices of call-and-response with Protestant traditions of 'lining-out'—the alternate repeating of lines of hymn texts common in western Scotland (Dargen 2006). Similarly, the music of the 'Polka belt' of the upper Mid-West reflected musical styles brought to the USA by successive waves of European immigrant farmers. Based substantially around German and Scandinavian stylistic practices, the music also shows influences of Irish, Scottish and English music (Leary 2006).

These and similar fusions had long interested collectors, who habitually interpreted them in light of their favoured narratives. A generation earlier, for example, the English folk music collector Cecil Sharp interpreted the heterogeneity of Appalachian music through the lens of looking for traces or 'survivals' of past cultures (Gold and Revill 2006). Lomax certainly recognised the heterogeneity of Appalachian music as one of its fundamental characteristics and, with Sharp in mind, noted that:

> Collectors have gathered scores of ballads dating back to the late Middle Ages from mountain singers. These ancient ballads served to link the pioneers with their British homeland and to keep alive ancient patterns of emotion and poetry which beautified their lives. The country singers, however, did not regard them as historical documents, but as dramas which exemplified traits of character, both for good and for evil, that they perceived in themselves and their neighbours... Folk singers seldom make distinctions between old and new ballads, or indeed between lyric songs, comic pieces and the ballads so cherished by scholars. (Lomax 1997: 2)

Rather than trying to strip folk songs back to their 'European roots,' Lomax seemed more interested in the extent to which folk music represented the new experiences of people living in America. In 1941, he had argued that:

> The American singer has been concerned with themes close to his everyday experience, with the emotions of ordinary men and women who were fighting for freedom and for a living in a violent new world. His songs have been strongly rooted in his life and have functioned there as enzymes to assist in the digestion of hardship, solitude, violence, hunger, and honest comradeship of democracy. (Lomax et al. 1949: xiii)

Subsequently, he maintained that Appalachian folk music was the first *British* folk music since it was the first tradition to blend English, Irish, Scottish and Welsh traditions. Intriguingly, he claimed that Appalachia had democratised British music by purging its aristocratic and medieval tone to create a new purer, hybrid form. Lomax viewed this mixing as positive evidence for the democratic qualities of the American 'melting pot' at work. In 1960, he stated:

> American folk songs are, above all else, American. They are a mixture of English, Scottish, Irish, French and African influence stirred together in a way that could happen only in this magnificently heterogeneous country. (quoted in Calkins, 1960: 205)

For Lomax, Appalachian folk music's heterogeneity was a starting point for a project that was centrally concerned with developing a highly politicised conception of the American nation. His own politics bore the imprint of the Roosevelt era, even though he classed people with strong left-wing sympathies, such as Pete Seeger and Ewan McColl, as amongst his closest associates. As such, Lomax saw folk music as both a conduit by which the people might speak to the centre and as a means by which the USA might discover its democratic identity. In 1981, Lomax recalled that:

> …the Roosevelt period was not only one of political development, when for the first time America became conscious of its social responsibilities to the whole population. It was also a time when a rising interest in American culture flowered and bore fruit… The developing concern about what our own American culture was actually like, about who we were as people peaked at this time. And the search for American folk roots was a part of this…. The Roosevelts and the bright, young, intellectuals of the New Deal and Congress under Roosevelt's baton put their arms around the whole of American culture—minorities, ethnics, blacks, poor whites, Indians, coal miners, unemployed…. And it was partially on the power of that discovery that we could fight World War II. That self-discovery poured energy right into the bloodstream of the people and helped us lick the fascists. (quoted in Cohen 2003: 93)

Yet despite his sense of patriotism and duty, Lomax was investigated as a 'communist sympathizer' on various occasions from 1942 onwards, arousing particularly hostile scrutiny when placed in charge of campaign music for Henry A. Wallace's 1948 Progressive Party's Presidential candidacy. Subsequently, Lomax was listed in the publication *Red Channels* as a possible Communist sympathizer and was blacklisted from working in US entertainment industries (Gioia 2006; Szwed 2010: 189–191). His decision to spend much of the 1950s in London was undoubtedly influenced by wishing to be beyond the reach of the anti-Communist witch-hunts of that era.

Whatever the reason, it proved an immensely prolific period for Lomax. He acquainted the British public with the diversity of American folk music styles, while at the same time collecting British and European folk music. From the latter, he compiled the 18-part *Columbia World Library of Folk and Primitive Music*—an anthology issued on newly-invented LP records that drew upon a network of folk musicians and folklorists in Britain and continental Europe, particularly Italy and Spain. Rather than simply constituting exercises in preservation, the accumulated materials gave Lomax *as collector* a sense of the dynamism, complexity and transformation of specific traditions within a broad spatial and temporal framework and, *as performer and populariser*, the potential for new emerging styles, forms and fusions that might capture the popular imagination. This was illustrated on his return to the USA in 1959 by the concert 'Folksong 59' that he produced in New York's Carnegie Hall. In addition to gospel, blues, bluegrass and folk revival music, the concert also featured a Black rock 'n roll group 'The Cadillacs.' This occasion marked the first time rock 'n roll and bluegrass were performed on the stage of Carnegie Hall (Cohen 2003: 140). Lomax was unapologetic in the face of protests from purists, arguing that Americans should set aside prejudices and embrace this music as their own (also Szwed 2010: 310–312). Although there are many moments in the 1950s and 1960s claimed by historians of popular music as being pivotal, there was symbolic significance in encouraging Americans to accept the emerging genre of rock 'n roll as an American tradition—as much part of the musical 'melting-pot' as the blues, gospel, cajun or bluegrass.

4 A Flawed Methodology

Lomax's writings reveal that he envisaged working towards a global theory of the folk by the mid-1950s. The regional music of Spain and Italy collected during the period 1952–1955, in particular, made him wary about the use of folk culture by totalitarian regimes and reinforced his sense that social science should find ways to allow all musical cultures to flourish equally. He therefore proposed a new science of musical ethnography based on the study of the musical styles or habits that should embrace the total situation in which human beings produce music. Its purpose was not only to achieve a new understanding of the nature of folk music, but also to show that ethnomusicology can contribute to understanding the relationships between cultural and social patterning. Lomax (1959: 928) proposed that: 'the new science of musical ethnography be based on the study of the musical styles or musical habits of mankind. I prefer the term "style" to "habit," because the former gives the sense of a dynamic current in culture, while the latter puts the accent on non-creative, mechanical activity.' He continued by arguing that the 'study of musical style should embrace the total human situation which produces the music,' which included:

1. The number of people habitually involved in a musical act, and the way in which they cooperate.
2. The relation between the music makers and the audience.
3. The physical behaviour of the music makers—their bodily stance, gestures, facial expressions, muscular tensions, especially those of the throat.
4. The vocal timbres and pitch favoured by the culture, and their relationship to the factors under 3.
5. The social function of the music and the occasion of its production.
6. Its psychological and emotional content as expressed in the song texts and the culture's interpretation of this traditional poetry.
7. How songs are learned and transmitted.

In contrast to other contemporary research on music such as Merriam (1964) and other mid-century American ethnomusicologists, Lomax argued that it was only when the behavioural patterns covered by these seven points were taken into account that formal elements (such as scales, interval systems and rhythmic patterns), could be properly understood as acting as symbols that stand for the whole. He concluded:

> A musical style is learned as a whole and responded to as a whole.… the very magic of music lies in the fact that its formal elements can conjure up the total musical experience. An Andalucian gypsy finds it difficult to sing well in his flamenco style unless he is in a bar with wine on the table, money promised, women to clap and dance the rhythms, and fans to shout encouragement. Yet a melody hummed at work in an olive grove conjures up this experience to his imagination. (Cohen 2004: 143)

Cantometrics took formal shape between 1963 and 1976 and drew on the assistance of numerous collaborators, most notably the anthropologist Conrad Arensburg and the composer and musicologist Victor Grauer. Arensburg, in particular, was co-director and a formative influence on the project through his belief in a holistic anthropological science that looked for unifying models to structure cultural data. He believed that anthropology, as a natural science, could build a science of observed behaviour based on the empirical study of human interaction—a form of interaction theory termed 'human ethology' (Arensburg 1972: 6). Cantometrics, Arensburg claimed, exemplified this way of doing anthropology, resulting in the successful discovery of pan-human, worldwide cultural correlations and highly specific cultural variations in social relationships, and measuring both musical and non-musical cultural data in behavioural terms (Arensburg 1972: 7; Arensburg and Kimball 1981).

Cantometrics was also shaped by the available funding, which came initially from the National Institute of Mental Health (NIMH), an agency of the US Federal Government. Perhaps not surprisingly, the NIMH's interest in Lomax's work centred more on social relations than musicology. His application to NIMH for a four-year grant under the name 'Folk Song as a Psycho-Social Indicator' proposed to extend phonotactics beyond its original Indo-European data set, develop content analyses of song texts, and begin a pilot study of dance style. However, what especially interested the NIMH was Lomax's theory that a stronger sense of self, rooted in a more authentic cultural identity, could alleviate juvenile delinquency in minority populations and ethnic groups (Cohen 2003: 239; Szwed 2010: 347).

As propounded in *Folk Song Style and Culture* (Lomax 1968), cantometrics essentially involved a complex and quantitatively sophisticated form of cultural and behavioural analysis. In brief outline, researchers analysed at least 10 songs for each of a sample of 233 different cultures and did so on the basis of coding sheets (Fig. 1), which covered musical and melodic structure, lyrical content, social context and participation, gesture and performance. As Lomax (1967: 220) notes:

> A rating sheet was designed by Victor Grauer and myself so that a trained observer might consistently record his judgement about many levels of the song performance. Consensus tests on this rating system indicate that agreement between judges on most parameters is 80 percent or better and that naïve judgement may be trained to achieve this level of consensus on at least half the system in a matter of a few days.

The sheets provided rating scales for each element, with between 3 and 13 points available on any given scale, depending on the perceived evaluation needs of the particular element (13 was then the limit that IBM computer coding cards could handle). Each song was coded separately, having a specific profile. These could be aggregated into data sets which told about the style of the folk music of particular

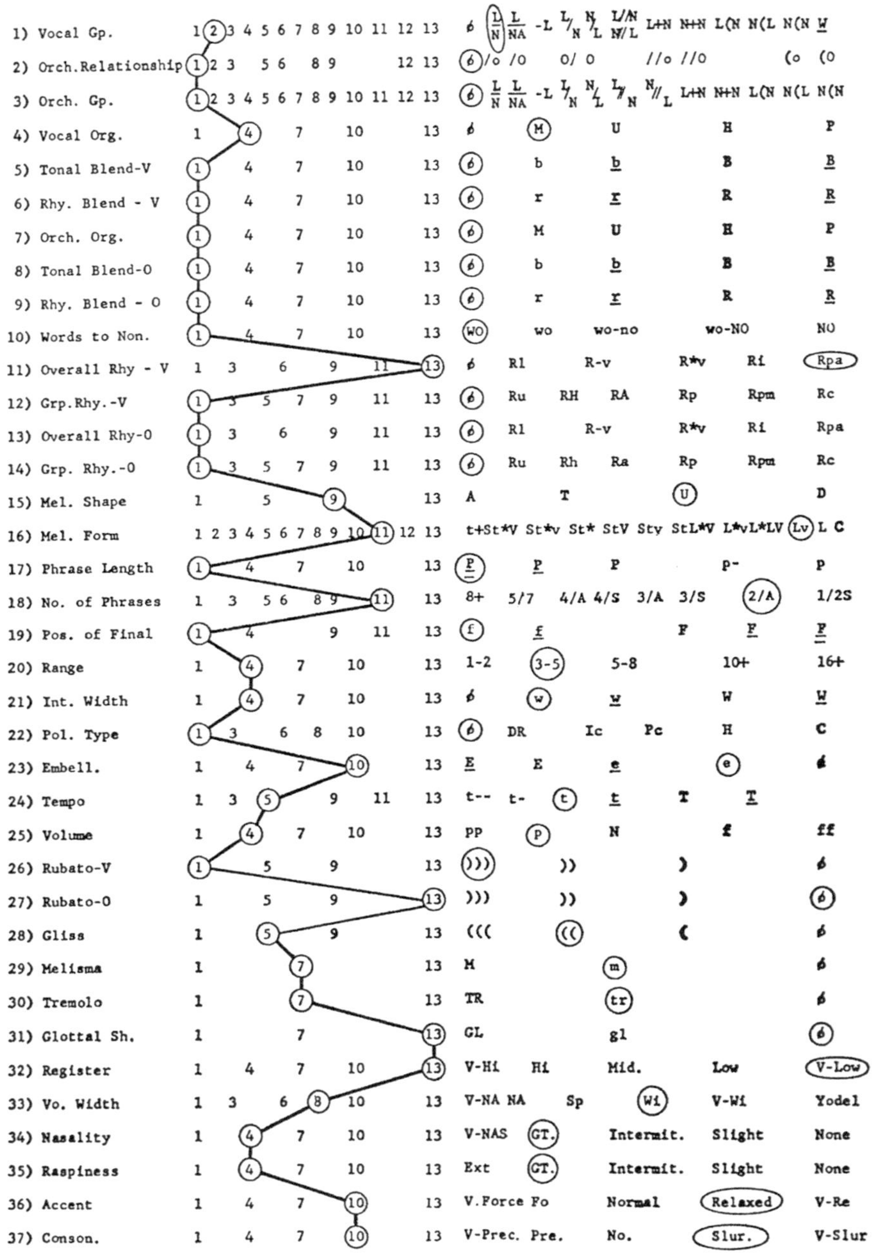

Fig. 1 Modal profile for North America, based on 374 songs. The profile is drawn on the most frequent coding points (*Source* Lomax 1968: 86)

areas and could be used in all manner of ways: for example, to yield information about particular elements of song, to produce world maps of say choral or solo singing, or simply to demarcate culture regions.

The researchers also collected data about subsistence type and social structure for each of the 233 cultures, which were cross-tabulated with the cantometric data. These were analysed and, using regions based on Murdock's (1967) *Ethnographic Atlas*, were used to generate 56 broad areas on the basis of musical 'style,' a characteristic that Lomax and his colleagues defined as that 'part of music that expressed the social or collective identity of its performers and listeners.' Through aggregation, the 56 culture areas were arranged into nine large, homogeneous, though interrelated, style regions. On this basis, it was possible to generate sets of topological maps to give visual expression to the basic principles of cantometrics. Figure 2, for example, shows the 56 culture areas of the world on one map, with the focus on North America. What this map purports to show is that the stylistic connections for American folk music are primarily within the bounds of the broader culture region, apart from significant links with Eastern Brazil and the Mato Grosso. Placing maps that focus on other culture regions alongside would give an immediate impression of homogeneity and difference.

Cantometrics attracted heavy criticism and had little impact on the methodologies used by other ethnomusicologists. Lomax's quest for universal, demonstrable

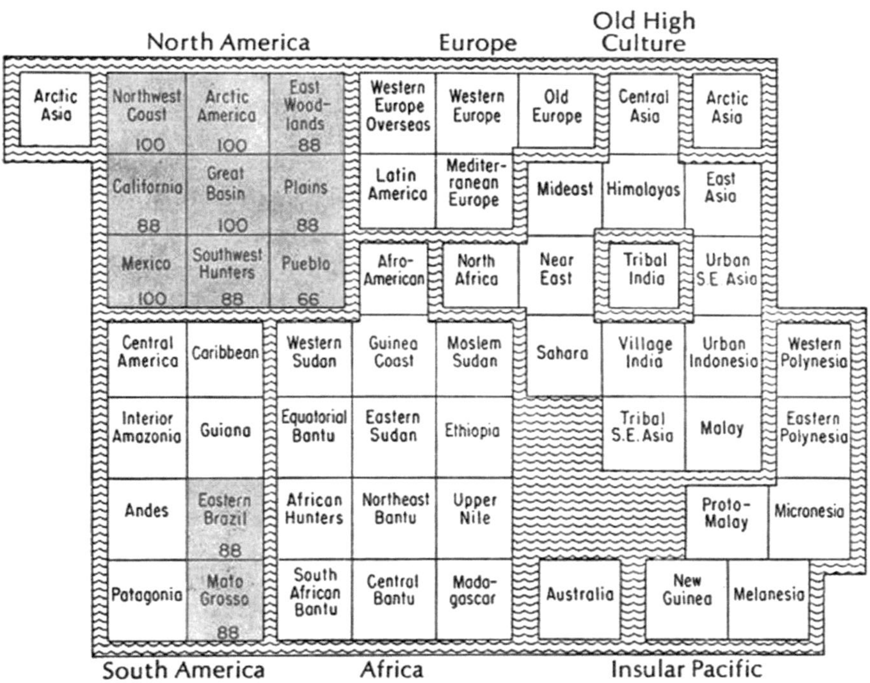

Fig. 2 Homogeneity mapping for North America (*Source* Lomax 1968: 85)

correlations fitted uneasily alongside an increasingly fashionable intensive ethnographic approach developed from the work of Franz Boas. However, the most damning criticisms were directed against the quality of the science. Panteleoni (1972: 243–244) accused Lomax of 'bias in approach and sloppiness in method.' *Inter alia*, he cited the following problems: the exclusion of data from China without discussion; omission of features that failed to yield the desired results; exclusive use of North American adjudicators; imprecise analytic categories and rating scales; implicit assumptions of European cultural superiority; small samples of songs from any given culture and the lack of a rationale for choice; and the preponderance of field recordings of male musicians by male researchers. Downey (1970) raised concerns that the *method* of coding influenced the *outcomes* of coding in the process, thereby posing a series of questions about reliability, bias and experimental control. Various commentators have pointed to ambivalence in Lomax's work about whether musical correlations derive from historical diffusionism or from parallel stages of development. From the first viewpoint, cultural similarities are spread through the influences of war, migration or expansion of trade. From the second, parallels derive from what in cantometric parlance might be termed societies that occupy corresponding evolutionary 'levels' or experience similar systems of subsistence. This contradiction is evident in Lomax's own statement that:

> First, the geography of song styles traces the main paths of human migration and maps the known historical distributions of culture. Second, some traits of song performance show a powerful relationship to features of social structure that regulate interactions in all cultures. (quoted in Cohen 2003: 244)

Ultimately the question of geography, which one might imagine would lie at the very centre of a project aiming to map culture areas, proved to be the least well theorised and most problematic of all aspects of the study. Victor Grauer (Cohen 2004: 240) noted:

> What really became a problem was that Alan was so focused on getting the correlations with the Murdock data to come out "just right" that other very important, meaningful and less problematic aspects of the project were neglected.

Nevertheless, Lomax continued to champion the validity of cantometrics as he developed his final project, the Campaign for Cultural Equity, and attempted to find funding for a computer-based GIS system called the 'Global Jukebox' (Filene 2000: 176; Szwed 2010: 384–385). Here Lomax adopted the language of environmentalism to counter what he perceived as 'pollution of the symbolic environment' and 'cultural greyout' resulting from the domination of modern media by international big business (Cohen 2003: 324–325).

5 Geography, Cantometrics and World Music

Though, as noted above, cantometrics developed entirely separately from the influence of human geographical research, criticisms of the project resonate with the powerful critiques directed at the quantitative and behavioural approaches within geography. It is sobering to learn that varieties of culture area theory often characterised within histories of geography as theoretically and methodologically conservative were being developed during the 1960s at the cutting edge of quantitative social science. In addition, while cantometrics substantially predates human geography's engagement with behaviouralism, parallels in criticisms are striking. These include: accusations of focusing on methodology to the exclusion of critical reflection on underlying assumptions regarding society, culture and behaviour; propagation of bourgeois values; and the adoption of atomistic models of human behaviour and deterministic models of social organisation in order to explain complex geographical processes, circulations and relations (see Gold 2009). In a sense, the maps themselves constitute eloquent critiques of cantometrics. The division of the world into equally sized blocks, each representing one of Murdock's 'culture regions' and closely parcelled together to form a schematic and abstracted representation of the earth's surface, suggests a deliberate break with qualitative accounts of culture. Yet the Murdock map of culture regions was a more representational and 'realistic' projection of the globe than that which resulted when the musical data was mapped on to it. For cantometrics, geography is objectified both as *a priori* knowledge and scientific abstraction.

Having said this, cantometrics and its subsequent developments were far from mere academic exercises in dispassionate and rationalistic scientific inquiry, since both were informed by strong, egalitarian and left-leaning politics. It is not surprising that a conception of world music should emerge in the United States in the context of a liberal democratic conception of American identity and a background of anti-fascism built, however optimistically, around a notion of the 'melting pot' and a plurality of cultures. There are parallels here with Herder's eighteenth century efforts to build cosmopolitan politics which defended the rights of small German states against the power of Prussia.

There are also contrasts between the folk collecting strategies of Lomax and previous collectors. Whereas Cecil Sharp searched for survivals in culture in order to locate and define historically-isolated outliers of English culture, Lomax, like Bartók in his later period, was well aware of the heterogeneous nature of folk music. For Lomax this did not threaten the integrity of American national identity but provided a primary building block for a democracy built on the idea of the melting pot. He went further than this, of course, and developed his idea of the melting pot beyond the USA into a global cartography of cultural hearths, diffusions, crossovers and hybrids. In doing so, Lomax retained the basic outlook and values of New Deal politics, continuing to see his role as, on the one hand, showcasing and providing a voice for the disenfranchised and, on the other, facilitating cultural bridges between apparently disparate and divided peoples through recognising the 'essential unity' of humanity. In this sense, cantometrics and other aspects of Lomax's work drew more generally on the Enlightenment ideals evident in Herder's original theorisation of folk culture.

Thus, for Lomax, the delimitation of specific cultures became the building block for a profound ecumenism. In a curious sense the foundational atomism of interaction theory, based on the idea that particularity and cultural meaning derive from the aggregation and collection of interactions into increasingly complex structures at progressively greater spatial scales, provides an appropriate theoretical ground. This is because interaction theory, at least as formulated by Arensburg, begins with events as abstraction and then traces the aggregation of these into socially meaningful organisational structures and networks. In the process, culture emerges from abstraction at higher levels of interaction.

Like earlier collectors, Lomax has been heavily criticised by recent theorists for the rigidity and dogmatism of his underlying philosophies and classificatory systems. As with other folksong collectors of his day, Lomax believed that vibrant and lively folk cultures would reinvigorate their respective national cultures and counteract the dominance of what were believed to be degenerate and decaying commercial cultures—a point reminiscent of the heterogeneous creative and regenerative potential of folk music discussed earlier. However, with this in mind, it is important to return to the question of how folk collectors negotiated the relationships between purity and heterogeneity when defining the spaces of culture, region and nation or, as with cantometrics, global spaces of local distinctiveness. It seems evident that Lomax, like Sharp and Bartók (Gold and Revill 2006), remained locked into a search for points of certainty and security to anchor his otherwise fluid sense of the folk. As such, that process of anchorage was achieved in three distinct ways.

First, Lomax believed that when and wherever folk musicians found music related to their own 'tradition' it would remind them of their own personal and community folk history and thus of being "at home." Hence, he felt that folk traditions were inherently conservative because to change was to challenge long-standing senses of self and community. Such thinking mirrors notions of cultural hearths familiar from North American cultural geography and was an idea shared by Sharp in his search for 'survivals in culture' or Bartók in his concern for finding a creative centre for folk culture. Yet while trying to ground music's mobile, fleeting and transient qualities firmly in ideas of place, his approach also differed considerably in a number of respects. Most important were his emphasis on the performative qualities of folk music, on its fundamental heterogeneity and on the cultural equality of world music traditions—ideas which have only recently taken a hold in ethnomusicology.

Secondly, the long-term collaboration of Arensburg and others at Columbia underscored a view that cantometrics should engage centrally with current social science debates. Lomax has been castigated for the apparently pompous pseudo-scientific language of cantometrics and the western, masculinist assumptions which critics detect as underlying its methodology and findings. Yet it is reasonable to assume that Lomax genuinely believed that what he was doing was fair and objective. This apparent dissonance is familiar in the world of folk collecting where many collectors have tamed the disturbing heterogeneity of music by resorting to an authenticity founded in the judgement of the collector (arranger, composer) as a sympathetic but dispassionate expert. In the context of the American New Deal liberalism championed by Lomax and Arensburg, the dispassionate expert was undeniably a key figure in the forging of an equitable democracy.

Thirdly, points of certainty were found by reference to a value system that defined 'beauty.' Lomax (1967), for instance, argued that the purpose of cultural analysis is to examine the ways in which value judgements (the good) materialise in culture (in what is considered beautiful) as part of the environmental and social adaptation of human societies. In this formulation, the study of aesthetics (or study of the beautiful) is our means of accessing the truth of the lived realities of societies in their specificity. Here Lomax reworks an idea from survivals theory with the sophistication of modern anthropology and some reference to his undergraduate years studying philosophy. It suggested that folk music is beautiful because it is true to a people's history and experience, given that this experience is distilled through generations of adaptive decision-making (see Boyes 1993).

6 Conclusion

Given Alan Lomax's sensitivity to other cultures, one might justifiably enquire why someone who was in many respects so enlightened in his approach to world folk musics could champion the rigid and partial classificatory system associated with cantometrics. Yet it is arguable that all attempts to create world music require the diversity of musical forms, practices and cultures to be subject to some form of conceptual homogenisation or overarching theory. Certainly the major criticisms of the recent world music phenomenon focus on this issue, highlighting concerns over issues such as the commodification, packaging and parcelling together of music created for a multiplicity of social, religious and ceremonial situations into a catalogue of bland style choices cut off from their social, religious and political contexts and meanings.

Cantometrics was undoubtedly open to some of these accusations. Though Lomax argued strongly for the cultural specificity of regional, local and national musical styles, his exposure to applied anthropology and transactional analysis enabled him to find a means of moving beyond the apparent limits of locally-generated meanings and bridge cultures by finding cross-cultural unity in interactional processes at higher levels of organisational complexity than that of musical events themselves. If for Lomax music presented universal truths, they were truths of structure rather than content. Perhaps this begins to help us understand the apparent contradiction evident in the cantometrics project between the integrity of local styles and the fetishisation of abstraction and correlation.

References

Andriolo, K. (1979). Kulturkreislehre and the Austrian mind. *Man NS, 14*, 133–144.
Arensburg, C. M. (1972). Culture as behaviour: Structure and emergence. *Annual Review of Anthropology, 1*, 1–26.
Arensburg, C. M., & Kimball, S. T. (1981). Cultural holism through interactional systems. *American Anthropologist, 83*, 562–581.
Aubert, L. (2007). *The music of the other: New challenges for ethnomusicology*. Aldershot: Ashgate.
Bhabha, H. K. (1994). *The location of culture*. London: Routledge.
Biddle, I., & Knights, V. (Eds.). (2007). *Music, national identity and the politics of location: Between the global and the local*. Aldershot: Ashgate.
Bluestein, G. (1972). *The voice of the folk: Folklore and American literary theory*. Amherst: University of Massachusetts Press.
Bohlman, P. V. (Ed.). (2012). *The Cambridge history of world music*. Cambridge: Cambridge University Press.
Bohlman, P. V. (1988). *The study of folk music in the modern world*. Bloomington: Indiana University Press.
Bohlman, P. V. (2002). *World music: A very short introduction*. Oxford: Oxford University Press.
Born, G., & Hesmondhalgh, D. (2000). Introduction: On difference, representation, and appropriation in music. In G. Born & D. Hesmondhalgh (Eds.), *Western music and its others: Difference, representation, and appropriation in music* (pp. 1–58). Berkeley: University of California Press.
Boyes, G. (1993). *The imagined village: Culture, ideology and the english folk revival*. Manchester: Manchester University Press.
Calkins, C. (1960). Getting to know folk music. *House Beautiful (April), 141*, 205.
Carney, G. (Ed.). (1998). *Baseball, barns and bluegrass: A geography of American folklife*. Lanham, MD: Rowman and Littlefield.
Carney, G. (Ed.). (2003). *Sounds of people and places: A geography of American music from country to classical and blues to bop* (4th ed.). Lanham, MD: Rowman and Littlefield.
Carney, G. (1990). Geography of music: Inventory and prospect. *Journal of Cultural Geography, 10*(2), 35–48.
Clayton, M. (2003). Comparing music, comparing musicology. In M. Clayton, T. Herbert, & R. Middleton (Eds.), *The cultural study of music: A critical introduction* (pp. 57–68). London: Routledge.
Cohen, R. D. (2003). *Rainbow quest*. Amherst: University of Massachusetts Press.
Cohen, R. D. (2004). *Alan Lomax: Selected writings 1934–1997*. London: Routledge.
Collins, S. (2004). *America over the water*. London: SAF Publishing.
Connell, J., & Gibson, C. (2003). *Sound tracks: Popular music, identity and place*. London: Routledge.
Dargen, W. T. (2006). *Lining out the word: Dr. Watts Hymn singing in the music of Black Americans*. Berkeley: University of California Press.
Downey, J. C. (1970). Review of 'Folk Song Style and Culture'. *Ethnomusicology, 14*, 63–67.
Filene, B. (2000). *Romancing the folk: Public memory and American roots music*. Chapel Hill, NC: University of North Carolina Press.
Francmanis, J. (2002). National music to National Redeemer: The consolidation of a 'folk-song' construct in Edwardian England. *Popular Music, 21*, 1–25.
fRoots. (2011). Minutes of meeting between the various "world music" record companies and interested parties, Monday 29th June 1987. Available on-line at http://www.frootsmag.com/content/features/world_music_history/minutes/page03.html. Accessed 11 Mar 2011.
Gilroy, P. (1993). *The Black Atlantic: Modernity and double consciousness*. London: Verso.
Gioia, T. (2006). The red-rumour blues. *Los Angeles Times,* 23 April.

Gold, J. R. (2009). Behavioural geography. In R. Kitchin & N. Thrift (Eds.), *International encyclopaedia of human geography* (Vol. 1, pp. 282–293). Oxford: Elsevier.
Gold, J. R., & Revill, G. E. (2006). Gathering the voices of the people? Cecil Sharp, cultural hybridity and the folk music of appalachia. *GeoJournal, 63*, 55–66.
von Herder, J. G. (1778, 1779). *Stimmen der Völker in Liedern*. [1880] Leipzig: P. Reclam.
Kluckhohn, C. (1936). Some reflections on the method and theory of the Kulturkreislehre. *American Anthropologist, 38*, 157–196.
Leary, J. P. (2006). *Polkabilly: How the Goose Island ramblers redefined American folk music*. Oxford: Oxford University Press.
Lomax, A. (1959). Folk song style. *American Anthropologist, 61*(6), 927–954.
Lomax, A. (1967). The good and the beautiful in folksong. *Journal of American Folklore, 317*, 213–235.
Lomax, A. (1968). *Folk song style and culture*. Washington, D.C.: American Association for the Advancement of Science.
Lomax, A. (1997). 'Liner Notes' to 'Southern Journey,' Volume 2: Ballads and breakdowns: Songs from the Southern Mountains. Washington, DC: Rounder Select.
Lomax, J. A., Lomax, A., & Seeger, R. C. (1949). *Our singing country: Folk songs and ballads*. New York: Macmillan.
Merriam, A. (1964). *The anthropology of music*. Evanston, IL: Northwestern University Press.
Murdock, G. P. (1967). *Ethnographic atlas*. Pittsburgh: University of Pittsburgh Press.
Nettl, B. (1983). *The study of ethnomusicology*. Champaign: University of Illinois Press.
Nidel, R. (2004). *World music: The basics*. London: Routledge.
Panteleoni, H. (1972). Review of 'Folk Song Style and Culture'. *Yearbook of the International Folk Music Council, 4*(1), 158–161.
Pickering, M., & Green, T. (Eds.). (1987a). *Everyday culture: Popular song and the vernacular milieu*. Milton Keynes: Open University Press.
Pickering, M., & Green, T. (1987b). Towards a cartography of the vernacular milieu. In M. Pickering & T. Green (Eds.), *Everyday culture: Popular song and the vernacular milieu* (pp. 1–38). Milton Keynes: Open University Press.
Radano, R. M., & Bohlman, P. V. (Eds.). (2000). *Music and the racial imagination*. Chicago: University of Chicago Press.
Revill, G. (2005). Vernacular culture and the place of folk music. *Social and Cultural Geography, 6*(5), 693–706.
Richards, S. (1992). *Sonic harvest: Towards a musical democracy*. Charlbury: Amber Lane Press.
Skultans, V. (1998). *The testimony of lives*. London: Routledge.
Szwed, J. (2010). *The man who recorded the world: A biography of Alan Lomax*. London: William Heinemann.
Taylor, T. D. (1997). *Global pop: World music, world markets*. London: Routledge.
Young, R. (1995). *Colonial desire: Hybridity in theory, culture and race*. London: Routledge.

Cultural and Participatory Mapping

Ann Kingsolver, Manuel Boissière, Michael Padmanaba, Ermayanti Sadjunin and Sasikumar Balasundaram

Abstract Anthropology, as a discipline, has been closely allied with geography for over a century, and mapping is an important practice in each of anthropology's four subdisciplines: archaeology and biological, cultural, and linguistic anthropology. This chapter focuses on cultural mapping, or representations of how humans understand social and physical environments and relationships, and participatory mapping, a particular technique of inclusive map-making in which researchers and community members-as-researchers create maps collectively. Such maps are often used to document understandings of space that contrast with official maps of state understandings of, for example, resources and rights. In participatory mapping, the questions about what is to be mapped are established collectively as well as the mapping process itself (often, as mentioned, done as part of a larger social justice project). This chapter discusses the history of, and variation in, cultural mapping and then goes on to provide several examples of cultural and participatory mapping. Manuel Boissière, Michael Padmanaba, and Ermayanti Sadjudin describe the participatory mapping process in which they engaged, with many others, in Mamberamo, Papua, Indonesia as part of a

A. Kingsolver (✉)
Anthropology Department, University of Kentucky, Lexington, USA
e-mail: ann.kingsolver@uky.edu

M. Boissière
Centre de cooperation International en Recherche Agronomique pour le Développement, Center for International Forestry Research, Bogor, Indonesia
e-mail: m.boissiere@cgiar.org

M. Padmanaba
Center for International Forestry Research, Bogor, Indonesia
e-mail: m.padmanaba@cgiar.org

E. Sadjunin
Conservation International, Jakarta, Indonesia
e-mail: ermayanti@conservation.org

S. Balasundaram
Anthropology Department, College of William and Mary, Williamsburg, USA
e-mail: sbalasundaram@wm.edu

S.D. Brunn and M. Dodge (eds.), *Mapping Across Academia*,
DOI 10.1007/978-94-024-1011-2_15

long-term project on biodiversity and natural resource management. Residents of six villages, working from base maps on the same scale as state maps, corrected and expanded information about rivers and resource diversity on the maps as well as adding livelihood activities and sacred places. These maps were used in a regional workshop, including government representatives, on land use planning. Sasikumar Balasundaram illustrates, in another example, how children in long-term refugee camps in southern India map their current circumstances and imagined futures.

Keywords Cultural anthropology · Social justice · Participatory approaches · Resources · Future

1 Introduction: A Penchant for Mapping in Anthropology

Anthropology is a social science with four subfields: cultural anthropology, linguistic anthropology, biological anthropology, and archaeology. The focus of the discipline is a broad one: studying human diversity and experience from the way the human body has evolved over time, adapts to different environments, and forms and recognizes sounds in diverse languages to the interaction between humans and our physical surroundings and social contexts through material culture and cultural interpretations that in turn shape actions. As sociology, anthropology, and geography have changed through time, their overlap has increased greatly. The distinction between sociology's emphasis on urban, "developed" societies and anthropology's emphasis on rural, "developing" societies has long since blurred with current academic programs and organizations focused on topics including urban anthropology and rural sociology, for example. Anthropology and geography have always been very related disciplines. Franz Boas, the founder of the first academic program in cultural anthropology in the United States and the teacher of Margaret Mead, was trained as a geographer. Place and space are important considerations in cultural anthropology. Rodman's (1992: 640) point that "the meaning of place too often seems to go without saying" in contemporary anthropology was well-taken, and cultural anthropology and cultural geography have become very intertwined. As Mitchell (2000: 294) defines cultural geography, it is "the study of how particular social relations intersect with more general processes, a study grounded in the production and reproduction of actual places, spaces, and scales *and* the social structures that give those places, spaces, and scales meaning." A conversation about cultural and participatory mapping is definitely informed by cultural geography; we just happen to be writing about it from an anthropological perspective in this chapter.

All four of anthropology's subfields utilize mapping methods. Linguistic anthropologists may map languages and dialects (cf. Barni and Extra 2008). Archaeologists increasingly use ground penetrating radar to map earlier human imprints on the environment (structures or other evidence of human occupation) without having to disturb subsequent layers, and have long used Geographic Information Systems technology to map archaeological sites. Using GIS data and other forms of documentation, Steven Wernke (2007) was able to map land tenure over a period spanning from Incan to Spanish control of the Colca Valley in Peru, for example. Urban and Schortman (2013), also archaeologists, mapped power relations through looking at how territory may have been maintained nearly 1000 years ago through the construction of monumental architecture and related social networks. Biological anthropologists might use maps to posit theories of human evolution through time and space, or—in this creative partnership with a musician—to create an aural map of infectious disease evolution (Achterberg and Pierce 2011). Cultural anthropologists use maps in many ways to consider how humans conceptualize our physical and social environments, including kinship relations, for example, and social networks. Tobin et al. (2010), for example, mapped the social networks of individuals in chronic disaster settings in which periodic volcanic eruptions cause displacement and looked at what kinds of networks best mediated vulnerability, and Vásquez-León (2009) mapped the relationship between the social networks of Hispanic farmers and farmworkers, water scarcity related to climate change, and livelihood vulnerability in Arizona.

Of the many ways that anthropologists use maps, this chapter focuses on cultural mapping. Culture is a set of learned and shared interpretations of the world that can shift over time as each new generation shapes and is shaped by culture. It has often been discussed as coinciding with national identity, but culture is much more complex and nuanced than that, with global influences in every locality. Political and ethnic boundaries are themselves a human cultural production, reinforced through cultural (and rather arbitrary and fluctuating) imagination of who belongs where. Cultural mapping is a broad technique that may be used in two general ways by anthropologists: to keep track of the social scientist's own findings—for example, in mapping the results of interviews about whether residents are aware or not of an environmental hazard in their community—or as a method for understanding how those being interviewed are themselves mapping whatever is being studied. This may or may not correspond to an observable landscape. A cognitive map might be a narration of understandings of power relations in a community over time. Such mapping processes can render what has been a silenced or marginalized experience of space very visible to members of a dominant group. Polanco (2012) worked with residents of Union, Virginia to map racialized understandings of history and place, and Hargrove (2009) mapped the political economy of racism in "heritage tourism" development in and around Charleston, South Carolina. DeVault (2014: 781), a sociologist, mapped "the invisible work that deaf people have to do in hearing spaces". Elicitation of narratives or drawn maps of how individuals navigate particular spaces, for purposes of comparison (to map power and/or social relations), is a common technique for social scientists. A researcher might ask people to draw their usual route through a farmer's market as a cultural mapping technique, for example. Sopranzetti (2014) described mobility itself as a form of political mobilization by tracing the routes of motorcycle taxi drivers during political protests in Thailand, Wendland (2012) compared the different "moral maps" of Malawian and international medical students (or "clinical tourists") in a teaching hospital in Malawi, and this technique has even be used in virtual space by Takhteyev et al. (2012), mapping Twitter networks.

This chapter provides some historical context for cultural mapping and then provides examples, especially of the process of participatory cultural mapping: a technique in which the researcher and those collaborating with the researcher in a local context together constitute the questions that cultural mapping is used, in a group setting, to address. Anthropologists today tend to use participatory mapping to create alternative, community-constituted representations that may counter dominant or official representations—those alternative cultural maps might be part of a larger social justice project in which voices that have been marginalized socially and economically are made central so that their interpretations are not silenced. One example in this chapter is of participatory mapping of natural resource use by Indonesian residents whose environment has been threatened by a development project; such a cultural map can become evidence that can be used to demonstrate differing forms of value represented by a single territory. The other main example in this chapter is of cultural mapping by young Sri Lankan residents of a long-term refugee camp in India as they convey, through drawings, the everyday violence of their surroundings and imagine an environment in which they see themselves thriving.

Anthropology's earliest use of cultural mapping techniques was far from these current social justice examples; it was associated with the oppressive practice of colonialism, since many of the earliest anthropologists were making their cultural observations as colonial officials. European anthropologists in the nineteenth century helped to map colonial territories, with their arbitrary lines and brightly colored "otherness" (Mbodj 2002), and promoted the self-defined superiority of upper-class Victorian culture on maps used in colonial and missionary classrooms across empires by drawing world maps with Europe always at the center. As Said (1979) pointed out, this created the "orientalizing" dichotomy between Western and Eastern cultures, as well as a reference point for telling powerful histories of agentive explorers and those `passively' explored and mapped. Colonial maps of trade routes omit cultural encounters and contributions going in many directions and tell a story of `bringing' culture to the colonized when in fact the colonized were equally shaping the colonizers, culturally (Wolf 1982). Although anthropology began in the mid-nineteenth century—its history intertwined with a colonial vantage point, Greek writers had long before promoted the assertion that Africans needed to be mapped and were not themselves mappers of the world (Mbodj 2002). Through ethnographic listening among or as members of groups formerly colonized (who had actually been culturally mapping the world long before the colonizers), anthropologists have shifted in perspective a great deal since the discipline's nineteenth-century origins. Postcolonial anthropologists (cf. Asad 1973; Harrison 2009) have been among the strongest critics of the colonial vantage point, and cultural anthropologists now tend to place at the center of their analyses (and maps) the perspectives of those who have been marginalized within powerful state discourses.

Looking back at those disciplinary origins, though, British social anthropologists used maps to document political, religious, kin and ethnic identities as perceived by states and by various constituencies within states. Franz Boas, who mentored the first cultural anthropologists in the U.S. in the early twentieth century, and Bronislaw Malinowski, a functionalist anthropologist doing fieldwork off the northeast coast of Australia in the same period, established methodologies in cultural anthropology that included cultural mapping—paying close attention to the way that individuals conceived of, and communicated, their social and physical environments. Malinowski (1961) famously mapped the *kula ring*, the circulation of ceremonial shell armshells and necklaces in opposite directions between a string of islands, arguing that the symbolic movement of these objects imbuing power to those possessing them along their routes, functioned to encourage interaction between otherwise isolated (he assumed) groups of people living on dispersed South Pacific islands.

Early twentieth century cultural anthropologists tended to associate specific theories and concepts with geographic regions of the world, associating discussions of colonial land tenure and labor systems with Latin America, for example, or matrilineality with western Africa (see Fardon 1990). Culture areas, locations assumed to share a practice, language, or identity, were not seen as necessarily coinciding with state borders or national cultures, and cultural anthropology as a discipline set up area studies courses in the curriculum to train students in culture areas: Peoples and Cultures of Africa, for example, or The Middle East. Discussions of regional culture or national culture have been problematized by postmodern and postcolonial anthropologists, who foreground for analysis the arbitrary construction of national boundaries and notions of cultural homogeneity within those borders (see, for example, Vila 2000 and Williams 1996). Anthropologists (informed by cultural relativism, or an emphasis on attempting to understanding a cultural concept or practice within its own context) are usually very comfortable analytically with seeing multiple, simultaneous mappings of reality from different vantage points. Culture has itself been defined as a "mental map which guides us in our relations to our surroundings and to other people" (Downs 1971: 35), and anthropologists are interested in the ways in which these mental maps converge and diverge, overlap and conflict. There are cultural and political conflicts over who can claim the same territory as a homeland, as on the West Bank. Sometimes these are high-stakes conflicts, as in the long civil war in Sri Lanka, and sometimes they are the subject of lighter contestation, as in discussions of whether one is occupying the tomato, vinegar, or mustard-based barbecue zone of South Carolina, USA (those food preferences have been mapped for the state).

Through close ethnographic fieldwork, cultural anthropologists (increasingly, indigenous to the region of study) have documented many different ways of mapping social and environmental resources besides the state-sanctioned cartography recognized in colonial classrooms. Landscapes can be narrated (see the work of cultural geographers, Blunt and Rose 1994) in ways that do not require paper or pen. Indigenous Australian elders, for example, have narrated and painted symbolic maps of the landscape with embedded memories of seasonal access to water and other resources; other indigenous nations were mapping landscapes through oral and artistic traditions long before their territories were "on the map" in Western terms.

There are story maps, song maps, danced maps, and walked maps. Amazonian rubber tappers, for example, have drawn maps of their walking trails in the forest between tapped trees in the shape of trees themselves, as a mnemonic (Mendes 1989).

Cultural anthropological research can include eliciting these alternative mappings of social and physical environments. Kingsolver (2011), for example, in fieldwork in her home community in rural Kentucky in the 1980s, would drive elders around the county with a tape recorder running (with their permission), so they could narrate the landscape. These phantasmagorical taxonomies of community and space were related to the directions often joked about as unintelligible to 'outsiders,' as in "go to where the old barn that used to belong to the Williams family burned down, and turn left." Kingsolver found that the narrations of memory and landscape revealed significant histories not otherwise recorded. A woman in her eighties pointed to a pasture, for example, and described the entire community inhabiting that hillside before the 1918 influenza epidemic killed nearly everyone living there. That spatial memory of a critical event is similar to the much larger-scale cultural memories associated with Ground Zero in New York City and Wounded Knee Creek on the Lakota Pine Ridge Reservation in South Dakota. Narrated mapping is a way to recover marginalized histories, as in sites on the Underground Railroad in the U.S. or the history of Koreans who were killed along with Japanese citizens in the nuclear bombing of Hiroshima (see Yoneyama 1999). Importantly, as mentioned earlier, silenced histories can be made visible in the landscape, as in markers on the sites of lynchings and disappearances. There are many different mental maps that individuals use to navigate through the same landscape, including sacred spaces as well as abhorrent ones.

Feminist cultural anthropologists have done structural analyses of how space can be mapped in gendered ways, as in private and public space, or spaces that may culturally be occupied only by men or women at specific times (see Hoffman 1976). Many cultural anthropologists interested in structural inequalities have looked at the powerful ways in which maps can both render visible those inequalities and also reinforce them. The Appalachian Regional Commission (2015), for example, publishes maps each year showing the counties within the thirteen Appalachian states in the U.S. with a modernist classification scheme of economically distressed (belonging to the lowest 10 % of counties nationally, in terms of economic indicators), at-risk, emerging, transitional, competitive, and counties with economic attainment (belonging to the nation's top 10 % of counties, according to economic indicators). Those in the economically distressed counties have argued that at times such classification can reinforce stereotypes about communities, perpetuating a "culture of poverty" perspective and distracting the viewer from a structural inquiry into what might have caused such disparities within a region (for example, extractive industries or absentee land ownership). Lewis (1966), the author of the culture of poverty argument, was himself a cultural anthropologist. Other cultural anthropologists have critiqued that model as ethnocentric (Miller 2005: 132), and have looked at the ways economic and other types of marginalization are related to the way power is structured and access to resources is controlled. Those power structures can, in turn, be mapped, as mentioned earlier. Thomas-Houston (2005) has written about how the layout of roads and cul de sacs in Oxford, Mississippi, marginalizes African American communities and shapes political participation in that city, for example.

Mapping power spatially, in terms of inclusion and exclusion, is a useful method for community activists and for cultural anthropologists. When Ann Kingsolver visited—with a group from the Society for the Anthropology of North America—a Vietnamese American community rebuilding after Hurricane Katrina in New Orleans East, residents said that they were receiving no help from FEMA (the U.S. Federal Emergency Management Agency) because they had been "off the map" to start with, and FEMA was only assisting with the reconstruction of what had been on the official maps before the storm. Inequalities can be mapped effectively in studies of environmental injustice and displacement, as in the shifts of Native American reservations on the landscape over time in the U.S. as valuable resources (uranium, for example) have been discovered on what was initially thought of as marginal land to which Native American nations' territories have been relegated (see Norgren and Nanda 1996). Social scientists and community activists doing environmental justice research have looked at the locations of toxic waste dumps and other radioactive or polluting sites on maps and found consistently that they are placed in areas where the residents are the least likely to protest because of class and racialized marginalization, for example, and the lack of access to political capital (cf. Bullard and Waters 2005). Cultural anthropologists, by listening carefully to the ways in which individuals and groups situate themselves, can map the various vantage points in contested spaces and the power relations between them. What does it mean to people to live in a neighborhood that some other people label "unsafe" or "undesirable," or to live in a "no-fly zone," a security zone, an economically distressed or contaminated region, or a food desert, for example?

2 Participatory Mapping

Participatory mapping is a technique used by cultural anthropologists, ethnobotanists, development professionals, community activists, and others to create alternatives to maps that reinforce powerful discourses and marginalize spatially those who are marginalized in other ways as well. In state or business planning processes, for example, an area may be seen as "blank" on the map when, to others, it may be very much mapped as home, a workspace, and/or a sacred space. Participatory mapping provides a way to give agency to local residents in a planning process, and to put maps into conversation with each other. As the Mapping for Rights (2015) web resource notes, the "processes used to create the maps can be as valuable as the maps themselves." Participatory mapping processes can take lots of different forms—drawings to which everyone adds, or narration, for example. Vajjhala (2006) points out that international, state, and NGO development processes now mostly call for citizen participation in development planning, and participatory mapping is an effective technique for providing community input since all development projects have a spatial component.

The two examples that follow show how groups often marginalized in state planning processes—villagers in Indonesia and children in a long-term refugee camp in India—can inform public discussion when their own views of the landscape are placed at the center rather than at the margins of analyses and planning processes.

3 Participatory Mapping in Mamberamo (Papua, Indonesia)

Local people are not generally involved in land use planning, when in fact they represent the main managers of large territories. In Papua, Indonesia, one village territory can cover up to 1000 km^2. Manuel Boissière, Michael Padmanaba, and Ermayanti Sadjudin have been participants, with others, in a long-term, interdisciplinary research project on biodiversity and approaches to natural resource management in Indonesia. They have studied how local perceptions and priorities can be reflected and included in decision made by local government institutions on land management. In Papua, decentralization and special autonomy make land use planning a key instrument for decision-making, especially regarding the implementation of development programs.

The study took place in six villages of Mamberamo Raya Regency (the regency is a political subdivision of the province, administered by an elected head of Regency and local parliament). It has a territory of 2.8 million hectares, with 23,000 inhabitants divided into 59 sparsely located villages. Villagers in Mamberamo are cultivators (mainly sago, bananas and sweet potatoes), fishermen and hunter-gatherers. About two-thirds of the regency are part of the Mamberamo Foja Wildlife Reserve, known for its rich biodiversity. Many villages are located inside the protected area.

The different steps are explained here for how participatory maps of important resources and sites, clan boundaries, and present and future land use were developed, representing the villagers' own vantage points. This was important in the context of possible large governmental logging concessions based on governmental maps that did not agree with local knowledge of forest resources and watersheds. The researchers invited regional governmental involvement in the participatory mapping process so that government staff would have the opportunity for more direct and equitable conversations with villagers about land use issues.

3.1 Working from the Base Map

The official land use plan (LUP) uses maps at the scale of 1 : 50,000. In this project, participatory maps were developed in each of the villages following the same scale, to be able to compare them with the official ones. A base map was provided that showed the main rivers and tributaries, and the positions of villages, roads and other features visible on a satellite image. Most villagers participating in the mapping activity were literate and could provide direct input. Indonesian was used as the communication language, as it is widely spoken in Mamberamo because of the presence of schools in which Indonesian is the language of instruction.

The first step in the participatory mapping process was for the villagers to recognize, add, and correct the rivers' name on the maps. They would look at the tributaries closest to their own village and then progress by identifying rivers upstream and downstream from the village, along the main rivers. The participatory mapping team included two groups of villagers, male and female, as the knowledge of a territory is not equally shared among genders. Women have a more intimate knowledge of the landscape near the village and gardens, while men know more distant places better (Fig. 1).

Once a map was drafted with the correct rivers' positions and names on it, villagers added important sites (i.e., gardens, cemeteries, sacred places, and the sites of former villages). They also added where significant natural resources might be found; they identified locations for what they saw as the ten most important plants and animals for local livelihoods.

Fig. 1 Participatory mapping in 2012 in the village of Bagusa, Mamberamo Raya District, Papua Province, Indonesia; participants are drawing and naming all the small rivers in their territory (Courtesy of Manuel Boissière.)

3.2 Ground Check

The men's and women's maps were then merged to get an idea of the overall village's communal knowledge. The research team then checked the position of each of river, resource and important site on each map directly in the field, using GPS, accompanied by local guides (normally representatives from the landowner clan). Based on the ground check results, the map was corrected in the village and a clean version was provided to the villagers.

3.3 Land Use Maps and Clan Boundaries

Current land uses were discussed during meetings of a group of villagers, mostly local decision-makers (e.g., the village head, customary leaders, clan heads, women and men elders). From these discussions, the group added to the new map what areas are devoted to agriculture, fishing, hunting, and reserves of natural resources.

Future possible uses of land and villagers' expectations in terms of development, conservation, and governance were discussed once the map of current land use was finished. The areas identified in the current land use map were updated and some new features were added as appropriate (for example, an airstrip, a small town, or roads). For example, some villages anticipated population growth and considered that more agricultural products (hence larger areas for gardens or sago groves) would be necessary in the future.

The last step undertaken by the mapping team, if villagers asked for it, would be to prepare a map of clan boundaries. The residents of some villages wanted this map, to use it as a tool for negotiation with a logging company. Other villages did not want to map their clans, as they worried about possible conflicts with other villages that might result from this.

3.4 Giving Back the Results: Workshop

A "restitution" workshop was organized at the capital of the regency, Kasonaweja, in March 2012. The aim of the workshop was to give back the project results to the regency government and villagers, and to launch discussions on future land use planning. Consent was asked from each of the villages concerning the utilization of the maps and other reports coming from this research. The participatory mapping research team explained the pro and cons when sharing these maps with others parties, but after discussion, residents of all the villages participating in the mapping project agreed to share their maps with the local government during the final workshop.

3.5 *Utilization of the Maps*

The final printed maps were given back to villagers and local government representatives during the workshop. About 150 people participated, most from civil society (villagers and NGO representatives), but also staff from the regency, provincial and national government, and from the private sector. Representatives from each village discussed with local government staff members, in small groups, case-by-case, the needs and priorities of local people and local government. They used the maps for the discussion. Villages' representatives presented their maps. They were confident in explaining them.

Because of their scale being similar to the official land use maps, and the participatory approach used, the maps became a powerful tool for negotiation on land use planning, development strategies, and collaboration between the government and the villagers (whose perspectives had been underrepresented in regional and national development discussions). The next step was to train regional staff teams to replicate these methods in the remaining 53 villages of the Regency.

The local government expressed great interest about using our methods to cover the remaining villages in the Regency. Some discussions were initiated during the workshop concerning concrete development issues: for example a proposition was made to enlarge a navigable channel in the mangroves for the village of Yoke. This would shorten their long and costly trips to the capital of the Regency. Villagers used the map and their knowledge of natural resources to make their point about what route would result in the least environmental damage. More negotiation is expected from the use of these maps in the future, and will hopefully lead to better informed and more inclusive decisions.

Local knowledge valorized through this participatory mapping process has been used not only in regional development discussions, but also in a larger project on how people living in tropical forest communities are adapting to climate change and what political, economic, and cultural factors complicate resettlement from flood-prone zones. The residents of Yoke, for example, did not want to move their community permanently because that might mean losing their designation as a village, which had significant implications (Boissière et al. 2013).

4 Children Mapping "Exile" and "Home" in Long-Term Refugee Camps in India

The cultural mapping among Sri Lankan young people in long-term refugee camps in Tamil Nadu, India, discussed by cultural anthropologist Sasikumar Balasundaram in this example includes the ways in which children map both their present reality and an imagined future. He discusses how drawing can be used as a tool for children to map the present (exile) and future (home) in relation to experiences of place and agency. Balasundaram analyzed drawings solicited through an NGO program, in which Sri Lankan children in long-term refugee camps in southern India that had been built as a result of the ethnic conflict in Sri Lanka describe the relationship between the spaces of exile and home (geographical, cultural, and political) and the meaning and consequences of occupying those two spaces. Because many of their families had been occupying the refugee camps for more than 25 years, for many of the children a home that represented something other than exile existed only in their imaginations. For further background on the camps and ethical considerations of doing research with children, whose views he feels are too often excluded from both social science research and the public sphere, see Balasundaram (2014).

4.1 Field Work

During his dissertation fieldwork (2010–2011) among Sri Lankans in long-term refugee camps, Balasundaram used several participatory research methods to gather data from various age groups. Considering the ethical dilemmas related to using traditional research techniques such as interviews with children under the age of 18, the Jesuit Refugee Service, one of the two NGOs working with refugee children in all the refugee camps in Tamil Nadu, and Balasundaram reached a consensus to employ drawing as a tool for children to represent their everyday life in camps as well as map their dreams about the future. Balasundaram agrees with anthropologists, sociologists, and psychologists who believe that children are capable of understanding the complexity of social and political lives and predicting possible futures (see Coles 1986; Montgomery 2001; Kingsolver 2011). Children explain their views in their own ways. Thus, the responsibility lies with researchers to understand and interpret their meanings and represent their views in ways commensurate with those of any other perspectives being represented. During research in refugee camps, Balasundaram found that the ways children, youth, and adults understand camp life are not significantly different from one another despite the age gaps. His comparison of analyses of the children's drawings with analyses of conversations with adults reveals the falsity of the notion that children are not capable of understanding complex social and political realities and/or lack an ability to envision a future for their communities as well as themselves.

4.2 Findings

One of the major findings of Balasundaram's (2012) dissertation research was that the physical structure of the refugee camp, particularly the 10 × 10-feet spaces in which families may find themselves housed, often within warehouses without windows, ventilation, or toilet facilities, has been a primary cause for the social and cultural problems faced in everyday life in long-term camps. Infectious disease, domestic violence, alcoholism, early marriage, elopement (traditionally only caste and regional ethnic endogamous marriages are approved by Sri Lankan Tamil parents), increasing numbers of premarital and extramarital relationships (non-marital relationships are culturally taboo), sexual abuse, and various forms of child abuse are some of the most common social and cultural issues directly associated with the physical structure of the refugee housing. The camp is a congested settlement for the refugees and they are often set up in abandoned government warehouses located on the outskirts of Indian villages in Tamil Nadu. Many of them are single-standing large buildings with a lack of ventilation and are portioned into 10′ × 10′ spaces using transparent linen cloths or saris (a long traditional cloth worn by women in South India and Sri Lanka). This warehousing has been a causal factor in sexual exploitation, sexual abuse, and health risks (especially respiratory diseases) for the refugees. Approximately 25–100 families may live in a single building, with only one door. The majority of the refugee camps were established 25 years ago when tens of thousands of Tamils from Sri Lanka left the country to escape from state-sponsored violence and the civil war between the Liberation Tigers of Tamil Eelam, the rebels fighting for an independent state ("Eelam") for the Sri Lankan Tamils, and the government of Sri Lanka.

For several reasons, including the restrictions on conducting research in camps, there are not many studies of Sri Lankan Tamils in Indian refugee camps. Balasundaram wanted to undertake research that included the perspectives of refugees of all ages in the camps. He was particularly interested in the perspectives of children, as they have been traditionally excluded in policy research. Since he was sensitive to their double vulnerability as refugees and as children, he did not include them as research subjects in his dissertation research, but he wanted to include their perspectives and decided to learn about their views through their drawings as a form of cultural mapping.

4.3 *Children's Drawings*

In collaboration with the Jesuit Refugee Service, he announced a drawing competition for the children under the age of 15. Based on the advice from the JRS staff, who have been working with children in camps for the last 20 years, and to ensure parental consent regarding children's participation in the project, the drawing competition was identified as one of the best options for children to articulate their viewpoints on the experience of living in camps and their opinions about the future. Participation was voluntary, and Balasundaram did not directly solicit the drawings or collect them. The reason for having a competition was not to promote competitiveness among children or evaluate their talents individually. It was a way to increase the number of participants in the process. The majority of the children were awarded with a small amount of cash as it was the usual practice of the NGOs to compensate and motivate children to participate in conversations related to everyday life in exile. The children were accustomed to being consulted by the NGO about their views, and participated in a children's parliament in which they discussed how to address social problems in the camp, for example. The theme of the drawing competition was "camp life and hopes for the future."

Fig. 2 "The brown tree is our life in the refugee camp; the green tree is what our hope is." Photograph of a drawing by a Sri Lankan child living in a long-term refugee camp in southern India, 2011 (Courtesy of Sasikumar Balasundaram)

Interestingly, even though no specific guidelines were provided for children to follow in the competition, almost all drawings had been structured in a particular style (Fig. 2). Each drawing is a map of both the present and future. Typically, the first half of the page is a map of the present, and the second half of the page is a map of the future. Drawings made by refugee children in many parts of the world show that children include images of war and violence (e.g., Coles 1986); however, Balasundaram did not find a single drawing in this project that depicts war—that is mainly because all Sri Lankan children in Indian camps were born and raised in exile. The map of the present in every painting includes the physical landscape of the camp housing and represents the economic, social, and cultural problems associated with that. In contrast, the map of the future represents an ideal life and an avoidance of all social and cultural problems faced in the refugee camps during exile. One child described her map of the present and future as a drawing of "hopelessness and hope" (Fig. 3). Thus, Balasundaram calls the drawings maps of the present and future or maps of hopelessness and hope. For the children, the map of hopelessness represents overcrowding, suffering, control, militarization, violence, sickness, sadness, dirtiness, ugliness, discomfort, abuse, stigma, dependence, and loss. There are specific representations of adults with bottles of alcohol, adults beating children, pregnant women crying, etc., in crowded conditions, either barracks or warehouse style. In contrast, the map of hope depicts happiness, belonging, comfort, freedom, cleanliness, healthiness, freedom from violence, peace, security, self-sufficiency, and pride. In these drawings, there tend to be free-standing houses, farm animals, trees, running water, children with places to play, and lots of plants

Fig. 3 "My future dream." Photograph of a drawing by a child in a long-term refugee camp made in 2011, mapping the spatial conditions of the camp (barracks-style, on the right) and the lively space the child dreamed of inhabiting one day (Courtesy of Sasikumar Balasundaram.)

and flowers. In many of the drawings, the present is drawn in black and white and the future in color. Just like the vast majority of the adults interviewed, through their drawings, the children concluded that restructuring the physical structure or a complete freedom from living in warehouse refugee camps would ensure a life with dignity.

Children used their drawings as cultural maps to describe the present situation in exile and future hopes for the refugees. Children's maps of the present and future serve two main purposes: (1) they provide an opportunity for them to express their perspectives on the refugee situation, and (2) they provide decision-makers the information necessary to understand issues related to camp life from the perspectives of children. The most important point that we all could learn from these children's drawings is that children are equivalent to adults in many ways in regards to understanding and mapping the complex cultural and political life of exile and an imagined home, or place of belonging.

5 Conclusion

These two examples illustrate the ways in which cultural analyses benefit greatly from the incorporation of mapping techniques, and participatory mapping is a way to bring to the center of analyses the perspectives of those who have often been marginalized in state discourses. There are many ways to map the significance and accessibility of landscapes; those who study culture focus on the way that accessibility and significance can differ according to who is drawing the map.

Participatory GIS is a cultural mapping technique used increasingly by geographers and anthropologists that shows promise as a way to bring into the same frame different ways of understanding and valorizing the same space. As Boissière et al. (2013), showed, the process of different stakeholders coming together to do the participatory mapping as well as the maps resulting from that process that are in a synchronized format with governmental maps can lead to officials taking the views of constituents more seriously. Some of the newest forms of cultural mapping discussed in this chapter—drawings and aural mapping, representing the often unseen and unheard and the vantage points of species other than humans—are also the oldest. Cultural mapping can be a form of collective knowledge production, and there is much to be learned from indigenous epistemologies and precapitalist as well as postcapitalist perspectives on relatedness in space and time as the conversation goes forward. Although it is a technique that might easily be used, for example, to make product advertising campaigns more effective (like focus groups), in this chapter, we have emphasized the social justice applications of cultural mapping.

Further Reading

Crawhall, N. (2008). The role of participatory cultural mapping in promoting intercultural dialogue: 'We are not hyenas'; A Reflection Paper. Concept paper prepared for UNESCO, Division of Cultural Policies and Intercultural Dialogue. http://portal.unesco.org/culture/en/ev.php-URL_ID=37746&URL_DO=DO_TOPIC&URL_SECTION=201.html

Elwood, S. (2006). Critical issues in participatory GIS: Deconstructions, reconstructions, and new research directions. *Transactions in GIS, 10*(5), 693–708.

Marshall, D. J., & Staeheli, L. (2015). Mapping civil society with social network analysis: Methodological possibilities and limitations. *Geoforum, 61*, 56–66.

References

Achterberg, J. T., & Pierce, A. J. (2011). Creation of an aural map: Convergence of music and biological anthropology methodologies. *Anthropology News*, January: *52*(1), 3–6.

Appalachian Regional Commission. (2015). *County economic status in fiscal year 2016 (map).* http://www.arc.gov/research/MapsofAppalachia.asp?MAP_ID=105. Last accessed 8 June 2015.

Asad, T. (1973). *Anthropology and the colonial encounter*. New York: Humanities Press.

Balasundaram, S. (2012). "Freedom from camps: Housing and power in the experience of Sri Lankan long-term refugees in India." (Unpublished Ph.D. Thesis, Department of Anthropology, University of South Carolina.).

Balasundaram, S. (2014). Children matter: Including camp children's perspectives in refugee research. *Practicing Anthropology, 36*(3), 38–42.

Barni, M., & Extra, G. (Eds.). (2008). *Mapping linguistic diversity in multicultural contexts*. Berlin: Mouton de Gruyter.

Blunt, A., & Rose, G. (Eds.). (1994). *Writing women and space: Colonial and postcolonial geographies*. New York: The Guilford Press.

Boissière, M., Locatelli, B., Sheil, D., Padmanaba, M., & Sadjudin, E. (2013). Local perceptions of climate variability and change in tropical forests of Papua. *New Guinea. Ecology & Society, 18* (4), article 13.

Bullard, R. D., & Waters, M. (2005). *The quest for environmental justice: Human rights and the politics of pollution*. Berkeley, CA: Counterpoint Press.

Coles, R. (1986). *The political life of children*. New York: Atlantic Monthly Press.

DeVault, M. L. (2014). Mapping invisible work: Conceptual tools for social justice projects. *Sociological Forum, 29*(4), 775–789.

Downs, J. F. (1971). *Cultures in crisis*. Beverly Hills, CA: Glencoe Press.

Fardon, R. (Ed.). (1990). *Localizing strategies: Regional traditions of ethnographic writing*. Edinburgh: Scottish Academic Press.

Hargrove, M. D. (2009). Mapping the 'social field of whiteness': White racism as habitus in the city where history lives. *Transforming Anthropology, 17*(2), 93–104.

Harrison, F. V. (2009). *Outsider within: Reworking anthropology in the global age*. Urbana: University of Illinois Press.

Hoffman, S. M. (1976). Discussion of the film, *Kypseli: Women and men apart, a divided reality. Annals of the New York Academy of Sciences, 268*, 382–384.

Kingsolver, A. E. (2011). *Tobacco town futures: Global encounters in rural Kentucky*. Long Grove, IL: Waveland Press.

Lewis, O. (1966). The culture of poverty. *Scientific American, 215*, 19–25.

Malinowski, B. (1961) [1922]. Argonauts of the Western Pacific. New York: E.P. Dutton & Co.

Mapping for Rights. (2015). *Participatory mapping*. Rainforest Foundation UK http://www.mappingforrights.org/participatory_mapping (Last accessed 18 June 2015).

Mbodj, M. (2002). So geographers in Africa maps with savage pictures fill their gaps': Representing Africa on maps. In G. C. Bond & N. C. Gibson (Eds.), *Contested terrains and constructed categories: Contemporary Africa in focus* (pp. 37–58). Boulder, CO: Westview Press.

Mendes, C. (1989). *Fight for the forest: Chico Mendes in his own words*. New York: Monthly Review Press.

Miller, B. D. (2005). *Cultural anthropology* (3rd ed.). Boston: Pearson.

Mitchell, D. (2000). *Cultural geography: A critical introduction*. Oxford: Blackwell Publishers.

Montgomery, H. (2001). *Modern Babylon? Prostituting children in Thailand*. New York: Berghahn Books.

Norgren, J., & Nanda, S. (1996). *American cultural pluralism and law* (2nd ed.). New York: Praeger.

Polanco, M. B. (2012). `Not to scale': Mapping race in a Virginia historically black community. *Transforming Anthropology, 20*(2), 118–130.

Rodman, M. C. (1992). Empowering place: Multilocality and multivocality. *American Anthropologist, 94*, 640–656.

Said, E. W. (1979). *Orientalism*. New York: Vintage.

Sopranzetti, C. (2014). Owners of the map: Mobility and mobilization among motorcycle taxi drivers in Bangkok. *City & Society, 26*(1), 120–143.

Takhteyev, Y., Gruzd, A., & Wellman, B. (2012). Geography of twitter networks. *Social Networks, 34*, 73–81.

Thomas-Houston, M. M. (2005). *`Stony the Road' to change: Black Mississippians and the culture of social relations*. Cambridge: Cambridge University Press.

Tobin, G. A., Whiteford, L. M., Jones, E. C., & Murphy, A. D. (2010). Chronic disasters and social networks. *Anthropology News*, October: *51*(7), 21–22.

Urban, P., & Schortman, E. (2013). Monumentality, territoriality, and networks during the Middle Preclassic in northwest Honduras. *Archaeological Papers of the American Anthropological Association, 22*(1), 87–106.

Vajjhala, S. P. (2006). `Ground truthing' policy: Using participatory map-making to connect citizens and decision makers. *Resources for the Future, 162*, 14–18.

Vásquez-León, M. (2009). Hispanic farmers and farmworkers: Social networks, institutional exclusion, and climate vulnerability in southeastern Arizona. *American Anthropologist, 111*(3), 289–301.

Vila, P. (2000). *Reinforcing borders: Social categories, metaphors, and narrative identities on the U.S.-Mexico frontier*. Austin: University of Texas Press.

Wendland, C. L. (2012). Moral maps and medical imaginaries: Clinical tourism at Malawi's College of Medicine. *American Anthropologist, 114*(1), 108–122.

Wernke, S. A. (2007). Negotiating community and landscape in the Peruvian Andes: A transconquest view. *American Anthropologist, 109*(1), 130–152.

Williams, B. (Ed.). (1996). *Women out of PLACE: The gender of agency and the race of nationality*. New York: Routledge.

Wolf, E. R. (1982). *Europe and the people without history*. Berkeley, CA: University of California Press.

Yoneyama, L. (1999). *Hiroshima traces: Time, space, and the dialectics of memory*. Berkeley, CA: University of California Press.

Mapping Religiously, or Religiously Minding the Map?

Lillian I. Larsen

Abstract This essay explores the programmatic pedagogy that has long linked religion and maps. Through comparative juxtaposition of contemporary surfaces, it underscores the degree to which disciplinary constructs continue to shape cartographic representation of religiously defined space. It then examines this contemporary phenomenon in light of the historical role that religion has played in iteratively naming particular geographic landscapes authoritative. Extending the work of Catherine Delano-Smtih on "Maps as Art and Science" (1990; Cf. *Maps in Bibles*, 1991), it demonstrates the particular merits of melding "change" with "tradition" in a manner that troubles common cartographic nomenclature. Following Harley (in *Cartographica* 26:1–25, 1989), it simultaneously affirms the importance of deconstructing authoritative surfaces with more mindful utilization of contemporary cartographic tools. As religiously defined landscapes are here 're-drawn' in ways that combine critical historical engagement with sophisticated digital artistry, both the most malleable, and the most stable maps are rendered less emphatic. Re-shaped and re-purposed, these surfaces no longer serve as static harbingers of hegemonic 'truth'—or malleable registers of cultural idiosyncrasy—but symbolically supple interfaces that temper certainty. Absent an ideological agenda aimed at culturally affirming inherent authority, they effectively demonstrate the usefulness of inverting established pedagogies. As the traditional tactics that have dictated 'mapping religiously' are re-deployed, they serve as tools for 'religiously minding the map.' Persuasive influence, however, is contingent. Even as 'the authority inherent in all mapped surfaces' is re-defined, effective re-configuration remains grounded in shared cartographic assumptions.

Keywords Pedagogy · Religion · Bible · Historical GIS · Maps in Bibles · Maps of Paul

---

Lillian I. Larsen and with Cartography by Steve Benzek

L.I. Larsen (✉)
Religious Studies/MSGIS, University of Redlands, Redlands, USA
e-mail: lillian_larsen@redlands.edu

S.D. Brunn and M. Dodge (eds.), *Mapping Across Academia*,
DOI 10.1007/978-94-024-1011-2_16

1 Introduction

The authority implicit to canonical texts, and per Harley (1989), "all mapped surfaces," makes the disparate defining emphases used to distinguish diverse maps of religion, something of a case study for exploring interpretive patterns in both religion and cartography. It is demonstrably the case that religious truth as map, and maps of religious truth, continue to exert, at once, implicit and explicit influence on both contemporary and historical understanding of conceptual landscapes. In fact, one might argue that maps—whether ancient or contemporary, overtly ideological or scientifically 'objective'—have played a singularly decisive role in communicating and/or masking a religiously prescriptive worldview.

Within this frame, comparative work with both contemporary and historical cartographic refraction serves well to illustrate the degree to which: (A) disciplinary definitions have served to shape visual representation of religious space, spatial processes, and spatial patterns; (B) disparate digital refractions of a common religious landscape, invite consideration of the malleability of contemporary cartography, when viewed in light of the pedagogical role religion has played in rendering mapped historical landscapes, remarkably stable; (C) in melding these motifs, supple contemporary surfaces can be mindfully molded to pedagogically re-shape and re-purpose the symbolic stability of historical maps, marking a new frontier of mapping religion.[1]

2 Maps of Religion

As digital tools have rendered religious landscapes almost infinitely malleable, three recently mapped refractions suggest that how one defines religion has everything to do with how its contours are presented on a map. One surface, produced by Buzzfeed, purports to represent the "religion of members of the U.S. House of Representatives" (Buzzfeed News 2013; Schwartz 2015) A second visualization, published by the Pew Research Center on Religion and Public Life, draws on the same body of data, but defines this congressional landscape in inverse terms (Pew Forum 2012; Schwarz and Schaul 2015a, b). A third, likewise captures the character of the American religious landscape, but refracts neither the ubiquity nor diversity of its composite constituencies. Instead, it marks emergent sites of interreligious initiative (Pluralism Project 2012–2016). As each map defines depictions of a common physical landscape, its conceptual contours offer dramatically different answers to the same question.

[1]As mapped sequences eloquently communicate the rhetorical character of both text and cartographic contexts, they elucidate the degree to which how one defines religion—and represents religious 'truth'—continues to be shaped by cultural and social norms, which have, themselves, rendered both religion and maps authoritative in character.

2.1 *Defining American Religion*

Given a sample size of 435 elected leaders, Buzzfeed's documentation of '31 religions' in the U.S. House of Representatives, appears conceptually plausible. Overall, tallies suggest a foundational core large enough to offer a significant base for sophisticated parsing. Closer scrutiny, however, provides a provocative portrait of both surveyor and subject. Five of the 'six' primary constituencies—here listed as discrete 'religions' (Fig. 1)—are more accurately characterized as 'Christian.' In fact, of the 31 'religions' named, 26 are included in this broader category. This leaves Judaism—the sixth constituency on the Buzzfeed list—something of an anomaly. With nothing that overtly marks any distinction between this tradition and the preceding five, 'by definition' it might easily be mistaken for a subset of the religion, which historically, it spawned/generated.

Religion	Number	Percent
Catholics	135	31
Baptists	66	15
Methodists	45	10
Anglicans/Episcopalians	35	8
Presbyterians	28	7
Jews	22	5
Other	104	24

Relatively scaled, the size of respective constituencies appears unremarkable—until one notices that the tally of Jews in the House of Representatives is on par with the number of Presbyterians. In turn, if constituent numbers are more conventionally defined, commensurate measures of 'religious' density leave 22 Jewish representatives juxtaposed with a ledger that includes over 300 Christians—of various stripes.[2] More detailed parsing of this Jewish constituency renders contours defined in still sharper relief. If one distinguishes various Jewish strains—for example, Orthodox, Conservative, Reform—in ways commensurate with those

[2]Understood as a measure of 'religious' sensibilities and assumptions, its odd equivalencies underscore the degree to which maps retain unexpected capacity to effectively render patent otherwise unregistered, and often unrecognized, assumptions. As elucidating is subsequent analysis that visually parses this data along party lines. Buzzfeed reports that the "majority of Catholics and Jews are Democrats while the remaining 'religions' [emphasis mine] on the top-six list tend to be represented by Republicans." Given that 21 out of 22 Jewish members are Democrats, the 'Jews' are here named the "most partisan" constituency; although Mormons remain a close second, with seven out of the eight Mormon members of Congress, identifying as Republican. States retaining a single seat—Idaho and Utah—derivatively bear the designation of representation by "one religion"—here, Mormonism (www.buzzfeed.com).

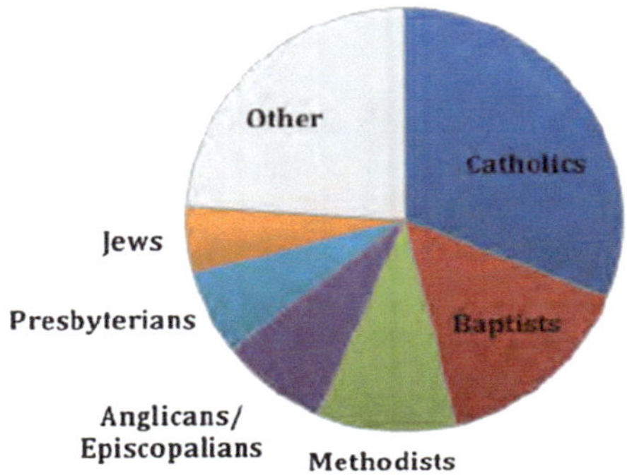

Religion	Number	Percent
Catholics	135	31
Baptists	66	15
Methodists	45	10
Anglicans/Episcopalians	35	8
Presbyterians	28	7
Jews	22	5
Other	104	24

Fig. 1 Religious affiliation of a sample of 435 elected leaders (Pie chart and table)

used defining Christian constituencies, any substantive vestige of diversity disappears. Rather than the small United Nations suggested by defining categories that premise 31 religions in Congress, one is instead faced with a governing body that remains resoundingly, and monolithically Christian.

2.2 *Defining Religion in America*

Alternate parsing of the same data is presented by the Pew Research Center on Religion and Public Life. Emphatically inverting the definitions that structure the constituencies refracted in the Buzzfeed tallies, Pew introduces its discussion with a series of "firsts"; each signaling something of a watershed moment in American religious and political history. Here, the monolithically Christian Congress similarly presented in the Buzzfeed map is alternately defined as a body that includes "the first Buddhist to serve in the Senate, and the first Hindu to serve in either chamber."[3] These individuals join two Muslim Representatives, already listed on the congressional roles. With the more substantive Christian and Jewish constituencies discussed above, what remains an "unprecedented level of diversity" in the House, is reflected not in 31 'religions', but five.[4] This fledgling range of representation, however, is noteworthy. Such radically revised numbers simultaneously call attention to the hegemonic political influence of Christian constituencies, and mirror "a gradual increase in religious diversity…in the country as a whole" (Pew Forum 2012).

[3]The latter of these preserves a seat formerly filled by a Sikh representative, who served during the 1950s and 1960s.

[4]Aggregation of "the first member of Congress to describe her religion as 'none'" remains an open question in both the Buzzfeed and Pew tallies.

2.3 Re-defining Religion in America

While the respective religious landscapes presented by Buzzfeed and the Pew Forum offer inverse refractions of common parameters, the presuppositions that structure a third set of visualizations define the American religious landscape in alternate terms.[5] Philosophically grounded in an initiative that has sought to physically document "how a 'Christian Country' has become the world's most religiously diverse nation," the mapped landscapes produced by the Pluralism Project at Harvard University, refract American "religion" on a surface that shares the same borders as the constituencies chronicled by Buzzfeed and the Pew Forum. However, as religiously delineated, this "portrait" draws on an alternate body of underlying data. Explicitly "selective," the resulting map eschews categorical definitions altogether. Instead, its geographical surface solely displays "emergent" loci of "interfaith infrastructure." Here, Christianity—by virtue of its dominance—is rendered 'virtually' invisible (Pluralism Project 2012–2016).

In broader representation, the Pluralism Project has historically emphasized the specific physical locations of discrete religious strains. As refracted, each "mapped surface" is organized by city. Each city, in turn, is communally defined by intimate detail that invites comparison between the monolithic face of a "Christian country"—implicitly represented in the popular parameters that inform the Buzzfeed map—and the rich spectrum of diversity manifested in constituent religious presence 'on the ground.' Such discrete maps document, at once, Christianity's continued dominance, and the degree to which the "changing contours of American religious demography…[remain] both visible and invisible in American public life" (Pluralism Project 2012–2016).

2.4 Synopsis

Like maps, for both the casual reader/viewer, and the serious student/scholar, the variable character of religious "truth" is often difficult to identify, and thus to engage. The perspective afforded by examining this melding of maps and religion, however, underscores the importance of recognizing the role religion may play in defining mapped landscapes. It simultaneously emphasizes the inverse role maps can play in shaping and articulating religious understanding. While some might

---

[5]Explicitly addressing the degree to which, "over the past five decades, immigration has dramatically changed the religious landscape of the United States," the Pluralism Project seeks to render visible the daily encounters with people of different religious traditions that take place in cities and neighborhoods across the United States. Devoted to documenting this emergent tapestry, an expanding network of affiliates has joined forces in efforts aimed at sketching "the contours of [a] multi-religious society, …explor[ing] new forms of interfaith engagement, study[ing] the impact of religious diversity in civic life, and contextualiz[ing] these findings within a global framework" (Pluralism Project 2012–2016).

argue “a world of difference…[separates a] modern map” from those created by individuals for whom “belief governed…description of the natural world” (Scafi 2006: 15), these maps suggest that “belief”—as variously defined—remains as proximate a parameter in guiding contemporary cartographic configuration as its historical counterparts. The malleable parameters that variously shape definition of both the American and World religious landscape become alternately interesting, however, when viewed in conversation with the historical role religion has played in rendering particular landscapes remarkably static.[6]

3 Mapping ‘Religiously’

The defining character of such malleable religious landscapes is inversely elucidated in a set of five maps that have been included in print, Protestant bibles since the 16th century (Delano-Smith 1991). In her detailed survey of “Maps in Bibles,” Catherine Delano-Smith introduces this corpus as one element in a larger Reformation-era reading program—explicitly aimed at guiding untrained readers, encountering vernacular translations of ‘scripture’ for the first time. Across subsequent centuries, the same core content was published and re-published. Over time, its contours likewise exerted singular influence in effectively defining the biblical landscapes; the maps were created (and published) to elucidate. In contrast to the supple surfaces examined above, for almost half a millennium, these maps have remained remarkably stable. The pedagogical predispositions that inform their static character invite readers, cartographers and scholars of religion, to ‘think about what they do.’

3.1 Maps in Bibles

In her seminal work, Delano-Smith attaches the history of “Maps in Bibles” to a lengthy, firmly established trajectory of using manuscript illustration and illumination “to communicate religious messages” (Delano-Smith 1990, 1991). She notes that from antiquity through the early medieval period, depicted subject matter was

[6]Apparent visual anomalies underscore the constructed character of religious truth. Likewise, mapped landscapes render patent the fluid character of static source material. They call attention to the pedagogical power of historical influence on the contemporary cartographic moment—whether articulated in words, surfaces, or symbol. Here, one must take seriously the complementary (and arguably commensurate) degree to which, in an era of increasing malleability, both cartographic and religious constructs have been imbued with a seemingly fixed character. In contexts increasingly defined by diversity of practice and presentation, this invites consideration of the terms that render particular maps and religions—and, as importantly, particular maps of religion—authoritative, while leaving others, virtually, ‘off the map.’

primarily drawn from Hebrew Scripture (Delano-Smith 1990: 66).[7] The 12th and 13th centuries broadened this emphasis to include portions of the Christian canon (Delano-Smith 1990: 66; Levin 1985; Weitzmann 1947, 1959).[8] With the onset of the Reformation—and the introduction of the printing press—the atlas of included content grew narrower, its focus more explicit. To introduce "uninstructed" readers to the bible, *sans* ecclesiastical intermediaries, primary representation was reduced to an essential core of five maps (Delano-Smith 1990: 67–69).[9]

This distilled corpus was comprised of a map of the "Israelite's Exodus from Egypt to Canaan," first printed in Zurich in 1525, and included in Lutheran Bibles printed in Antwerp from 1526 onwards (Delano-Smith 1990: 67).[10] The remaining 'set of four' included:

> Calvin's map of Eden (first published in 1554, included in bibles from 1560); the division of Canaan (first recorded as a separate map in 1559 [re-using tribal boundaries shown on earlier maps of the Exodus]); the Holy Land in the time of Christ; and the eastern Mediterranean (showing places visited by Paul and the Apostles, first recorded in a New Testament of 1549). (Delano-Smith 1990: 67)

By Delano-Smith's (1990: 67) tallies, the same five maps account for roughly 80 % of the maps included in 16th century bibles.

In contrast to the manifestly malleable, contemporary religious landscapes examined above, Delano-Smith observes that with few exceptions, "identical material appeared on version after version and copy after copy of each map." As woodblocks were "reused, closely copied, hired, perhaps even pirated," they were left unaltered. "Apart from translating place names" to facilitate linguistic access for vernacular readers, even "formcutters or artists with very different styles" used common configurations (Delano-Smith 1990: 73). In fact, per Delano-Smith, "four of the five maps "were used as a more or less complete set by at least 24 different publishers for 48 different editions of the complete bible in nine languages in many countries for half a century." Albeit copied, 'updated' and recopied, "each remained to all intents and purposes identical" (Delano-Smith 1990: 73).

---

[7]In her focused study of "Maps as Art and Science," Delano-Smith (1990: 66) delineates the content of this corpus in some detail. She suggests that "whatever the medium, manuscript or wall paintings or other," one encounters representations drawn from Hebrew Scripture. These include "the Creation cycle (Eden, Adam and Eve, the Flood, Noah's Ark); the Exodus cycle (the Israelites in Egypt, their Red Sea crossing, their forty years of wandering in the desert with associated events); the King [Monarchy] cycle (David, Solomon with his temple and its furniture, etc.); the Prophet cycle (Ezekiel with his vision of the rebuilding of the Temple, Daniel with lions, the young men in the fiery furnace, and his apocalyptic vision of the Four Beasts)."

[8]Delano-Smith notes the particular emphasis here placed on cartographically friendly narratives, like those included in the Gospels and Acts. She simultaneously signals sustained interest—even at this early point—in the narrative episodes related to the Apostle Paul.

[9]Delano-Smith (1990: 67) notes that "there was little that was entirely new about any of the maps in the set." For example, although these maps came to be associated with Calvinist editions, exegetically, they likewise incorporated aspects of Luther's teaching.

[10]Delano-Smith (1990: 67) observes, "an Exodus map of some sort was also included in almost every illustrated Calvinist (Genevan) bible from 1559 onwards."

3.2 *'Biblically' Mapping*

While in contemporary parlance such design choices might be reduced to something of a historical curiosity, Delano-Smith notes the degree to which the maps' stable character functionally served to reinforce particular exegetical principles.[11] Mapped, and re-remapped, each surface affirmed, and re-affirmed "a cardinal feature" of Reformation theology. By virtue of its status as "Holy Scripture…the written text…was clear" (Delano-Smith 1990: 75).[12] Introducing any representation that "render [ed] the gospel uncertain," fundamentally compromised its authority. Because "the power of the Word…lay in its unambiguity…the strength of the Scriptures" was embodied in its "directness." Additional "elaboration" risked confusing and/or compromising the clarity of an "essential message" (Delano-Smith 1990: 75). As graphic manifestations of "the simplicity and plainness of…the written word," selected maps were multi-purpose. Each served not only to "innocent[ly] aid…the reader's understanding," but also to provide "strict guidelines" for the "proper" interpretation of Scripture—in conformity with authoritative Protestant precepts—and consonant with "Protestant pedagogical strategy" (Delano-Smith 1990: 78).

Albeit "metaphorical in form, the bible maps…claimed geographical accuracy." Through effectively visualizing "the 'truth' of the text," they derivatively established a "hermeneutic connection between geography and history." This was manifested in co-existent and complementary tendencies that paired "change and tradition" (Delano-Smith 1990: 73). As perceptions of geographical 'accuracy' changed, the maps incorporated the nomenclature of these cartographic adjustments, but retained—and traded on—static claims of exegetical exactitude. For example, Delano-Smith observes that, "early on, when most people's sense of distant places would have been rather vague, it was sufficient to outline a country, and label it and its towns, mountains and rivers, to convey the notion of geographical reality." Over time, as heightened geographical literacy allowed for "incorporation of first-hand geographical knowledge into maps," the same content was re-configured "to satisfy a more critical readership" (Delano-Smith 1990: 73). By incorporating established 'historical' content, contemporary cartographic nomenclature, effectively re-affirmed "the reality of the outlines and places depicted" (Delano-Smith 1990: 73–74). For example, as mundane, "regional maps were given gridlines, co-ordinates, and degrees of latitude and longitude," elements like the "the scale bar, compass rose, graticule, and graduated margin," were also incorporated into bible maps (Delano-Smith 1990: 74).

---

[11]Delano-Smith (1990) acknowledges the considerable investment required to create and print any 16th century mapped reproduction of a biblical landscape. She simultaneously argues that re-use of this common core of maps cannot be explained solely on economic grounds.

[12]Emblematic is William Farel's premise that "he defeats the gospel who renders it uncertain" (*Aux Lecteurs Crétiens*, Proposition 7).

3.3 *Synopsis*

The effectiveness of the Protestant's pedagogical strategy is confirmed by the number of contemporary bibles that carry some permutation of the same five maps that have appeared in print editions since the 16th century. That the set can now be found in versions produced and published by Catholic and Protestant (and Orthodox) presses, signals the scale of programmatic success. Equally suggestive, is the degree to which overtly religious renderings have influenced broader, non-religious representation of the biblical landscape.[13] As increased access to digitally produced maps has facilitated a paradigm shift that parallels the textual popularization precipitated by Protestant printing of vernacular bibles, with inverse proportionality, technology has often simply rendered "the scientistic rhetoric of" maps "more strident" (Harley 1989: 2). However, by tempering increasingly versatile tools with the lessons of history, the same positivist ideologies can be as readily re-directed. Through defining and shaping biblical, popular, and scholarly cartographies, in more nuanced, and critically productive ways, the didactic structures that insured the success of Protestant pedagogies, in fact, provide a rich sandbox for more 'mindful' meldings of 'art and science', 'tradition and change.'

4 Mindfully Mapping

Among the five maps included in early Protestant bibles, those related to the apostle Paul have proven themselves particularly persuasive. Iteratively re-produced, 'updated', and re-published for over half a millennium, both historical and contemporary manifestations offer provocative loci for exploring the multi-faceted character of pedagogically packaged, religious influence—in space, and over time. Like the broader corpus of maps discussed above, as digital tools have rendered mapped interfaces increasingly malleable, cartographic representations of the "journeys of Paul" have remained remarkably stable. While these static surfaces register the stubborn influence of Protestant map-making priorities, contemporary cartographies invite re-presenting Paul's 'travels', in ways that re-define conceptual frontiers, and pedagogically re-purpose the aims of putting 'religion' on a map.

4.1 *The 'Journeys' of Paul*

Early prototypes that ground subsequent depiction of the 'journeys of Paul' are simply identified as maps of the "Eastern Mediterranean." Rather than tracking

[13] Here recent maps produced by National Geographic are particularly provocative. See further discussion below.

Paul's movements, included detail instead emphasizes locations of "the places visited by Paul and [/or] the Apostles" (Delano-Smith 1990: 67, 71, Fig. 2D). As this cartographic register develops, emergent surfaces absorb iconographic detail consonant with the cultural and cartographic trends of a given historical period. For example, a late 16th century map includes decorative cartouches embellished with period drawings that visualize narrative detail associated with Paul's movements, on the road and at sea (Fig. 2). An early 17th century map introduces route markings that faintly trace a biblically sequenced itinerary of investment (Fig. 3). Derivative maps are characterized by more clearly defined trajectories. These are, in turn, accompanied by detailed additional visuals, drawn from narrative incidents associated with Paul's life and travels (cf. Vischer 1642).

Full sequencing of the cartographic emendation that has shaped contemporary presentations of the "*Peregrinatio Pauli*" remains a work in progres (cf. Larsen and Benzek 2014, 2016). However, spanning religious, scholarly and popular divides, maps of "Paul's Journeys" might be classed as a genre unto themselves. In contemporary manifestation, some maps are color-coded to aid visual accessibility (Fig. 4). Others depict each 'Journey of Paul' as a discrete entity accounting for a particular segment of Paul's life and work. One of the most recent renditions—a compilation published by National Geographic in 2012—combines color-coded itineraries with Renaissance portraits of the 12 male apostles (Toddhunter and Johnson 2012).[14]

Half a millennium post introduction, these maps remain standard fare in the ancillary material included in both popular and scholarly bibles. The same presentations are readily available in well-respected reference works. For any who wish to embark on a more tangible venture, a simple internet search produces, not only maps of Paul's missionary 'adventures', but also links to travel sites where one can follow 'in the footsteps' of 'the Apostle' on organized sightseeing tours that encompass the Mediterranean world. Whether recorded as individual itineraries or a composite image, inclusive or devoid of ancillary illustration, incorporated into popular or scholarly, print or digital media, Paul's movements are presented as secure entities. Evocatively underscoring the perennial patterns of "incremental change" and "enduring tradition" emphasized in Delano-Smith's (1990, 1991) work on 16th-century bibles, even as seismic shifts have re-shaped historical understandings of Christian origins—and derivatively, Paul's role in the trajectories that

[14]This map, with its accompanying article, explicitly links Paul's journeys with contemporary pilgrimage. It also tacitly acknowledges growing historical awareness that Paul was certainly not the sole, or perhaps even primary, proto-Christian traveler. While content included in the primary surface mirrors contemporary convention in its delineation of color-coded itineraries, a small pull-out map traces the less familiar path of Thomas's 'apostolic' journey to India.

Fig. 2 "Peregrinationis Divi Pauli Typus Corographicus" (Ortelius 1598)

Fig. 3 "Peregrinatio Pauli" (Mercator 1607)

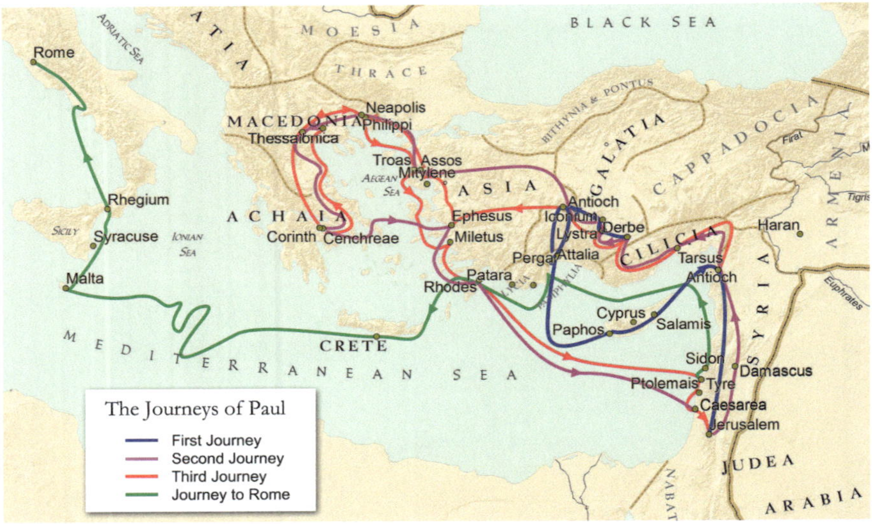

Fig. 4 Contemporary map of the journeys of Paul (Larsen and Benzek 2014, 51, 2016, 132)

mark its progression—these established itineraries remain routine reference points for 'secure' delineation of this storied 'Apostle's' actions and activities.[15]

4.2 *Deconstructing Paul*

Although the detailed itineraries that structure familiar maps of Paul's "peregrination" suggest an exhaustive trove of information pertaining to Paul, like most ancients, Paul comes to us piecemeal. Texts written "by," "in the name of," and "about" Paul surface both within and beyond the canon. Countering stable mapped trajectories of the apostle's 'storied' travels, these sources mark alternate ways of defining Paul's itinerancy, and record a shifting network of Pauline associates and communities. Examined in conversation with conventional 'maps of Paul', from both a religious and cartographical perspective, what is perhaps most interesting about included geographical content is not a delineated course of the apostle's

[15]In measuring the success of the Protestant pedagogical program, one cannot ignore the degree to which the clear sequencing of this itinerary has increasingly come to define not only Paul's life and investments, but derivatively, the westward, Eurocentric movement of emergent Christianity, itself. Whether encountered in the classroom, online, or in the appendices to a garden-variety Bible, a trajectory that begins in Jerusalem and ends in Rome, raises few questions. Instead, each mapped and re-mapped iteration reassuringly renders historically hypothetical constructs more authoritative. In turn, broad evidence for less linear, diasporic refractions of communal formation and contested individuation, becomes more difficult to discern.

'journeys', but rather the variable character of the locative detail with which he is represented (Larsen and Benzek 2016).

Returning to the underlying source material with new questions in mind, one encounters a less than clearly defined textual and cartographic landscape. On a variety of grounds, it has long been argued that only seven of the thirteen canonical letters attributed to Paul were authored by Paul himself.[16] Authorship of the six additional letters ascribed to Paul is 'disputed.' Among critical scholars, it is generally held that these documents are pseudonymous. They are thought to have been written by associates or a school of Paul, and distributed across two generations that span roughly half a century.[17] The Acts of the Apostles does not overtly make any claim to Pauline authorship. However, nearly three-quarters of its 28 chapters address some aspect of Paul's life, work and travels. While scholars affirm that this text is clearly composed by a figure sympathetic to Paul, the story itself appears to present an independent narrative tradition, composed long after Paul's death.[18] Whether the author of Acts had access to Paul's 'undisputed' letters and/or additional source material related to Paul, remains a topic of ongoing debate. It is, nonetheless, Acts' relatively late, secondary story-line that supplies core details and overall structure to the mapped profile that, in popular imagination, has come to define Paul as a historical figure.

4.3 *Reconstructing Paul*

When one seeks to re-construct the "journeys" of Paul, attending to the relative historical reliability of available sources, even the simplest presentations yield refractions that counter the well sequenced trajectories of established portrayal (Fig. 5) (Larsen and Benzek 2014: 56). In turn, introducing more sophisticated symbology—for example, replacing clearly demarcated lines with dispersed, proportional pie charts—readily registers the disparity that characterizes geographical

[16]Roughly dated to the final full decade of Paul's life (49–60 CE), this undisputed corpus includes: First Thessalonians, Philippians, Philemon, Galatians, 1 Corinthians, 2 Corinthians, and Paul's culminating address to the Romans.

[17]The earlier subset of 'disputed' letters includes 'Paul's' second letter to the Thessalonians, as well as texts addressed to the Colossians and to the Ephesians. On linguistic and theological grounds, this body of deutero-Pauline material is dated to the decades immediately following 'the Apostle's' death. The subsequent subset of letters, often termed 'the Pastorals', includes two letters addressed to Timothy, and one to Titus. Loosely dated from the late 1st through mid-2nd century of the Common Era, these documents appear to be the work of a second generation of followers, still asserting Paul's authority, and writing in Paul's name.

[18]Along with its prequel, the Gospel of Luke, Acts has conventionally been dated to the final decades of the 1st century. In alternate consideration, this two-volume compendium is assigned a chronological frame that roughly aligns with the Pastorals. For a detailed summary of these debates, and comprehensive bibliography (see Pervo 1987, 2006, 2010; Penner and Vander Stichele 2003).

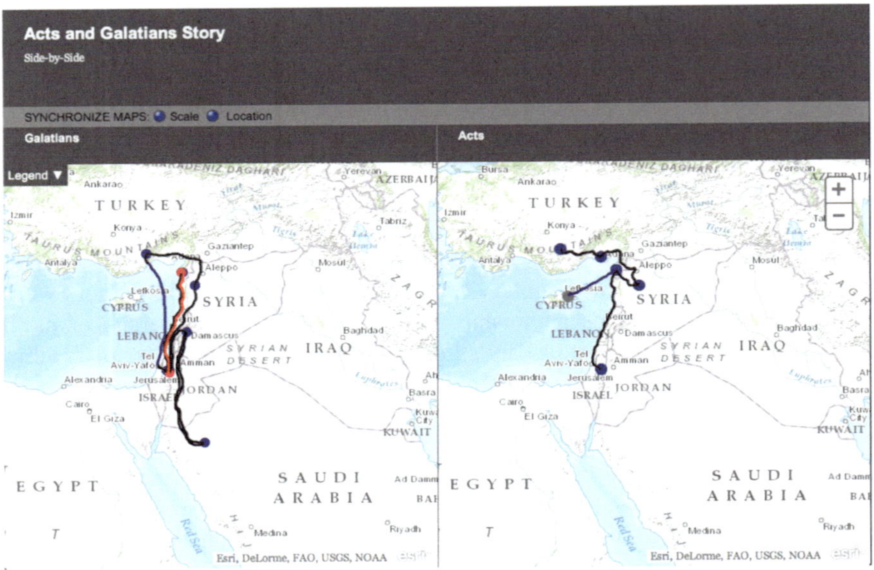

Fig. 5 Student map comparing Acts and Galatians

data drawn from, and delineated by inclusion in respective textual source layers (Fig. 6) (Larsen and Benzek 2014: 55).

Introducing marked contrasts in color and size to visually distinguish between the relative frequency and distribution of places mentioned in letters written by Paul, and those penned by later followers, makes temporal distinctions clearer, even as more evocative cartographic nomenclature—for example, a 'scratchy' pen and ink graphic—de-stabilizes any residual notions of certainty. Re-presenting, instead, the inexact and fragmentary character of extant historical data, sketchy loci of concentration emphasize—rather than efface—marked disparity in the geography of respective source layers (Larsen and Benzek 2016). Simultaneously, translucent layers created by 'sketchy' pen strokes, readily accommodate overlapping digital renditions of respective source strata on the same surface.

Viewed singly, the surfaces that emerge from discretely plotting respective layers of Paul-related source material are strikingly sparse, and decidedly non-linear (Fig. 7) (Larsen and Benzek 2016). Their impressionistic concentrations register the variation that distinguishes geographic data included in the "undisputed" letters of Paul (Fig. 7a), from that named by later letter writers (Fig. 7b). Clear contrasts in the geographic density and character of respective portions of Acts, likewise underscore this document's less than uniform character. As recounted (and mapped), Paul's investments (7.54–28.31) (Fig. 7c) stand in sharp relief relative to the limited scope and detail assigned a broader range of communal leaders/apostles (1.1–7.53) (Fig. 7d).

Fig. 6 Pie-chart map of places mentioned in undisputed (lilac) and disputed (maroon) letters of Paul (Larsen and Benzek 2014, 55, 2016, 136)

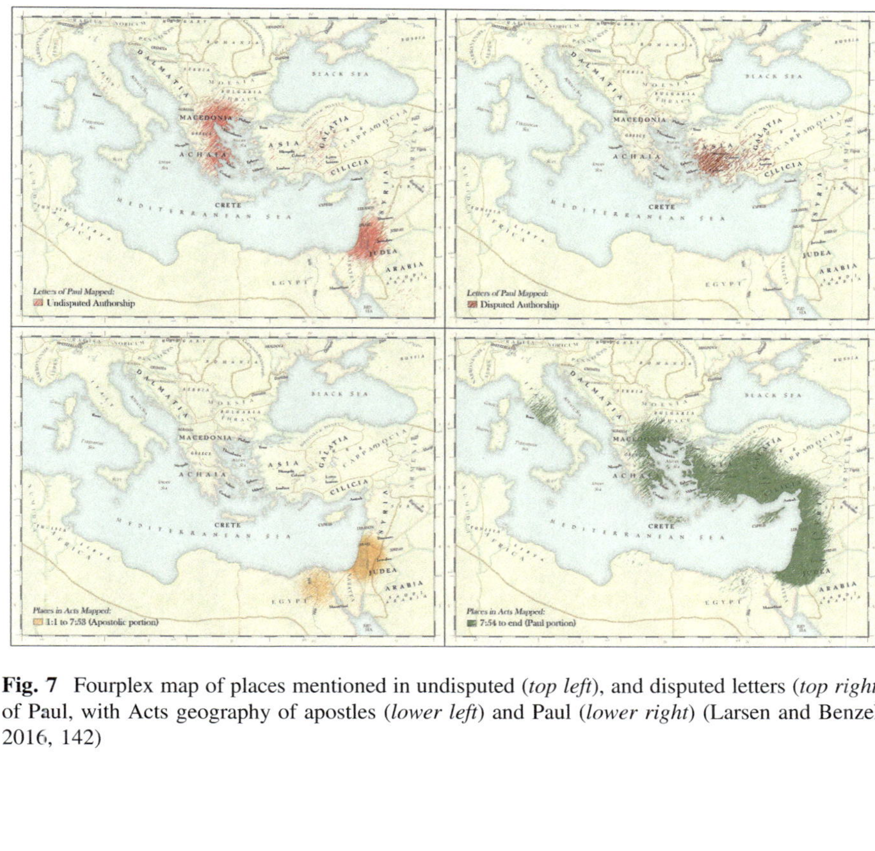

Fig. 7 Fourplex map of places mentioned in undisputed (*top left*), and disputed letters (*top right*) of Paul, with Acts geography of apostles (*lower left*) and Paul (*lower right*) (Larsen and Benzek 2016, 142)

As translucent color gradations facilitate the inclusion of chronologically distinct geographies on the same map surface, the contrasts apparent in composite images are compelling. Juxtaposed combinations highlight patterns both in data excised from, and included in the most familiar maps of Paul. While 'scratchy' registers of iterated pen strokes mark areas of concentration, they visually mirror (and reinforce) the historically nebulous nature of the source material in question. Although these contrasting geographies offer more historically reliable refractions, the look of hand-drawn pen and ink serves to remind that even apparent loci of convergence (and divergence) remain, essentially, impressionistic (Fig. 8) (cf. Larsen and Benzek 2016, 143).

Trading on the Protestant pedagogical premise, these maps simultaneously subvert its programmatic emphases. As non-seamless surfaces 'authoritatively' elucidate—rather than efface—the disparate geographies of temporally discrete source layers, their contemporary melding of 'art and science' remains rhetorically persuasive. By didactic design, however, the goals have shifted. Included refractions communicate not a single, clearly delineated, authoritative reading, but instead the diversity that grounds the contours of critical debate. The aim is no longer doctrinal certainty, but instead, engaging viewers in 'thinking about' Paul—and religious reification, more generally—in complex ways. As mindful deployment of digital tools renders the same surface in richly variable ways, semi-transparent gradation encourages—and facilitates—iterative interrogation of the historical and narrative elements refracted in disparate strata of source material. Through dis-assembling and re-assembling ancient spheres of influence—almost as one

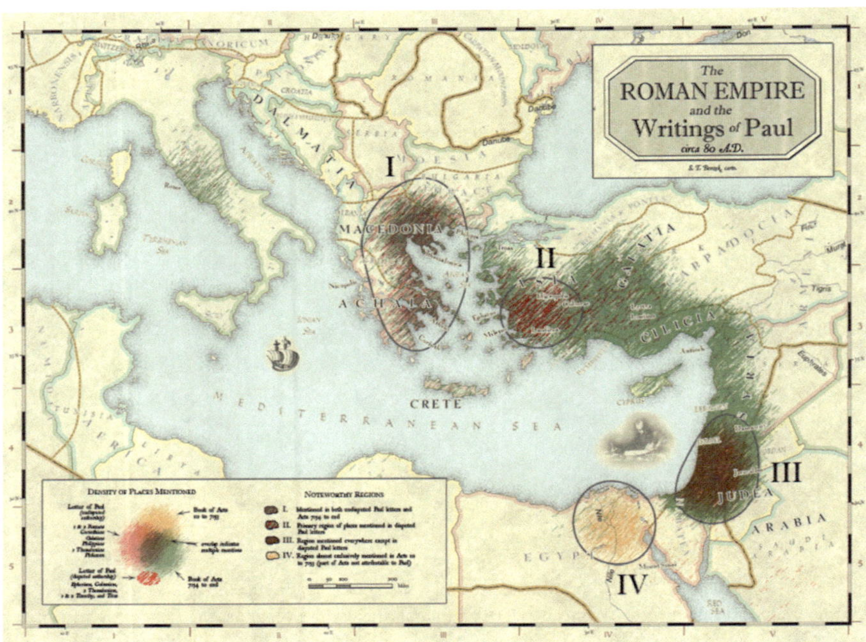

Fig. 8 Composite map of places mentioned in Pauline Letters and Acts

would a puzzle—the land–(and sea-) scapes Paul, and later associates, ostensibly traversed and inhabited, offer provisional perspectives, qualities and/or settings. These can be judged more or less plausible by virtue of their historical weight, consonance and/or disparity, relative to elements in emergent, and/or related, iterative layers.

4.4 Synopsis

Within the field of critical scholarship, the conclusions affirmed in these alternate Pauline cartographies are not new. However, the correspondence that translates sound critical assessment into effective visualization is conceptually promising. As virtual gaps in the historical record replace seamless apostolic 'journeys', one is left with snapshots of Paul, more evocative than prescriptive. Because iterative re-configuration of any set of maps can readily render alternate 'geo-textual' juxtapositions, nodes of intersection and/or divergence are no longer vehicles for inculcating defined doctrinal delineation, but instead elastic permutations that raise new questions, and invite closer investigation. Rendered virtually and visibly patent, challenging geographical variation, at once, elucidates the fragmented character of the sources, and lends insight in addressing broader debates.

Emergent configurations may bear variously little or greater resemblance to more familiar maps. However, both discrete iterations and composite images invite pondering the true scope of this apostle's itinerancy, and derivatively, the shape of his own, and his successors' communal investments. Fissured surfaces, simultaneously, serve as provocative reminders of how little we actually know about Paul. The sequential topographies of Acts remain valuable, not as authoritative refractions of Paul's 'Journeys', but as conceptual depictions of a worldview imagined, and arguably embraced, by at least one nascent Christian community seeking to follow (and/or regularize) Paul's footsteps, long after his death.[19]

Given that within each digital interface, underlying (and overlapping) data can be repeatedly re-configured, and re-framed, it is not difficult to imagine that 'mindful' mapping of a variety of combinations—gender, travel, migration, economic demographics and constituencies—might yield additional insight into the historical character of particular individuals, religious texts, and/or the broader fabric of emergent communities and textual traditions. At points of confluence,

[19]Brock (1982: 9) emphasizes the "pernicious influence on the writing of almost all subsequent ecclesiastical history" exerted by Eusebius' "picture of the history of the Christian church…being inextricably interwoven with the history of the Roman Empire." The narratives recounted in Acts, and derivatively familiar maps of Paul, have arguably been similarly determinative in shaping Western perceptions of the geographical spread of early Christianity. Here, Harley's (1989) seminal discussion of the rhetorical character of all mapped interfaces offers rich food for thought.

discussion can move forward—to a greater or lesser extent—on more historically sound footing. Simultaneously, as cracks and irregularities in a heretofore seamless landscape are rendered visible, emergent apertures afford additional opportunities for ongoing investigation and experimentation.

Perhaps predictably, the malleable capacity that powers iterative cartographic visualization—like the subtle surety communicated by static maps—retains derivative effect. As versatile interfaces invite movement from passive encounters with maps as authoritative end products, to active use of maps as tools for critical inquiry, ancillary engagement moves readers from passive encounters with authoritative text, to active interrogation of the historical processes endemic to its formation and interpretation. 'Mindfully' operationalized, the authority inherent to all mapped surfaces (Harley 1989: 3), effectively inverts longstanding pedagogical priorities. In fact, reflexively redeployed as a catalyst for recognizing critical shifts in both textual and visual paradigms, emergent 'uncertainty' may become a given map's most instructive aspect (Larsen and Benzek 2014, 2016).

5 Mapping Religion

When gauging the degree to which cartographic configurations have been defined by determinative predisposition, even a cursory survey of the maps found in 'world religions' textbooks is provocative. In fact, the persuasive presuppositions that texture these explicitly pedagogical surfaces suggest that ideological goals have shaped not only contemporary understandings of American Congressional geography, and historical refraction of biblical terrain, but broader representation of religion as a whole.[20]

Patterns that link religious definition with mapped visualization are particularly patent in textbook reflections of indigenous practice.[21] In many surveys, any record of presence is absence. When included, the disparate tenor of verbal and visual nomenclature is both predictably and unexpectedly revealing. As products of an

[20]Thanks to Jack Hawley, for affording access to his rich library of 'World Religions' textbooks, during graduate studies at Columbia University. Professor Hawley's generosity fostered what remains a fascination with pedagogical and editorial investments, and ongoing critical exploration of 'how' religion is taught, both in classroom and civic settings.

[21]In classroom settings, the shifts in verbal and visual nomenclature employed across a spectrum of World Religions textbooks remain a significant register of the degree to which how one "defines religion" is manifested in mapped representation. Even cursory survey of the terminology used in a text's 'Table of Contents', relative to included maps, is often revelatory.

earlier era, a number of the first world religions textbooks register indigenous constituencies using value-laden language.[22] Simultaneously, the same early textbooks, are the few (of any era) to meaningfully demarcate discrete indigenous communities on a "Religions of the World" map. Although more recently published texts deploy less overtly valued nomenclature in discussing "indigenous" practice, corresponding cartographic imagery is more troubling. Encoding legacies that have defined indigenous peoples by the 'absence of religion,' many contemporary surfaces either leave populations 'off the map' (Brodd et al. 2013, 2016), or mark (and efface) their 'presence' with blank space (Hitchcock with Esposito 2006: 8–9). Such a symbolic register tacitly mirrors foundational narratives of European encounters with communities that retained "no [religious practice] at all" (Eck 2001: 36).

Landscapes linked to 'primary' religious practice are alternately revelatory. Here, verbal nomenclature remains relatively stable; nonetheless, disorienting visual disparities invite further exploration of the historical role 'mapping religiously' has played in molding contemporary cartographies. For example, maps included in a number of relatively recent textbooks equate the 'world' presence of Judaism solely with the modern State of Israel (see, for example, Partridge 2005: 34–35; Brodd et al. 2013: 6–7). The same cartographic configurations cover a sizable swathe of the globe's surface with multi-hued affirmation of Christianity's monolithic influence. Whether amalgamated, or discretely delineated (Protestant/Catholic/Orthodox), the numerical measures that substantiate such display, render Judaism's global diasporic presence—conspicuously absent (Oxtoby 1996).[23] Cartographically displaced by the religion(s) it spawned and shaped,[24] its singular influence is masked by Christianity's statistical dominance. As defined, both hegemony and erasure mirror the exigencies of history. Resultant surfaces affirm that "…maps are always a reflection of the culture[s] in which they are produced" (Harley 1989: 14–15).

[22]Now in its 13th edition, and respectively authored and edited by Noss and Noss (2011), *A History of World Religions* (original published in 1949 as *Man's Religions*) might be named the oldest 'living' World Religions text. Until his death in 2010, David Noss served as posthumous editor of his brother's foundational text. After editing/'curating' his brother's volume for almost half a century, David Noss noted the degree to which shifts in defining nomenclature, even over the span of five decades, were dictated by market driven forces—that is, by publishing houses focused on retaining a loyal faculty readership—rather than authorial preference (Spring 2002, Personal Correspondence).

[23]It is interesting that the Noss brothers' early map is one of few to globally register both Indigenous and Jewish presence. Lewis Hopfe's, *Religions of the World* (1979), now in its 11th edition (2009), is similarly noteworthy for including a chapter on "Native American Religions."

[24]This slim geographical footprint is inversely misrepresentative, relative to the statistical disparities encoded in the contemporary Buzzfeed maps discussed above. In fact, the strategies deployed in parsing data so as to suggest a relatively sizable subset of Jewish representation in Congress, here render Judaism's 'world' presence almost singularly negligible. Each suggests something of a 'mapped misnomer.'

It is perhaps tacit recognition of the problematic character of 'mapping' the complexity of a 'world' religious landscape that has led some authors and editors to eschew inclusion of maps altogether.[25] Others populate their pages with geographically limited landscapes that present foundational loci of particular historical eras, or targeted arenas of teaching and practice. Among textbooks that seek to incorporate the multifaceted character of 'world religions' on a single surface, results are mixed. Beyond five or six 'dominant' delineations, the verbal nomenclature used in categorizing depicted strains varies broadly. The range of traditions included, or left 'off the map', is equally fluid. Although Hinduism, Buddhism, Judaism, Christianity, and Islam are persistently present, each is differently parsed.[26] The scope and logic that governs inclusion/exclusion of a broader range of traditions is often opaque.[27]

In the classroom setting, calling attention to emergent patterns that connect how one defines religion, with how religion is pictured on a map, has proved compelling. Recurring motifs invite further consideration of the role 'mapping religiously' has played in molding both global and local landscapes. When students are tasked with 're-drawing the map,' their collaborative deliberations effectively underscore the degree to which the frontiers of 're-presentation' remain rooted in 'mapping mindfully' (Fig. 9).[28] Equipped with tools that facilitate both de-constructing and re-constructing 'world religious' contours, even fledgling re-imagination invites conceiving surfaces not solely as proprietary zones of passive conversion (or perennial conflict), but also as loci of sustained engagement.[29] Trade, travel, and migration corridors, surface as links that render geographical centers (and derivatively, communities) visible. As convergent and dispersed loci

---

[25]Among the most notable is Huston Smith's (1995) iconic, *Illustrated World Religions: A Guide to Our Wisdom* Traditions. As interesting, are Neusner's (1994/2003), *World Religions in America*, and Oxtoby and Segal's (2007), *A Concise Introduction to World Religions* (Segal's focused re-working of Oxtoby's earlier compendium). Respectively, Neusner and Segal's volumes constitute the rare 'World Religions' text to be edited by a Jewish scholar.

[26]For example, as in the congressional maps with which discussion began, the geographic landscape that aligns with particular sectors of Christianity is often closely delineated. Less proximate (and/or less familiar) traditions are respectively represented as singular entities, or sometimes, capriciously constructed amalgams.

[27]Inversely consonant with the static contours that have defined popular maps of Paul, the diversity that characterizes both historical and contemporary configurations of 'world' landscapes must give pause.

[28]A recent grant, awarded by the Wabash Center for Teaching and Learning in Theology and Religion, will support extending the strategies developed in re-framing 'maps of Paul', to 'Re-drawing the Map of World Religions.' Dispersed monies will fund a series of workshops that involve students and faculty in de-constructing, researching, and re-constructing, a geographic, critical, and spatially informed, 'World Religions' curriculum. See http://www.redlands.edu/news-events/news-landing-page/2016-news/february/support-from-wabash-center-will-support-curricula-on-re-drawing-the-map-of-world-religion/.

[29]A number of World Religions texts register such melding with nested surfaces distinguished by contrasting color or symbology.

Fig. 9 "Re-drawing the Map" student workshop (2016)

mark shared space, their nodes trace instrumental networks shaped by cross-cultural confluence and diffusion. The complex constituencies that emerge from these configurations, belie easy categorization (Fig. 10).

5.1 *Synopsis*

The variability that characterizes both established and 're-drawn' surfaces, underscores the importance of not only 'religiously minding the map,' but also re-thinking the pedagogical goals of 'mapping religiously.' Here, simply replacing passive perusal of persuasive panoramas, with active manipulation of malleable and exploratory models, relativizes 'the authority inherent to all mapped surfaces' (Harley 1989). Similarly, experimental approaches that emphasize—rather than efface—the melded character of both dominant and diffused belief and practice,[30]

[30]Whether ancient or contemporary, 'secular' or 'religious,' cosmic or mundane, discrete or global, the determining influence of clearly delineated contours can be traced across disciplinary and temporal divides. Iteratively imbued with mutually reinforcing innate authority, there is little question that 'scientific' maps, wrapped in religious 'truth,' continue to claim a static office that belies their pliable character. However, as the deceptively simple lines of digitally generated maps produce an illusion of surety that easily effaces the borders which separate representation from

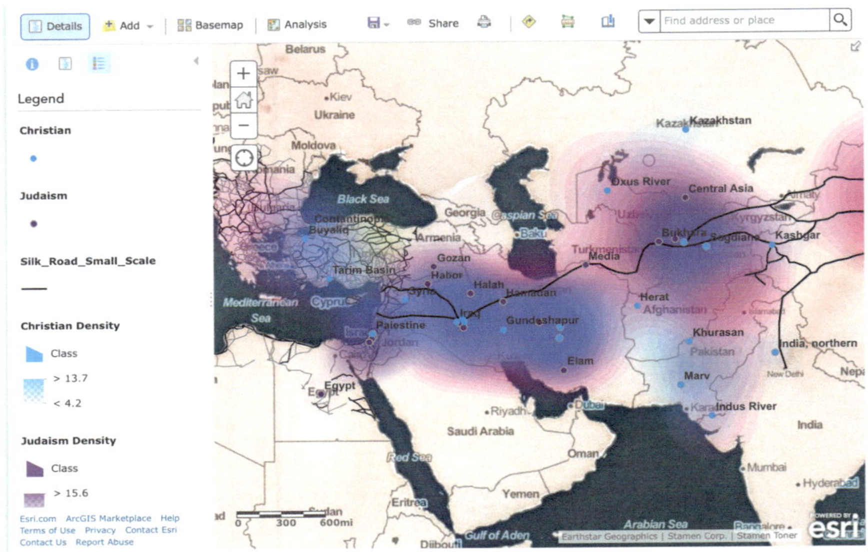

Fig. 10 Larsen/Benzek, "Re-drawing the Map" Judaism and Christianity on the Silk Road (2016)

trouble religious constructs long defined as 'stable' repositories of truth. As re-distributed statistical measures disrupt hegemonic dominance, they produce tapestried landscapes, threaded with corridors of intersection and exchange. The cracks and irregularities that mottle less immediately accessible substrata, double as apertures. Whether diasporic or dense, each signals opportunity for further investigation.

6 Conclusion

Just as "all historical interpretive efforts and their methods and approaches illuminate some things, cast shadows over others, they foreground some things, render into the background certain others" (Wimbush 1997) so "…maps are always a reflection of the culture[s] in which they are produced" (Harley 1989: 14–15). However, at a moment when 'maps of religion' can be iteratively shaped, re-shaped, and/or rendered authoritatively static with the click of a mouse, both contemporary deployment and established patterns commend critical cognizance of the authority inherent to all mapped surfaces. By communicating foundational

(Footnote 30 continued)

reality, the emergent 'frontiers' of mapping religion render any suggestion of certainty suspect. They invite, instead, more 'mindful' melding of religion and maps.

values of the communities from which they derive, maps refract a reality that ostensibly links a created landscape with its locus of origin. Their surfaces simultaneously retain the capacity to "transform [this] reality, [by] highlighting some phenomena at the expense of others" (Scafi 2006: 28; Harley 1989; Wimbush 1997).[31]

If religion is to be identified and communicated in terms commensurate with contemporary critical engagement, the lessons that emerge from re-defining Congressional constituencies, and more mindfully mapping "Paul journeys," are provocative. That they effectively extend to re-configuring, not only proto-Christian landscapes, but also the map of world religions is, at once, sobering and promising. Brokering the programmatic pedagogy of melding "change" and "tradition," mindful utilization of contemporary cartographic tools invites re-presentation that de-authorizes "correct readings" with surfaces which foster radical re-evaluation. Absent an ideological agenda aimed at culturally affirming scriptural authority—or Christianity's hegemonic influence—both reading and mapping can be re-purposed. Persuasive influence, however, is contingent. Ironically attached to the success of the Protestant project, even critical re-configuration remains grounded in shared assumption. Effective re-deployment entails strategically identifying the stubborn, authority-laden, operatives that inform cartographic pedagogy, while turning these to new ends.

Further Reading

Delano-Smith, Catherine. "Maps as Art and Science: Maps in Sixteenth Century Bibles" *Imago Mundi* 42 (1990): 65–83. Delano-Smith examines the Protestant pedagogical project presented in the first generation of 'Maps in Bibles.' While primarily focused on cartographic questions, her close analyses remain foundational to re-thinking the structural complementarities and confluences that link geography, information and spatial literacies, with contemporary conceptualization of religious landscapes. Cf. C. Delano-Smith, *Maps in Bibles 1500–1600: An Illustrated Catalogue* (Geneva: Librairie Drox S. A., 1991).

Eck, Diana. *A New Religious America*. New York: Harper Collins, 2001. This volume presents in narrative form, what the Pluralism Project documents in digital nomenclature (www.pluralism.org). Eck's work, here and elsewhere, remains foundational to redefining the nuanced, and ever shifting spaces that anchor and contextualize, both historical and contemporary expressions of religious experience.

Harley, J. Brian, "Deconstructing the Map" *Cartographica* 26.2 (1989), 1–20. Harley's relativization of the authority inherent in all mapped surfaces marks an essential counterpoint to the roles routinely accorded cartographic refractions of ancient and contemporary religious landscapes.

Knowles, Ann Kelly, et al., eds. *Geographies of the Holocaust* (Bloomington: Indiana University Press, 2014). The work of Ann Kelly Knowles offers a particularly provocative set of models

[31]The common emphases (no pun intended) that link Vincent Wimbush's interpretive-historical insight with Harley's seminal discussion of the rhetorical character of all mapped surfaces, offer a provocative frame for re-examining the relationship between religious text and map, and map as religious text.

for essential consideration of why spatializing religious and cultural categories matters. While this most recent study examines those questions within a religio-historical frame, Knowles' broader portfolio addresses, and contextualizes a rich array of spatial categories in evocative and nuanced ways; Cf. Knowles (ed.), *Past Time, Past Place: GIS for History* (Redlands, CA: ESRI Press, 2002).

Larsen, Lillian I. and Steve Benzek. "Minding the Gaps: Exploring Ancient Landscapes through the Lens of GIS" *Transformations: The Journal of Inclusive Scholarship and Pedagogy* 25 (2014), 45–58. Re-framing the role accorded digital tools in re-thinking contemporary manifestations of ancient source material, Larsen and Benzek invite further consideration of the degree to which 'gaps' in the religio-historical record, can usefully foster and facilitate contemporary critical engagement with authoritative textual tradition; Cf. "Min(d)ing the Gaps: Digital Refractions of Ancient Texts" in *Ancient Worlds in Digital Culture* (Claire Clivaz, Paul Dilley and David Hamidovic (eds.) Leiden: Brill, 2016), 128–147.

Scafi, Allessandro. *Mapping Paradise: A History of Heaven on Earth* (Chicago: University of Chicago Press, 2006). Scafi presents a rich overview of the interface between religious sensibilities as a register of historical lived contexts. Eloquent prose and lavish illustration serves well to underscore the continuum of historical maps that imbue heavenly landscapes with this-worldly import.

References

Brock, S. (1982). Christians in the Sassanian Empire: A case of divided loyalties. In S. Mews (Ed). *Religion and national identity*. Oxford.

Brodd, J. et al. (2013/2016). *Invitation to world religions,* (2nd ed.). New York: Oxford University Press.

BuzzFeed News. (2013). *Religion of members of the US House of Representatives*. http://ak-hdl.buzzfed.com/static/enhanced/webdr01/2013/8/17/19/enhanced-buzz-wide-16883-1376780519-9.jpg

Delano-Smith, C. (1990). Maps as art and science: Maps in sixteenth century Bibles. *Imago Mundi, 42*, 65–83.

Delano-Smith, C. (1991). *Maps in Bibles 1500–1600: An illustrated catalogue*. Geneva: Librairie Droz S. A.

Eck, D. (2001). *A new religious America*. San Francisco: HarperCollins.

Harley, B. J. (1989). Deconstructing the map. *Cartographica, 26*, 1–25.

Hitchcock, Susan Tyler with Esposito, John. (2006). *National geographic: Geography of religion*. Washington, DC: National Geographic.

Hopfe, L. M., & Woodward, M. R. (2009). *Religions of the world*, (11th ed.). Upper Saddle River, NJ: Pearson Education.

Larsen, L., & Benzek, S. (2014). Min(d)ing the gaps: Exploring ancient landscapes through the lens of GIS. *Transformations: The Journal of Inclusive Scholarship and Pedagogy, 25*, 51.

Larsen, L., & Benzek, S. (2016). Min(d)ing the gaps: Digital refractions of ancient texts. In: C. Clivaz, P. Dilley & D. Hamidovic (Eds.), *Ancient world in digital culture*. Cambridge, MA: Harvard University Press.

Levin, I. (1985). *The Quedlinburg Itala: The oldest illustrated bible manuscript*. Leiden: Brill.

Mercator, G. (1607). Peregrinatio Pauli. Amsterdam. http://www.davidrumsey.com.

Neusner, J. (1994/2003). *World religions in America*. Louisville: Westminster John Knox Press.

Noss, J. B., & Noss, D. S. (2011). *A history of world religions (originally, Man's Religions, 1949)*. Upper Saddle River, NJ: Prentice Hall.

Ortelius, A. (1598). Peregrinationis Divi Pauli Typus Corographicus. Antwerp. http://www.davidrumsey.com

Oxtoby, W. G., & Segal, A. J. (2007). *A concise introduction to world religions*. New York: Oxford University Press.

Oxtoby, W. J. (Ed.). (1996). *World religions* (2 vols). New York: Oxford University Press.
Partridge, C. (Ed.). (2005). *Introduction to world religions*. Minneapolis: Fortress Press.
Penner, T., & Vander Stichele, C. (Eds.). (2003). *Contextualizing acts: Lucan narrative and Greco-Roman discourse*. Atlanta: Society of Biblical Literature.
Pervo, R. I. (1987). *Profit with delight: The literary genre of the acts of the apostles*. Minneapolis: Fortress Press.
Pervo, R. I. (2006). *Dating acts: Between the evangelists and the apologists*. Santa Rosa, CA: Polebridge Press.
Pervo, R. I. (2010). *The making of Paul: Constructions of the apostle in early christianity*. Minneapolis: Fortress Press.
Pew Forum. (2012). Faith on the Hill: The religious composition of the 113th congress. http://www.pewforum.org/2012/11/16/faith-on-the-hill-the-religious-composition-of-the-113th-congress/.
Pluralism Project. (2012–2016). America's interfaith structure: A pilot study. http://www.pluralism.org/interfaith/.
Scafi, A. (2006). *Mapping paradise: A history of heaven on earth*. Chicago: University of Chicago.
Schwarz, H. (2015). Map: What religion does your member of congress identify with? http://www.buzzfeed.com/hunterschwarz/us-house-of-representatives-members-religion#.yaE3E8zxn.
Schwarz, H., & Schaul, K. (2015a). The religions of the 114th Congress, mapped *Washington Post*. https://www.washingtonpost.com/news/the-fix/wp/2015/02/06/the-religions-of-the-114th-congress-mapped/.
Schwarz, H., & Schaul, K. (2015b). Religions of our representatives. http://www.washingtonpost.com/graphics/politics/114th-house-religions/.
Smith, H. (1995). *Illustrated world religions: A guide to our wisdom traditions*. San Francisco: HarperCollins.
Toddhunter, A, & Johnson, L. (2012). In the footsteps of the apostles. *National Geographic, 221* (3), 38–65.
Vischer, N. (1642). Geographische beschryvinge, van de Wandeling der Apostelen ende de Reysen Pauli. Amsterdam.
Weitzmann, K. (1947). *Illustrations in roll and codex*. Princeton: Princeton University Press.
Weitzmann, K. (1959). *Ancient book illumination*. Cambridge, MA: Harvard.
Wimbush, V. L. (1997). Interpreting resistance. Resisting interpretations. *Semeia, 79*, 1.

Maps, Mapping and the Visual Arts

Lou Cabeen

Abstract Maps, mapping and the visual arts have long been related disciplines, each offering distinct means and methods of recognizing and depicting space while sharing an implicit assumption of the importance of doing so. During the first half of the last century, these distinctions were paramount, and visual art became ever more abstract and distant from representation in any form. But since the 1970s those distinctions—like boundaries of many types—have become porous and blured as artists sought to re-engage visual representation. This is the time period in which artists revived their historic efforts to describe place, space and our experiences within it but needed to find new visual strategies that acknowledged the power of abstraction while clearly referencing the real. This search led many artists to embrace the abstractions of cartography, and the communicative power of mapping strategies, finding in maps and mapping not only a means of representating the real but also tools with which to expand the boundaries of where and when visual art could be experienced. This chapter, written by a practicing artist, focuses on two and three-dimensional visual art in order to provide a concise introduction to this change. It provides an overview of the key exhibitions, curators, writers and artists who pioneered this shift in visual art thinking and practice in the latter part of the last century—often noted as the shift from Modernism to Post-modernism. Three contemporary artists whose work represents distinct approaches to this newly charted interdisciplinary waters are discussed and future directions in this arena are charted via suggested readings.

Keywords Maps · Mapping · Visual art · Modernism · Post-modernism · Abstraction · Representation

L. Cabeen (✉)
University of Washington, Seattle, USA
e-mail: lcabeen@uw.edu

S.D. Brunn and M. Dodge (eds.), *Mapping Across Academia*,
DOI 10.1007/978-94-024-1011-2_17

1 Introductory Definitions

This chapter will discuss some of the relationships among maps, mapping and the visual arts that rose to prominence late in the 20th century and continue to inform contemporary practice. Although the visual art of past centuries has celebrated this connection (Alpers 1982; Fiorani 2005) the visual art practice of modernity did not (Smith 1996). However, as we shall see, with the end of that particular cultural era and the advent of post-modernism (Bennett et al. 2005; Perry and Wood 2004) artists began to see maps and mapping as potent allies in their effort to expand the reach of the visual and communicate to new audiences in new ways.

For the purposes of this review, "the visual arts" refers to static objects whose aesthetic purpose and effectiveness relies primarily on their being seen. Traditionally, the visual arts are distinguished from the literary and performing arts by their gestalt. That is, like maps, they are apprehended (if not comprehended) all at once, rather than over time (Kardon 1977). Consequently media such as film are beyond the scope of this chapter, even though the spatial aspects of cinema provide a potent area of investigation for other scholars. This essay will further narrow its focus to consider only works created from the latter half of the last century to the present. These decades are considered by most commentators to have marked a definitive and permanent change in the nature of visual culture in Eurocentric society, and the history of this time period has shaped all current cultural practicioners (Smith 1996).

A further distinction that will be helpful is that which exists between studio art and art history. Art history is an academic discipline that studies art objects with an emphasis on their historical development. It is a systematic discipline, establishing the stylistic connections that link groups of works, and the political, intellectual, social and economic contexts that informed them. For example, the authors cited already in the chapter are art historians who decipher, evaluate and catalog visual art that already exists and offer cogent and thoughtful analysis of it (for additional insights into the practice of art history and criticism see Paglen 2008: 30). Art history is the academic bedrock of my own education, and while I lean heavily on art historical insights and methods, I must warn you that I am not an art historian.

I am an artist, a clinician if you will, whose observations on the following topic are rooted in practice rather than theory. I and my fellow artists make the objects that art historians place in context. Consequently my perspective is not that of the forest, but rather that of being among the trees. In speaking of the issues that follow —what has led—and what still leads—individual artists to use maps or mapping practices in the making of their art—I speak as someone who has in the past and still is living those questions, not merely observing and reporting on them.

2 Space and Spatial Processes in Visual Art

In the visual arts the word "space" covers a lot of ground. There is the space that may be depicted within the artwork. There is the space of the art object itself. There is the space in which the art is experienced—be it a gallery, a museum, an open field or the viewer's mind. These aspects of space in the visual arts have multi-layered histories beyond the scope of this chapter. However, an overview of the highlights of those histories follows. A more complete discussion of the development of these themes in the Western art canon can be found in Arnason (1986) and Kleiner (2010: Chapters 23–25).

2.1 *Depicting Space*

The historic goal of the Eurocentric artist was to create convincing representations of space using both linear and atmospheric perspective (for additional information see www.oxfordartonline.com: perspective and perception). These techniques were perfected during the early Renaissance and continue to be taught and appreciated today, as a visit to almost any undergraduate introductory drawing course will reveal.

The invention and development of photography in the course of the nineteenth century, however, provided an alternative method by which to document the visual world. Artists working in traditional media subsequently began to focus their attention on depicting the perception and experience of three-dimensional space rather than its direct representation. The Impressionists provide a classic example. These artists working in France from the 1860s onward sought to reveal space as it is seen in a passing glance, to create with paint on canvas an impression of a space and the figures or objects within it (Kleiner 2010). Their paintings, when successful, provide us with a concrete manifestation of fleeting visual perception. Close examination of these paintings reveal the means by which they were made—the space of a park, or restaurant, or dance hall dissolves into daubs of paint on an utterly flat surface. The impression that a specific space has been depicted in the painting only returns upon a more distanced view.

2.2 *Crafting Space*

The literal representation of three-dimensional space was further problematized by the Cubist experiments of the early 20th century (Kleiner 2010). The Cubists, most notably Pablo Picasso and George Braque began to depict space in their paintings as broken up, or "cubed" rather than as a seamless whole. They painted a table for example not from one point of view, but rather from multiple points of view all of which co-existed at the same time in the space of the painting. The paintings appeared to be entirely abstract, but in fact the jumble of shapes they contained were all derived from the experience of moving through space—the shifting views one has of a table as one walks around it and sees it in profile, from the top, in three-quarter view, and so on. This "breaking up of [the painting's] space," to use basic art-education speech, was seen to be a more honest way to depict three-dimensional reality in two-dimensional art.

Which brings us to consider the space of the artwork itself. Much of the time, this space is referred to as "the surface" or "the picture plane." The shift in emphasis from the surface of a painting being a vehicle on which to document three-dimensional space to the surface of a painting being seen as a space in and of itself is the ongoing legacy of the Cubists' experiment. Carefully scanning a painting or print for details, moving visually across its surface, is to experience "the space" of the painting no matter what imagery (or lack of imagery) exists there. The viewer, in discovering those details and making associations between them, becomes an active participant in defining the space of the artwork and in the construction of its meaning.

2.3 *The Space of Experience*

The viewer's agency in creating meaning brings us to the issue of the space in which the experience of art is generated. Does the experience of art reside in the object the artist makes? This was a primary belief in the Middle Ages when the aura of the artefact was sufficient to draw pilgrims long distances in order to be in its presence. Or is the art generated in the artist's mind and transferred to the art object, which then exists as a kind of Rosetta stone for people to decipher? This idea rose to prominence in the Renaissance, as artists were eager to leave the guilds of tradesmen and artisans and sought acceptance into liberal arts academies as intellectuals. The spaces in which art was encountered further augmented these experiences, the mystery of the cathedral lending credence to the aura of the altarpiece, and the intrigues of court life reinforcing the need to appropriately interpret the palace murals. Both of these models assume that the meaning of the art object and its aesthetic impact were received by the viewer either by the experience of being sufficiently devout in its presence, or by the intellectual exercise of successfully decoding its message. Both models became increasingly problematic as photographic reproduction began to offer viewers the opportunity to see a depiction of an art-work separate from the context of cathedral, palace or the modern equivalent—the museum (Benjamin 1968).

Consequently, an alternative relationship between viewer and art object began to emerge based on the realization that art is, in fact, an experience. As an experience, art is generated in the space between the viewer and the object. This model assumes active viewers who will engage in the construction of (their own) meaning when confronted with the art object. It has been the primary assumption about the space in which art "happens" since at least the 1970s when the artists now known as Conceptualists posited that the experience of art so thoroughly resides in the mind of the viewer that reading the artist's stated intention would be sufficient to create it (Arnason 1986; Perry and Wood 2004).

I am an artist who came of age in the 1970s, as assumptions about the "spaces" of art were being whole-heartedly re-examined. Trained by professors who had risen to prominence through their adherence to the art as Rosetta stone model, I took the half guilty-half defiant pleasure of the undergraduate in exploring the creative ferment happening outside the confines of the academy (art as experience). This is the time that saw the rise of artists turning to maps and mapping as potent allies in depicting space in new ways, in crafting new spaces in which to make art available to audiences and in creating new conceptual spaces in which it could be experienced and understood. There follows an outline of this process (for a contemporaneous discussion of this development see Smith 1981).

3 Maps

3.1 *Modernism's Last Gasp*

The Cubists set Eurocentric art on its Modernist course towards ever more reductive abstraction. By the middle of the last century, Minimalism, the ultimate expression of Modernist thought, rejected all references to the real in favor of artworks that were entirely self-referential, that is art works that were about and only about the nature of their materials (Perry and Wood 2004). This resulted in an art-landscape populated by two-dimensional works such as Robert Ryman's *Classico III* composed of white paint on squares of white paper. This art-for-art's sake philosophy and practice began to be seen as less and less relevant in the face of the social upheavals of the 1960s and 1970s. Arnason (1986: 560–561) describes the situation with precision:

> … [T]he fundamental modernist credo … held that the aesthetic object possessed meaning only insofar as it asserted its independence of all but its own inherent nature and process of creation …. [But] in an impure world … purity of form could no longer suffice. It began to seem as if large-scale abstraction were less a symbol of the heroic and the sublime than a neurotic escape from life.

And so the next set of questions for artists was framed—to develop ways of making art that would re-engage life and its contingencies.

Artists sought a way to reconcile the visual beauty and intellectual power of abstraction with a dynamic (re)engagement with the real without resorting to direct representation. They came up with a variety of solutions, but a common denominator among many of them was an embrace of maps and mapping—those fellow (visual systems that use two-dimensional abstract forms to describe) three-dimensional space. The how and why of this process is discussed below, followed by two artists who are now considered to provide classic examples of the successful resolution of this tension between the real and its representation.

3.2 Return of the Real

In 1994, New York Museum of Modern Art curator Robert Storr would put his finger on the (im)pulse of artists using maps by saying that:

> … [T]he particular opportunities maps provide visual artists—and their special appeal to modern sensibilities—result from their being the *ultimate pictorial coincidence of exacting representation and total abstraction.* (Storr 1994: 13, added emphasis)

Maps provided a system of abstract visual forms that were understood to both reference the real world, and remain entirely abstract at the same time. The range of expressive potential that existed between the common understanding of the map as a signifier of "real" space and its actual existence as a flat surface on which to display abstract forms provided artists with a means to resolve the "crisis of representation" that had arisen with the end of the Modernist agenda of total abstraction. And although we are now aware that maps are just as subjectively edited by their makers as any other human product, artists in the last quarter of the last century often used them as Storr describes—as visual documents understood to be abstract representations of an objective reality.

The abstract visual language of maps would seem to provide a distanced, objective evocation of space in keeping with Modernism's detachment. However, maps, being familiar visual components of everyday life, in fact served as a vehicle through which the viewer's own experience of space could be tapped, providing an additional layer to the newly valued relationship between artist, object and viewer, and a perhaps unintentional bridge between not only abstraction and the real, but also abstraction and private emotion. Roberta Smith, in her 1981 publication *4 Artists and the Map* noted that artworks incorporating maps and mapping were not stylistically similar. Nor were they linked by the conceptual goals of their makers. Artists were redeploying existing maps to make art in wildly diverse forms and formats and to serve widely disparate intellectual agendas. Smith did, however, articulate the underlying shift in focus that had led artists to use maps in the generation of these new subjects and forms of visual art:

> There is a shared attitude inherent in the use of the maps, which, if general, is also fundamental: it emphasizes what is given over what is felt, gives precedence to selected facts of the world over the expression of the artist's personal emotion, and opts for a shared, legible vocabulary of form over a private invented one … [This] establishes an immediate relationship to the viewer, calling forth [the viewer's] own knowledge and sense of the world. (Smith 1981: 7)

In other words, the map, being an artifact of daily life, calls forth our remembered experience of space and our experience of moving through it. Smith speaks above in the context of her time. More recent commentators, notably cartographic critic Denis Wood and art critic Lucy Lippard have identified the ideological biases that shape this vocabulary, and identified many of its component parts. Lippard (1997: 78) gives voice to the problematic nature of this 'legible vocabulary':

> The 'naturalization' of maps—the myth that maps show the world the way it really is—veils the fact that maps are cultural and even individual creations that embody points of view. They map only what the authors or their employers want to show; resistance is difficult.

However a more recent commentator, in an article discussing art works which specifically critique the ideological uses of conventional maps returns to Smith's point about the viewer's relationship to maps and map imagery:

> A map's beauty lies in its limits, in its ability to focus attention, to shape perception. A good map stimulates the imaginations of users who invent road trips, battles, and love stories as they follow the long spindly highway lines and pools of blue water. (Berwick 2010: 102)

3.3 *For Example…*

Two artists active in the 1970s and beyond provide particularly clear examples of the connections between maps and visual art at this moment of cultural change. Nancy Graves and Richard Long participated in most of the early art exhibitions that focused on maps and art. Their work continues to be exhibited, discussed and studied for its relevance to contemporary culture. I have chosen to discuss elements of their work that provide examples of the topics discussed above.

Nancy Graves (1940–1995) was a prolific artist, working in a wide range of media including painting, sculpture and printmaking (for an overview see Hughes 1987). For Graves maps provided a direct link between discredited Modernist abstraction and the then radical desire to return to representation. In fact, it is Graves' use of maps that articulated in visual form much of what continues to be stated about the connections and contributions of maps, mapping and visual art. Graves' paintings and lithographs can be understood within the established art world hierarchy of minimalist color-field abstraction—the "American sublime" as curator Robert Storr puts it (1994: 19). They also and at the same time represent the natural world as perceived by the advanced scientific technology of the time. Roberta Smith (1981: 11) points out that for Graves:

> … the map is both an image of nature … and an image of man's thoughts about nature. 'Mapping encompasses all our significant efforts,' Graves has said. 'It is the most advanced level of conceptual abstraction we have at the moment.'

Graves work in the 1970s is a direct illustration of this point. She immersed herself in this 'most advanced form of conceptual abstraction' by using satellite images—understood to be the most technologically advanced mapping available at the time—as the substrate of her drawings, prints and paintings. Using the forms and formats of weather maps, bathymetric charts, maps of the moon's surface, and images of Mars, Graves created abstract paintings whose subject was firmly rooted in nature, albeit in aspects of nature unseeable by normal means (Dreikausen 1985; for images see www.nancygravesfoundation.org/painting).

Many of the works are huge, and are composed of layers of imagery derived from different sources, or are composites of satellite images taken over time. Bright pinks, glowing yellows and crisp blues swirl through her canvases, following faithfully the lines, and shapes revealed by the mapping technology (see Padon 1996 for a more complete discussion of this aspect of Grave's work).

Unlike Graves, Richard Long (b. 1945) was not an artist in search of a subject—his subject was space, time and landscape. But he was an artist in search of new ways to communicate his subject to viewers. Maps provided the means to bring a record of Long's landscape based artworks inside, where gallery visitors could view them (Fuchs 1986; for imagery see www.richardlong.org/sculptures).

Long's art making took the form of walks in the English countryside. These walks would conform to a system of pre-established, self-imposed rules. Long then inscribed his walking routes onto Ordnance Survey maps, making a record of his personal experience of the landscape. At the bottom of the map he would print the title of the work. These titles described precisely the spatial task he had set himself in the landscape. For example, in *5 Fifteen Minute Walks on Dartmoor* we see a 16″ × 16″ Ordnance Survey map labeled SX57. On this map five short, straight lines are drawn, scattered seemingly randomly on its surface. The lines are labeled "start 1″, "start 2″ etc. The map provides documentation of the making of Long's work, and brings his personal, physical encounter with space and landscape into the white cube of the gallery (Smith 1981).

However, Long's use of the Ordnance Survey map encouraged another non-traditional (at the time) art experience for the viewer. The survey map provides a wealth of detail, since the scale is 2.5″ to the mile. Subsequently the abstraction of the seemingly random lengths of line reveals itself to be a direct statement of the terrain, and reflects the relative distances that can be covered in fifteen minutes depending upon whether one is walking along a paved road, across moorland, or slogging through a marsh. Viewers could decode Long's challenging approach to sculptural space via the familiar act of reading a map. The art existed, and continues to exist in the viewer's decoding of the information provided by the map, the title and Long's marking of his route(s).

4 Literature

4.1 Exhibition Catalogs

As early as 1977 Janet Kardon at the Philadelphia College of Art curated a major exhibition of twenty-six works by twenty-five artists titled simply *Artist's Maps.* The works were wildly diverse in their approach, and included works by both

Graves and Long. Kardon's goal as a curator was to bring together artists working with maps to provide an overview of what was at the time a dramatic departure from the traditional means of art-making. While she admits that there is no shared aesthetic among these artists using maps, she identifies the following conceptual links.

> Maps ... have an interesting relationship to drawing and painting. They have historically been a motif, a mode of measurement, and frequently a metaphor. Converging on maps are a startlingly rich group of concerns that have occupied recent art—perception, space/time relationships, location, topology, locomotion, systems, scale, language and the nature of ground. However used, maps, abstract configurations that miniaturize vast distance, when appropriated into the context of art re-expand according to the scale of the imagination. (Kardon 1977: 17)

A few years later in 1981 Roberta Smith curated a highly focused exhibition titled *4 Artists and the Map: Image/Process/Data/Place* at the Spencer Museum of Art at the University of Kansas. By confining the exhibition to the works of just a few artists Smith sought to more clearly identify the major intellectual themes that led visual artists to engage with maps and mapping. The four artists included were Jasper Johns, Nancy Graves, Roger Welch, and Richard Long each a representative of one of the four themes Smith identifies in her title. Smith makes important observations about these now well known and thoroughly documented artists, and she discusses early works of Richard Long not included in subsequent evaluations of his work. In addition, Smith offers a snapshot of art historical uses of maps by artists with an especially insightful discussion of Vermeer's *Officer with Laughing Girl* (1655–1660) in which the painter includes a recognizable contemporary map in the background of this domestic scene (see www.essentialvermeer.com: Fig. 7). Smith's insights into the evocative power of Vermeer's placing the map behind the girl and the window behind the soldier are instructive reading even now, 30 years later.

Robert Storr curated *Mapping* at the New York Museum of Modern Art in 1994. Thirty artists were included in this exhibition. In his essay, Storr emphasized the dynamic between power, representation and our perceptions of the real. This catalog documents the work being done by established artists working with "inscribed" maps or map-based imagery. The exhibition created controversy in the art world in no small part because of its title. Younger, lesser known and more politically radical artists had begun to create art work that mapped, or remapped the status quo in various ways, and they objected to the use of the active term "mapping" to refer to an exhibition of works which were, in their view, entirely passive (Weil 1995). It is, in fact, this more active and activist use of mapping that has shaped art's relationships to mapping in the 21st century as we shall see. Wystan Curnow (1999) expands upon this particular point well, and I will discuss his insights below.

World Views: Maps and Art held in 1999 at the Weisman Art Museum on the campus of the University of Minnesota, was curated by Robert Silberman. The exhibition included 15 contemporary art works in addition to three installations commissioned for this exhibition. The installations were by Ilya Kabakov, Laura Kurgan, and the artist co-operative known as the KNOWMAD Conspiracy led by

Mel Chin. The exhibition also included 32 maps, ranging in date from the 11th century to the present.

Silberman's (1999) essay, titled "Maps and art: The pleasures and power of worldviews," provides a brief overview of the history of western cartography as well as an excellent short history of late 20th century art and its embrace of maps and mapping. The catalog also includes an essay by Yi-fu Tuan entitled "Maps and art: Identity and utopia" in which Tuan compares and contrasts cartography and portraiture, pointing out "A common understanding of the map is that it is the portrait of a place" (Tuan 1999: 14). He then discusses maps as portraits that idealize, and are therefore utopian. The art installations made expressly for the show take up the problematic inherent in the utopian expectations placed upon contemporary cartography by exploring different aspects of cartographic limitation by focusing on aspects of human experience not recorded by traditional cartography.

In spring of 2003, Ronald Feldman Fine Arts presented a solo exhibition of the work of the artist team, Newton and Helen Mayer Harrison. Eleanor Heartney's review of this exhibition offers not only insightful descriptions of the works, but also a useful short course on the evolution of 'Eco' art—that is art that takes an activist stance on global issues and uses maps, mapping and other spatial notation systems to visually communicate. The Harrisons, whose work together began in 1969, are among the best known and thoroughly documented pioneers in this field (for imagery and documentation see theharrisonstudio.net). The 2003 exhibition included two of their major works, *Peninsula Europe* and *Vision for the Green Heart of Holland*. Both works are multi faceted installations that use maps, charts and explanatory texts. Their use of such imagery, Heartney points out, serves as:

> … a reminder that maps are not simple neutral topographical charts or location finders. They also function as political tools, shaping our perceptions of center and periphery, foreground and background. (Heartney 2003: 114)

More recent exhibitions of interest are documented online with catalog essays, images of the works displayed, videos, and links to artists websites. A few of particular interest include the following:

The Creative Compass in 2010 was hosted at the Royal Geographical Society in London. In addition to works from the Society's collection, the exhibition included a number of works by Agnes Poitevin–Navarre and Susan Stockwell. These artists made art works in direct response to the Society's extensive collection of maps and atlases, focusing on maps and mapping as a format through which to explore power, authority and hidden narratives. The exhibition included works by both artists which have been discussed and viewed in a number of subsequent venues. Poitevin–Navarre's *Proustian Map of London* in which she annotates the memories of over a thousand Londoners on wallpaper sized maps of the city makes visible the private, or 'hidden' narratives that construct an individual's sense of place. Stockwell's *Colonial Dress*, a dress patterned on those from the age of British Empire is constructed from maps of colonial expansion. Together with its companion piece *Money Dress*, a similar garment constructed from paper money Stockwell offers a quiet commentary on power, and its potential cost. A catalog is available from the Royal Geographic Society, and the text of Dr. Harriet Hawkin's catalog essay can be accessed at www.agnespnavarre.com.

Apamar: Charts, metrics and policies of space, in Barcelona in 2011 was curated by Maral Mikirditsian, Ramon Parramon and Lail Sole. The works by eight artists or artist cooperatives included wall-drawings, traditional maps annotated in non-traditional ways, video documentation of under-recognized urban spaces, and digital networks made in response to the spatial realities of specific geographies. This diversity of forms was reflected in the diversity of nationalities of the artists themselves, making this an international exhibition in every sense. Exclusively documented online, the curators state their intentions in this way:

> *Apamar* means to measure a field but also to know something very well ..... The participating projects in the exhibition ... work with the many underlying qualities of geographical maps. Experimenting with this seemingly conventional medium, while staying faithful to some of its main attributes such as accuracy, factuality and intentional objectivity, they create new models from a critical perspective with the objective of proposing alternative strategies [with which to understand and visualize space] (Mikirditsian et al. 2011)

We Don't Record Flowers Said the Geographer was held at the Betonsalon Art and Research Center in Paris, also in 2011. Curated by the art partnership known as Bo-ring, comprised of Virginie Bobin and Julia Klaring, the exhibition included works by 15 artists and artist groups who, according to curator Bobin

> ... act as geographers, explorers and cartographers ... traveling through both material and immaterial territories or terrains to gather data—images, facts, stories—which is then, often literally *mapped* and shared. (emphasis in original; Bo-ring 2011)

The artists include Ursula Biemann Swiss curator, artist and geographer and Trevor Paglen geographer, artist and activist. The Bo-ring curators were highly influenced by the work of these two scholars, especially Paglen's emphasis on the "transformable" nature of geography today and Biemann's observation that

> ... geography has become a predominant tool to think the post-modern world... [that is] a distinct mode of producing and organizing knowledge regarding the way natural, social and cultural conditions relate to one another. (Biemann and Bogoff 2003: 21)

Cultural geographer Caleb Johnston, now a lecturer at the University of Edinburgh, curated a more thematically focused exhibition titled *Counter Mapping* in 2011. The exhibition was held in conjunction with the Push Festival, an annual international music festival held in Vancouver BC. The exhibition centered on works by seventeen visual artists residing in Vancouver and was accompanied by works that were the result of collaborations between cultural geographers from the University of British Columbia and artists from Simon Fraser University School for the Contemporary Arts. These artists and collaborative groups were invited to create works that offered alternative ways to map the city of Vancouver BC. Johnston is clear about his intellectual intentions in this exhibition:

> From Mercator projection to Global Positioning Systems, maps have long remained the language of power. In Counter Mapping, we push back against this history. This exhibition

> began with a question: if maps are the instruments of power, how might we produce and circulate a poetic, even, radical cartography? … I am motivated by the desire to appropriate mapping technologies and disrupt the authority of the gridiron to flatten and deaden the complex social worlds of Vancouver. (Johnston 2011: 3)

The works included sound installations, an alternative guided tour to be used while riding the Canada Line train, and a video animation version of Ruth Scheuings GPS drawings discussed below. This exhibition includes no inscribed maps but rather art-works in a range of media that emphasize the lived experience of the inhabitants of a specific space and time. Although the catalog for this exhibition can be downloaded at the Push Festival website, the art works can be more fully understood by visiting Dr. Johnston's website where color imagery, sound tracks and videos of the works can be found (www.calebjohnston.ca/counter-mapping).

4.2 Books

Lucy Lippard is a prolific and influential writer on art and culture. She champions those forms of art making seen by many commentators as most relevant to the act of mapping: site-specific art, performance art, and installation. She is the go-to author in the postmodern moment to discover and learn about non-object makers in contemporary art. In *The Lure of the Local* from 1997, she explores art that is engaged with the issue of place: how it is established or re-established, experienced, remembered, obliterated, politicized, and celebrated. Always a champion of the arts as an agency of cultural critique, she makes an important distinction between the work of visual artists and cartographers.

> The 'naturalization' of maps—the myth that maps show the world the way it really is—veils the fact that maps are cultural and even individual creations that embody points of view. They map only what the authors or their employers want to show; resistance is difficult. They are 'powerful precisely to the extent that the author disappears.' Artists trying to combat and expose hegemony, on the other hand, put their names on their work and are vulnerable in their individualism; they lack the social power of the nameless mapmakers, who, like the image-makers of the mass media, determine how we see and are not called upon to take personal responsibility. (Lippard 1997: 78)

She also gives voice to an aspect of mapping common to both artists and cartographers as they ponder the nature of spatial representation, and seek to make work that recognizes

> … the importance of place not as a grand earth work carved in the desert but as a personal spiritual, mental form we each carry inside us, a sculpture carved by memory and exposed by simple conversation. (Lippard 1997: 81)

Lippard specifically credits cartography critic Denis Wood's (1992) *The Power of Maps* as a central influence in her 1997 text. Wood's book is, in fact, a demonstration of the close connection between maps and art as objects of visual culture, both of which are best understood via semiotic readings. Wood's deconstruction of the silent assumptions underlying the visual vocabulary of cartography ploughs ground familiar to visual artists, and finds an enthusiastic audience there. Wood's observations and arguments are further refined and distilled in his article entitled "Map Art" from the Winter 2006 issue of *Cartographic Perspectives*. In this article Wood focuses on the visual cartographic codes that serve the societal assumptions outlined in his 1992 work. He refers to these codes as "masks" behind which political, economic, and cultural assumptions hide. He points out that painting [read all visual arts] has shared these codes, or "masks":

> Maps pass as descriptions of the territory most readily when they appear to be describing an objective state of affairs … and they appear to be doing this when they wear masks of impersonal authority … This mask, for so long worn by painting, makes maps an irresistible target for contemporary artists who either take the map's mask off, or refuse to put it on. (Wood 2006: 8)

His article refers to a range of artists whose work with maps and mapping actively eschew this "mask," most especially Joyce Kozloff (for images of this artist's work see www.joycekozloff.net).

Although art making is informed by the contemporary cultural moment, artists also look to the past for role models and cautionary examples. The Dutch painter Jan Vermeer (1632–1675) is almost always cited as a forerunner of artists using maps because images of maps play such a prominent role in so many of his paintings (see www.essentialvermeer.com). Consequently, Svetlana Alpers' essay "The Mapping Impulse in Dutch Art" which originally appeared in David Woodward's (1982) *Art and Cartography* is instructive. Alpers places Vermeer's body of work in the larger context of other Dutch artists, all of whom saw painting as an act of "graphic description" by which one could record or evoke something seen. This northern European impulse created a strong cultural bond between Dutch draftsmen, painters and cartographers. This emphasis was in sharp contrast to the southern European desire to make pictures that were instruments of "rhetorical persuasion" (see Fiorani 2005). Alpers (1983) expands upon the northern European preference for visual description over rhetoric in her subsequent text *The Art of Describing: Dutch Art in the 17th Century*.

The essays in *Mapping*, a 1999 anthology edited by Denis Cosgrove, take up the process of mapping, rather than maps themselves. In his view, a rethinking of this process is in order due to the permeability of the planet's political, economic and industrial systems, and the radical changes in both information and representational technologies. His contributors

> explore some of the contexts and contingencies which have helped shape acts of visualizing, conceptualizing, recording, representing and creating spaces graphically—in short, acts of *mapping* (emphasis in original) (Cosgrove 1999: 6)

The anthology includes Wystan Curnow's chapter entitled *Mapping and the Expanded Field of Contemporary Art*. Curnow, a Professor of English at the University of Auckland, is a respected poet, curator and writer of art criticism. In this essay Curnow provides an overview of international exhibitions dedicated to art, maps and mapping from the late 1980s through the late 1990s. The chapter describes the distinction to be made between art that employs "inscribed maps" and art created as an act of mapping. As one particularly useful example of the distinction, Curnow describes at length the "counter-exhibition" *Mapping: A Response to MoMA* curated by Peter Fend in repost to Storr's exhibition at MOMA mentioned above. Curnow quotes Fend himself on the distinction between noun and gerund:

> [Storr's exhibition *Mapping* was] about paintings, sculptures, or drawings which happen to include maps. It has nothing to do with mapping … [which is] an action of charting or planning a domain or space, within which action would take place—real world actions, or at least performed actions. (Curnow 1999: 256)

The political implications of this distinction are present throughout the Curnow essay. Although Peter Fend continues to practice mapping as a central aspect of his creative endeavor, this early exhibition is rarely referenced. Curnow's essay provides useful insight into this particular aspect (or limitation) of the visual art world's understanding of the potential maps hold for constructing meaning.

Fiona Fiorani's (2005) *The Marvel of Maps: Art, Cartography and Politics in Renaissance Italy* is particularly useful in her well considered discussion of the rhetorical power inherent in large scale interior vistas of maps, especially those in the map galleries of Cosimo I in the Guardaroba Nuova in the Palazzo Vecchio. Her analysis of how the physical reality of walking among them choreographs the reception of their meaning in the viewer offers particularly useful insights to artists involved with active elements of mapping as contrasted to using inscribed maps.

Harmon's (2009) *The map as art* offers full color, well reproduced images of the work of 160 artists using maps and mapping in the creation of their work. Harmon offers a brief introduction, timeline and commentary on the individual works. Gayle Clemans provides five in-depth essays. While most of the art works rely on inscribed maps, or images of maps a notable exception is the work of Jane Lackey discussed below. Harmon's (2004) work *You are here: Personal geographies and other maps of the imagination,* is equally lavish in its color illustrations. It is an openly subjective and idiosyncratic view of the uses and abuses of map imagery in art, advertising and popular culture. It extends the field of mapping to include poems, lists, meditations and other textual works that seek to evoke the nature of specific locations.

GeoHumanities, an anthology edited by Michael Dear, Jim Ketchum, Sarah Luria and Douglas Richardson published in 2011, explores the intersection of geography and the humanities. Ketchum's observations on the contributions of the visual arts to this kind of interdisciplinary venture will be discussed below.

5 Three Contemporary Examples

I have chosen the following artists as contemporary examples for several reasons. They work individually, producing artworks that are viewed in traditional art spaces, the better to illustrate the impact of the histories discussed above. They share, with me, a desire to blur the distinction between public and private knowledge, and to work in the intersection between concepts of the subjective and objective. They seek to participate in the creation of more inclusive histories, more inclusive forms of understanding experience. But most importantly their work, like my own, specifically emphasizes embodied, tactile aspects of knowing the world. Each artist seeks to create a more inclusive map of whatever terrain she is investigating, and create an artwork that references the lived experience of inhabiting, visiting, or understanding that space.

5.1 Lou Cabeen

My own goal in this regard is to specifically imbue public forms of spatial knowledge with the subjective experience of that space. Consequently I use inscribed maps as the substrate for much of my work, specifically USGS topographical maps with their familiar color palette, cartographic codes and scale. I count on the viewer's assumption that these maps are objective, scientifically determined documents. This expectation is then confounded by my interventions with imagery or my disruptions of the surface itself. I use the maps both as visually sophisticated and pleasing objects in their own right, and as familiar objects of material culture whose use and function are known. Juxtaposing the familiar with the unfamiliar, or the mundane with the unexpected is a strategy that has allowed my work to disrupt the expectations of viewers and engage them with ideas that they might otherwise resist. A familiar material or form in an artwork can support us as we confront the unfamiliar, or perhaps even the unwelcome content represented there.

In the *Apparition Series* I record 25 apparitions of the Virgin Mary that have been reported within the continental US since 1945. The precise location of the sighting is found at the center of concentric rings of text. These are drawn in gold pencil and graphite directly onto the USGS maps. The text used on each map is a chant or prayer to the Virgin Mary, specific to that apparition story (Fig. 1).

In the two examples shown here, *Apparition Series: Jamaica NY and Apparition Series: French Village IL*, the luminous circles of text radiate from the spot where a distraught housewife and mother (Jamaica NY) or a disabled hospital volunteer (French Village IL) witnessed apparitions. The texts, drawn in concentric rings with graphite and gold pencil, are portions of Latin chants to the Virgin relevant to the site. The gold pencil leaves a distinct but translucent mark, one through which the cartographic markings show. This allows both the words of the chant and the carefully recorded, recognizable details of the local terrain to be seen simultaneously. In this way each of the twenty-five modified maps juxtapose the topography of our assumptions of the known with the numinous experience of vernacular devotion, leaving a record of the unseen (for images of the complete series see www.loucabeen.com). The *Apparition Series* is discussed by Wilson and Elwood (2014):

Fig. 1 Lou Cabeen, *Apparition Series: Jamaica NY and French Village IL* gold pencil and graphite on USGS topographical maps. Photo credit: Ricardo Martinez

> … [H]ow a mapmaker conceptualizes and approaches data influences the whole endeavor …. Cabeen is certainly working with data. She manipulates the precise location of apparitional sightings on the surface of USGS topographic maps, themselves a familiar product of centuries-long surveying techniques. These data capture. They hold within their bits and bytes: field notes and lines, the diverse knowledges of expertise and experience—spiritual, observational and scientific in one.

5.2 Jane Lackey

Jane Lackey maps the unseen, especially the unseeable but persistent dynamics of human interactions and communication. Lackey works with the most mundane of materials, placing and layering tape, stickers, dots, labels and thread on translucent paper to create the work discussed below. Shapes refer to architecture, lines denote connection, and dots can symbolize individuals or their interactions. A detail from *Survey* reveals Lackey's basic visual vocabulary (Fig. 2).

Harmon (2009) includes two images of *Survey,* a series composed of 14 separate compositions. Harmon describes Lackey's working method well:

> [In *Survey*], Lackey mapped discussions about creativity during an artist residency at the Camargo Foundation in Cassis France. As participants in this work/live community came together to discuss their projects in various disciplines, Lackey visualized the ephemeral

Fig. 2 Jane Lackey, *Survey*, detail showing the range of materials used in constructing this work. Photo credit: Tim Thayer

> connections between ideas and means of communicating them. In her series, identified pathways underscore the choices artist make in pursuing particular ideas over others, and the "you are here" dots mark interactions of particular significance. (Harmon 2009: 136–137)

Survey maps the conversations held between the scholars themselves during the residency. Of the fourteen works in the series, two reflect overall experiences of the scholar's presentations and interactions, *Matrix* and *Hybrid*. The other twelve, subtitled with initials, are "details" of these works reflecting the interactions of individuals (Fig. 3).

Lackey creates a visual record of her own and others' experience of a specific space—the seminar room—and the interactions of those people who move through it. The spatial experiences of individuals are mapped in purely abstract terms, providing a visual record of the spatial dynamics of a group for those viewers who can decode the abstraction. In this way Lackey's works are products of the desire to map, to remember our own experiences of space and to record the experiences of others. Lackey herself sums up her work as "… giving spatial dimension to human behavior … [Making] a schema of interaction, entanglement, memory, and contemplation mapped as a progression through the space of material" (Lackey 2012).

Like traditional maps, these works are the product of a visual vocabulary of abstract mark making whose complexity is immediately apparent, and which can be decoded with the use of Legends, which are made available to viewers of Lackey's installations (Fig. 4).

In *Panorama*, shown above, Lackey produced a drawing composed of 42 segments joined together to create a continuous surface 9 inches tall by 62 feet long. The individual drawings are mounted on prepared boards that can interlock to create the panoramic view of the work shown above. *Panorama*, like *Survey*, records the interactions of the whole Carmago Foundation residency community,

Fig. 3 Jane Lackey, *Survey*, mixed media. Photo credit: Tim Thayer

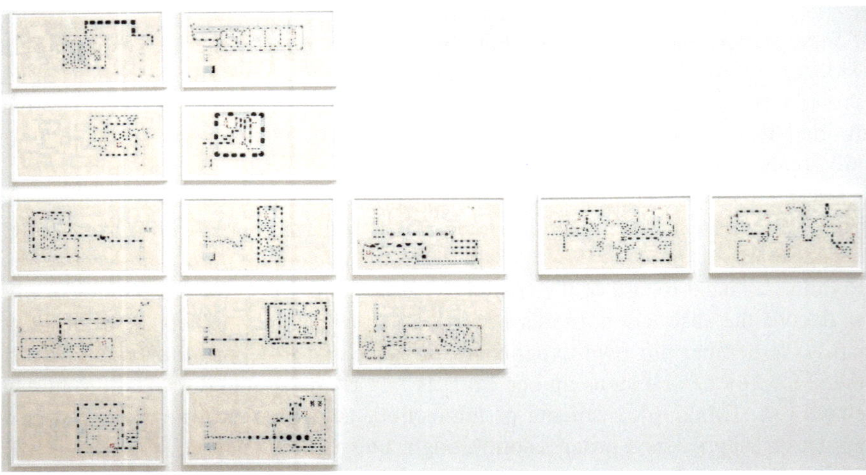

Fig. 4 Jane Lackey, installation image of *Panorama*, mixed media. Photo credit: Tim Thayer

not just the scholars. Each segment of the drawing refers to an individual—scholar, partner or staff—with the panoramic display revealing the interconnections of these diverse units. The size of the units and the overall piece was determined by the perimeter of Lackey's studio during the residency.

Although each individual work could be seen and appreciated in isolation, Lackey prefers to present these works as installations in which one moves along the entire perimeter of the gallery space in order to appreciate the understated, circuitry-like surfaces. This perambulation links the movement of the viewer with their visual appreciation of the work, and echoes Lackey's conceptual premise. Lackey herself describes the spatial aspects of her process in these terms:

> My work depicts interconnected relationships condensed to … material surfaces … [P]hysical movement is coupled with viewing as one progresses through an installation moving closer, bending, turning, circulating, looking at detail and making comparisons …. Each [piece] becomes a schema of imagined space involving interaction, entanglement, memory and contemplation. (Lackey 2012)

Although the system is consistent in Lackey's mind, we are left to construct our own meanings from her elegant and understated surfaces. Once again, abstraction is a reference for the real, and provides a means and a method for representing lived, carefully observed experience. And the mundane materials of everyday life, the tape, stickers and thread, communicate not only the complex but also the ethereal aspects of the seemingly commonplace (for additional imagery see www.janelackey.com).

5.3 *Ruth Scheuing*

Ruth Scheuing uses the most technologically advanced software systems to create her work. Scheuing's art is rooted in her fascination with the current interface between technology, textiles and topography. She gathers and manipulates spatial data from a variety of sources to create her imagery. This imagery becomes literally material—cloth—as Scheuing then uses the digital jacquard loom to manifest it in fabric—evoking the tactility of lived experience. While some are startled by such technologically driven imagery rendered in a textile, in fact there are strong connections between them. Textiles, like maps, are surfaces that are created on a grid. Looms are tools that function as binary systems, in that a thread is always either up or down. She revealed to me in a 2011 email that these connections fuel much of her interest in these technologies.

Scheuing uses imagery from Google Earth, NASA and GPS data as the visual basis for her works. This imagery makes evident aspects of nature that are unseeable in the normal course of one's moving through space. Although she chooses these technologically mediated images of topography because they are beautiful, she places them in the context of other imagery relevant to the site depicted in order to counter romanticized assumptions about place, history and terrain. That is, Scheuing seeks to confound our tendency to see unfamiliar terrain as somehow frozen in time, eternally exotic, and isolated from human experience or interaction.

Scheuing's methods are best seen in the works included in her *Silk Road* series, which examine the myth and metaphors of the Silk Road as well as the terrain it traversed (Fig. 5).

Sogdian Child's Coat on the Karakoram Highway; 38°56'8"N 75°25'34"E provides us with a stunning example. An image of a tribal textile, the child's coat of the title, is superimposed on the forbidding topography of the Karakoram Highway between the current nation states of China and Pakistan. The coat conforms to the specifics of the terrain even as the relative scales are wildly out of proportion. It is distinct in some places, and hard to distinguish from the mountains in others. Scheuing visually references the intimacy of place and people with this juxtaposition even as she reminds us of the larger world that traverses it via the golden line, which shows the route of the Silk Road that crosses both mountains and coat as it winds its way out of the piece.

Scheuing also creates work that draws on her fascination with the abstract mark making that results from her own GPS tracks. Scheuing uses these tracks as a kind of personal notation system, a record of her embodied experiences with the landscapes she knows intimately. Consequently, she incorporates these GPS tracks into artworks that reference the domestic environment and private life. Scheuing has created a series of white on white textiles framed with colored stripes in the size and textures of dishtowels (Fig. 6).

Fig. 5 Ruth Scheuing, *Sodigan Child's Coat on the Karakoram Highway*; *38°56' 48"N 75°25'34"E* jacquard woven textile. Photo credit: Michael Lawlor

Fig. 6 Ruth Scheuing, *Drive to Horseshoe Bay*, jacquard woven textile. Photo credit: Michael Lawlor

Each track is surrounded by familiar textile patterns that follow a cartographic system, even as they camouflage the terrain that surrounds it. Close examination reveals the name of her destination woven into the colored stripe of each piece (for additional imagery see www.ruthscheuing.com).

More recently, Scheuing has presented her now substantial collection of GPS track-drawings in animated digital form, without the context of text or textile. This work was part of the *Counter Mapping* exhibition discussed above, and is documented in the accompanying catalog. There Scheuing describes her working method and specifically her interest in GPS tracking. It is worth quoting at length:

> These lines were captured as tracks with a hand-held GPS. They represent simple daily activities, driving to work, shopping, visiting friends, walking around the neighborhood, or riding a bike around the park. The emerging patterns anchor my memory and trace narratives with simple lines. GPS tracks represent real lived patterns, and at the same time, they are evocative abstract shapes and scientific records of daily life. Although invisible in the work, my body is agent of these drawings … My work has long engaged in blurring the lines between global perspective and domestic spaces, between nature and technology, here and there. (Johnston 2012: 10)

In Scheuings work, as in the works of the other artists referenced here, we are shown unexpected representations of terrain. By pondering the surfaces, the materials, the patterns and the unmistakable abstractions of mapping, by seeking the familiar in these unfamiliar forms, we are led to make connections between our own embodied experience of time and place, and our internal spatial relationships with the places we think we know.

6 Conclusion

As we have seen, maps, mapping and the visual arts have long been related disciplines, each offering distinct means and methods of recognizing and depicting space while sharing an implicit assumption of the importance of doing so. During the first half of the last century, the distinctions were paramount, but since the 1970s those distinctions—like boundaries of many types—have become porous and blurred. The shared desire to describe place, space and our experiences within it have led artists to embrace the abstractions of cartography, and the recognition that all human products are mediated by power in some form have led geographers to grapple with post-modern critiques of such efforts. Although many artists, like those discussed above, draw on maps and mapping methods to create stand-alone artworks created by individuals, other artists are engaging in interdisciplinary dialogs, working in collaboration with geographers and others to delineate and represent the contours of our newly porous and mutable world.

An early field guide to this new interdisciplinary terrain is the anthology *GeoHumanities* where a number of contributors delineate strategies for what they refer to as transdisciplinary research. These include transgressing traditional disciplinary boundaries, making connections between layers of transdisciplinary data and imagining the world as well as describing it (Dear et al. 2011). All of the artists discussed in this chapter offer concrete examples of these strategies. From Nancy Graves crossing the intellectual boundaries between art and science with her use of satellite imagery as source material for her art, to Richard Long ignoring the boundary of studio walls and creating his art on the landscape itself, transgression of the traditional or expected has been of primary importance to contemporary artists. Ruth Scheuing's images—textile, caravan route and topography, the product of her expansive investigations into a range of academic disciplines—are presented visually in literal layers the better to reveal the connections among them. Scheuing not only describes the discrete components that inform our understanding of the historic Silk Road, she imagines a world in which their interdependence is also fully comprehended.

In *GeoHumanities* we also find encouragement for transdisciplinary researchers to allow their research methods, forms and formats to be 'content-responsive' that is to "allow one's subject matter to direct the project, its methods and the form it eventually takes …." (Dear et al. 2011: 57). This, too, is a primary strategy in the visual arts where the effective transgressions of traditional boundaries are understood to arise from the demands of the subject matter under consideration rather than on a whim. Whether one considers the Cubist desire to fully describe all the possible views of a table on the same canvas, or my own desire to juxtapose the quantifiable specificities of government issued maps with the subjective experience of the numinous, the creation of a new methodology must spring directly from the subject matter under scrutiny. Jane Lackey's materials—the most mundane of office supplies—reference and reinforce the routine nature of the interactions she charts even as her record of those interactions create complex and mysterious patterns that defy simple interpretation. And, although Lackey has a preferred method of presenting this work, the art is made in modules and can be displayed, seen and approached (both literally and figuratively) in a variety of ways. This also is a reflection of the nature of the interactions she documents. The content inspires even creates the form of her work.

As we have seen, these artists and many others have embraced maps and mapping as potent allies in the quest to reference and understand the world around us via abstract visual languages that can be understood on many levels simultaneously. Beyond the examples discussed here, a wealth of subjects is being examined via the trans/interdisciplinary ventures of art, maps and mapping. Groups such as the Center for Land Use Interpretation, made up of artists, geographers and scientists based in Los Angeles map prohibited sites and publish them as a 'Land Use Database' online (www.clui.org). This effort charts "anthropogeomorphology" or ways in which humans have changed the land/landscape (Smith 2011: 287). The Center for Urban Pedagogy is a team of artists, designers and activists who "help people understand and change where they live" (www.welcometocup.org). A recent project involved New York City trash disposal, and relied on maps to not only introduce community groups to the nature of the issue, but produced a map-based image describing the interconnected aspects of trash in urban life. The complexities of the network of government workers, neighborhood residents, sanitation workers, etc. are made visually legible via a map like graphic titled *Welcome to Garbage City* (Thompson 2008). These examples and others cited in the works below offer pathways for future research. Topics such as political realities and contexts, neuro-physiological connections, and psychological and sociological responses to space and its representation are just a few of the areas of inquiry that are fertile ground for such work.

Maps, mapping, and the visual arts offer interdisciplinary research multiple examples of methods and materials with which to further explore and understand the interconnected spatial realities we inhabit. Whether those spatial realities are internal, external, public or private artists will continue to engage any and all tools at their disposal to make manifest their insights into the world around them—maps and mapping primary among them.

Key Readings

Ashcroft, B., Griffiths, G., & Tiffin, H. (1995). *The post-colonial studies reader*. London: Routledge. The themes and issues raised by post-colonial theory inform much of the newest interdisciplinary work done between geography, cartography and the arts. For those not familiar with the core texts of this body of thought, this reader offers an excellent guide, especially Part IV: Postmodernism and post-colonialism and Part XII: Place.

Berwick, C. (2010). Remaking the map. *ARTnews*, *109*(9) 100–103. This review in the October 2010 issue of ARTnews is an excellent, concise introduction to some central projects and artists engaged in what Berwick describes as radical cartography. She defines radical, or counter cartography as "the new field … populated by geographers, cartographers, artists and designers who want to convey the kinds of information usually omitted in conventional maps."(100). Included in her overview are *An Atlas of Radical Cartography* from 2007, works by the Center for Urban Pedagogy, and writer Rebecca Solnit's *Infinite City* and *The Lost World* projects among others.

Smith, T. (2011). *Contemporary art: World currents*. London: Prentice Hall. This text picks up the story of the evolution of art making in the West where my chapter leaves off. Smith's discussion of contemporary art making in a global context is an excellent example of scholarship informed but not confined by post-colonial theory. Part Three, titled "Contemporary Concerns" is especially relevant in discussing works that further the issues outlined in this chapter.

Thompson, N. (2008a). *Experimental geography*. Brooklyn: Melville House. This exhibition catalog provides information about a number of art works, artists and art/geographer teams working in the field and essays by artist/scholars such as Trevor Paglen. The entries concerning the Center for Land Use Interpretation and The Center for Urban Pedagogy are especially pertinent.

Wood, D. (2010a). *Rethinking the power of maps*. New York: Guilford Press. As the title implies, this work is not a revised edition of Wood's 1992 *Power of Maps,* but rather a continuation of the ideas stated there. Wood's thesis in this volume is that "Maps *are* engines that convert social into social space, social order, knowledge" (Wood 2010: 6; emphasis in the original). Part Two of the text is devoted to counter-mapping.

References

Alpers, S. (1982). The mapping impulse in Dutch art. In D. Woodward (Ed.), *Art and cartography: Six historical essays*. Chicago: University of Chicago Press.

Alpers, S. (1983). *The art of describing: Dutch art in the 17th century*. Chicago: University of Chicago Press.

Arnason, H. H. (1986). *History of modern art*. New York: Harry Abrams.

Benjamin, W. (1968). *Illuminations* (pp. 217–251). New York: Schoken Books.

Bennett, T., Grossberg, L., & Morris, M. (Eds.). (2005). *New keywords: A revised vocabulary of culture and society*. Malden, MA: Blackwell Publishing.

Berwick, C. (2010b). Remaking the map. *ARTnews, 109*(9), 100–103.

Biemann, U., & Bogoff, I. (2003). *Geography and the politics of mobility, ex. cat.* Vienna: Generali Foundation.

Bo-ring (Bobin, Virginie and Klaring, Julia). (2011). We don't record flowers said the geographer. Available at: www.nousnenotonpaslesfleur.net

Cosgrove, D. (1999). *Mappings* (pp. 1–23). London: Reaktion Books.

Curnow, W. (1999). Mapping and the extended field of contemporary art. In D. Cosgrove (Ed.), *Mappings* (pp. 253–268). London: Reaktion Books.

Dear, M., Ketchum, J., Luria, S., & Richardson, D. (Eds.). (2011). *GeoHumanities: Art, history, text at the edge of place*. London: Routledge.

Dreikausen, M. (1985). *Aerial perception: The earth as seen from aircraft and spacecraft and its influence on contemporary art*. Philadelphia: The Art Alliance Press.

Fiorani, Francesca. (2005). *The marvel of maps. Cartography and politics in Renaissance Italy*. New Haven: Yale University Press.

Fuchs, R. H. (1986). *Richard long*. New York: Thames and Hudson.

Harmon, K. (2004). *You are here: Personal geographies and other maps of the imagination*. New York: Princeton Architectural Press.

Harmon, K. (2009). *The map as art: Contemporary artists explore cartography*. New York: Princeton Architectural Press.

Heartney, E. (2003). Mapping a better world. *Art in America, 91*(10), 114–119.

Hughes, R. (1987). *The sculpture of Nancy graves: A catalogue raisonne*. New York: Hudson Hills and the Fort Worth Art Museum.
Johnston, C. (2011). Dr. Caleb Johnston. Available at http://www.calebjohnston.ca/
Johnston, C. (2012). Counter mapping. Available at: www.pushfestival.ca/shows/counter-mapping
Kardon, J. (1977). *Artist's maps*. Philadelphia: Philadelphia College of Art.
Kleiner, F. S. (2010). *Gardner's art through the ages: The Western perspective* (13th ed.). Boston: Wadsworth.
Lackey, J. (2012). www.janelackey.com
Lippard, L. (1997). *The lure of the local; senses of place in a multicultural society*. New York: The New Press.
Mikirditsian, M., Parramon, R., & Sole, L. (2011). Apamar: Charts, metrics and policies of space. Available at: www.ACVic.org
Padon, T. (1996). *Nancy graves: Excavations in print*. New York: Harry N. Abrams Inc.
Paglen, T. (2008). Experimental geography: From cultural production to the production of space. *Experimental geography* (pp. 27–33). Brooklyn: Melville House.
Perry, G., & Wood, P. (2004). *Themes in contemporary art*. New Haven: Yale University Press.
Silberman, R. (1999). *World views: Maps and art*. Minneapolis and London: University of Minnesota.
Smith, R. (1981). *4 Artists and the map: Image/process/data/place*. Lawrence, KS: Spencer Museum of Art.
Smith, T. (1996). Modernism and modernity. In J. Turner (Ed.), *The dictionary of art* (pp. 775–779). New York: Grove.
Smith, T. (2011). *Contemporary art: World currents*. London: Prentice Hall.
Storr, R. (1994). *Mapping*. New York: The Museum of Modern Art.
Thompson, N. (2008b). *Experimental geography*. Brooklyn: Melville House.
Tuan, Y.-F. (1999). Maps and art: Identity and utopia. In R. Silberman (Ed.), *World views: Maps and art* (pp. 11–25). Minneapolis: University of Minnesota.
Weil, B. (1995). Mapping: A response to MoMA. *Art Monthly*, 184 (March pp. 33–34).
Wilson, M., & Elwood, S. (2014). Capturing (Mapmaking). In R. Lee, N. Castree, R. Kitchin, V. Lawson, A. Passi, C. Philo, S. Radcliffe, S. Roberts, & C. Withers (Eds.), *The SAGE handbook of human geography* (pp. 231–249). London: Sage Publications.
Wood, D. (1992). *The power of maps*. New York, London: Guilford Press.
Wood, D. (2006). Map art: Cartographic perspectives. *Journal of the North American Cartographic Information Society, 53* (Winter pp. 5–14).
Wood, D. (2010b). *Rethinking the power of maps*. New York: Guilford Press.
Woodward, D. (Ed.). (1982). *Art and cartography: Six historical essays*. Chicago: University of Chicago Press.

The Contested Maps, Multiple Worlds, and Negotiable Borders of Theatre

Edmund B. Lingan

Abstract In the art of theatre, many effects are created by way of a thorough understanding of the performance space, and also through the manipulation of the borders that divide the architecture of the theatre building from the temporary scenery of a production, the actors from the audience. These realms—the building, the scenery/stage, the auditorium—have been conceived of as "worlds", "continents", or "seas" that constitute the global geography of the theatrical experience. This essay demonstrates some of the ways in which maps, cartography, and geography have impacted the academic discipline of theatre studies, and it also considers how theatre artists' conceptualizations of the theatrical event as a universe that includes worlds informs the way they create their art.

Keywords Scene design · Stage management · Theatrical mapping · Performance space · Semiotics · Audience performer relationship · Visual literacy

For centuries, ideas about mapping and conceptualizations of geography have informed the way that theatre artists and theatre critics have created and thought about new works of art. In the field of theatre history, maps are usually employed to associate events and practices with specific geographical regions. Despite the usefulness of this function, some critics have identified traditional Western cartography as a contributing factor to the overemphasis of European traditions within the study of world theatre. Such critics have proposed new approaches to mapping the regions and migrations of theatrical practices, which challenge the idea that theatrical practices can be contained within artificially-created national borders and natural geographical barriers. In fact, the usage and creation of maps within theatre studies is an issue that is contested and currently a topic of discussion.

E.B. Lingan (✉)
Toledo, OH, USA
e-mail: Edmund.Lingan@utoledo.edu

S.D. Brunn and M. Dodge (eds.), *Mapping Across Academia*,
DOI 10.1007/978-94-024-1011-2_18

As far as the creative practice of theatre is concerned, cartographic activities and geographic thinking are an organic part of the creative process. In the art of theatre, many effects are created by way of a thorough understanding of the performance space, and also through the manipulation of the borders that divide the architecture of the theatre building from the temporary scenery of a production, the actors from the audience. These realms—the building, the scenery/stage, the auditorium—have been conceived of as "worlds," "continents," or "seas" that constitute the global geography of the theatrical experience.

This essay illuminates and demonstrates some of the ways in which maps, cartography, and geography have impacted the academic discipline of theatre studies, and it also considers how theatre artists' conceptualizations of the theatrical event as a universe that includes worlds informs the way they create their art. The first section of this essay considers the conventional use of maps within the study of theatre history, as well as a recent challenge to that practice that was made in a recent article in a widely-respected theatre history journal, entitled *Theatre Survey*. The second section of this article concerns how the notion of worlds and boundaries impact the creation of new works of theatre. As will be seen, both cartography and geography visualization are essential to the process of making theatre.

1 Conventional Map Use in Theatre Studies

Within theatre studies, the most conventional use of maps has been made by theatre scholars who seek to associate theatrical traditions with specific areas of the world. One of the most well-known examples of this sort of map usage is found in *History of the Theatre*, by Oscar G. Brockett and Franklin J. Hildy. This book, which is quite possibly the most popular theatre history text of the English-speaking world, features many maps. Chapter 1 concerns historical perspectives on the origins of theatre, and it contains a simple map entitled "ancient Egypt," which is a simple black and white map that shows the borders of the Mediterranean Sea, the Red Sea, the Nile, and dots indicating the locations of Memphis, Saqqara, and Abbydos (Brockett and Hildy 2008: 7), all of which are identified in the text as possible loci of theatrical activities in ancient Egypt. Throughout *History of the Theatre*, maps usually provide a visual representation of an area of the world that is associated with a specific theatrical tradition. The use of maps within this book may seem overly simple to some readers, and the reason for this simplicity is that most theatre historians view maps as little more than one of many tools that contribute to the overarching task of describing and critically responding to theatrical practices in a variety of places and historical eras.

2 Current Criticism of the Conventional Use of the Map in Theatre Studies

In his 2011 *Theatre Survey* article, Steve Tillis critiques the sort of mapping and geography that is used in *History of the Theatre*, and he proposes that a more complex "geographic understanding" might lead to "a far richer appreciation" of the migration of theatrical practices throughout history. Tillis laments that theatre historians have created few "theatrical maps" —by which Tillis means maps that concern geographic information that is directly linked to the practice of theatre (Tillis 2011: 301). Brockett and Hildy provide an example of such a map in *History of the Theatre*: in "Chap. 5: English Theatre to 1642," which includes a map that represents the locations of theatres built in London, England between 1574 and 1642 (Brockett and Hildy 2008: 119). This map constitutes what Tillis describes as a "theatrical map" because it locates specific sites where theatrical performances took place in early modern London.

Tillis argues that theatre is a transcultural human experience that "expresses itself in infinite variety," and he suggests that relating the multitude of theatrical forms of expression to various areas of the globe in which they are known to have appeared can counteract the influence of Western mapmaking, which has led theatre historians such as Brockett and Hildy to draw what are, in some cases, artificial distinctions between so-called "Western" and "Eastern" forms of theatre (Tillis 2011: 304). The result of resting upon these preconceived notions of mapping, which are based on an overestimation of the geographic significance of "mutable" nations, is the obscuration of the "differences and similarities among the theatrical practices of different lands" (Tillis 2011: 302). To deal with this issue, Tillis develops a set of theoretical concepts that might ameliorate the dearth of theatrical maps in theatre history and contradict the geographic simplification upon which many historians have based their categories of theatrical activity.

In his article, Tillis creates new maps that pinpoint areas where certain types of theatrical activities take place, and the geographic terminology he devises reveals the density and variety of theatrical activity in various areas of the world. These maps show the movement of theatrical traditions across borders, rather than compartmentalizing various traditions within national and geographic areas that have been established by other mapmakers. In order to create these maps, Tillis speaks of "regional theatre complexes" and theatre "border regions" (Fig. 1; Tillis 2011: 309, 321). Regional theatre complexes are "areas where various theatre forms have more than occasional interactions with regard to shared dramatic literature …, shared techniques of performance, shared artists …, shared audiences, or shared descent from a common 'ancestor' form" (Tillis 2011: 308). Border zones are areas "where two or more regional theatre complexes overlap," and in such areas the theatrical practices of the overlapping complexes "exert a notable influence on one another" (Tillis 2011: 309). Tillis's theoretical terms lay the groundwork for a new approach to mapping theatrical activities, and he demonstrates the effectiveness of this approach by designing a map that indicates the location of theatrical regions

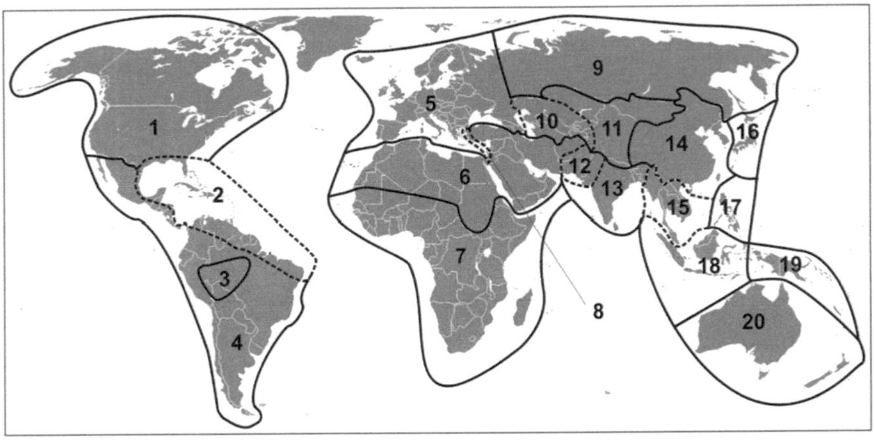

Fig. 1 Steve Tillis's map of theatre "regions" and "border zones," which was published in his 2011 article, "Conceptualizing Space: The Geographic Dimensions of World Theatre." Courtesy of *Theatre Survey* and Steve Tillis. Copyright: *Theatre Survey*

and border zones all over the world (Tillis 2011: 310). A quick glance over the map shows that Tillis strove to defy the national and continental borders that are represented in conventional Western maps of the world. In a caption under this map, Tillis explains that it represents "contemporary world theatre regions." Within the map, theatre regions are represented with solid black lines, and "border zones" with dashed lines (Tillis 2011: 310). The many openings in the dashed lines imply that the boundaries of the border zones are passageways through which performance traditions may travel from one nation to another.

By representing theatrical border zones in this way, Tillis uses cartography to question longstanding associations of theatrical forms with the national borders of Western maps. According to Tillis, this practice is essential to avoid the gross exaggeration of "the importance of European theatre" within theatre history, and it also acknowledges "the depth of heterogeneity" of theatre in Asia, the Americas, Africa, and other parts of the globe whose theatrical and performance traditions remain understudied (Tillis 2011: 304). The theatre regions and border zones on the map transcend the boundaries of continental, national, and oceanic borders. For instance, a theatre border zone that Tillis's map identifies as the "African-American Border Region" includes the northeastern coast of South America, the east coasts of Central America and Mexico, the West Indies, almost all of the state of Florida, and the coastal areas of the US states that surround the Gulf of Mexico. Above the African-American Border Region is a theatre region complex that Tillis calls the "North American Region (neo-Europe)," which includes the remaining portions of the US and all of Canada (Tillis 2011: 304). By stressing the mobility of theatrical traditions across national and geographic borders, Tillis questions the tendency to associate performance practices primarily with the regions of the world in which they were first discovered by commentators.

The use of maps by theatre historians such as Brockett and Hildy, as well as the critical response to such map usage made by theatre scholars such as Tillis, shows that maps play a significant role within the field of theatre studies, by helping scholars to develop a historical, visual, cultural, and spatial understanding of the development, transformation, and transportation of theatrical forms of expression. This, however, does not exhaust the significance of cartographic and geographic conceptualizations within the practice of theatre.

3 The "Worlds" of the Theatre

In the early-modern era, cartographers represented the Americas as a "new world" to be juxtaposed with the "old world" (Europe). Although this dichotomous view of the relationship between Europe and other continents contributed to cultural misunderstandings and separatism, it also heightened the public's cultural curiosity and interest in world exploration. Similarly, the theatre is often thought of as a space that includes multiple worlds, which are open for creative exploration. Shakespeare made one of the most lasting connections between worlds and stages in *As You Like It*, when he wrote "all the world's a stage, and all the men and women merely players" (II, vii). For theatre artists, a related, but distinguishable, principle also holds true: stages, theatre buildings, and auditoriums are spaces in which worlds are temporarily realized. The theatrical event might be said to be a universe that includes many worlds—some physical and some imaginary—and the boundaries between these worlds vary in their flexibility. Three regions of the conventional, Western theatrical event that have often been thought of as worlds are (1) the permanent theatrical building or structure that houses the event, (2) the stage upon which the performance takes place, and (3) the auditorium from which the audience views the performance. Before a visual theatrical concept can be effectively realized with lights, scenery, props, and actors, those who design, direct, and perform must take into careful consideration these three worlds of the theatrical event.

The borders between the "worlds" of each theatrical event are negotiable, and the relationship between the audience and the actor is perhaps the most negotiable of these theatrical world-borders. In some cases, the audience is separated from the performance by several feet of space; in others the actors and audience may interact directly with one another in a single, shared space. For David Wiles, author of *A Short History of Western Performance Space*, the relationship between "performers and audience" is one of the most important "relationships of performance" (Wiles 2003: 3). It is important to consider the border between the worlds of the audiences and the actors, because this spatial relationship contributes to the psychological response of the audience to the production. According to William Faricy Condee, author of *Theatrical Space: A Guide for Directors and Designers*, "the edge of the stage and the first row of the auditorium" is a place "where … two worlds meet", and he identifies this flexible boundary as "a key area in the audience's relationship to the performance" (Condee 1995: 39).

Because the audience-actor relationship is crucial to the success of failure of any production, it has occupied the minds of theatre artists for centuries. Some audience-actor configurations create distance between the audience and the world upon the stage, while others allow a sense of intimacy that allows for actors and audience members to feel a sense of connection with one another. Generally speaking, large and clear divisions of space between the world of the actor and the world of the audience encourages more emotional distance from a performance than a configuration in which the audience sits very near—or perhaps even within—the action of the play.

In Western culture, the proscenium or picture-frame theatre is one of the most commonly-used audience-actor configurations. In this configuration, the audience sits in a series of rows and views a stage and set that are encased within a proscenium arch that neatly and clearly divide the audience from the imaginary world of the theatrical production. The sense of division between the stage and the audience is often increased by decorative elements adorning the opening of the proscenium arch. Of course, curtains can be used to hide the scenery from the audience until the opening moment of the play, and this effect also accentuates the sense of division between audience and performance in the proscenium arrangement. Because of the distance placed between actors and audiences in proscenium spaces, the actors are compelled to use larger gestures and louder voices than would be needed in a closer configuration. Thus, there is a correlation between acting style and the distance placed between actors and audiences. It is no surprise, therefore, that the proscenium stage arrangement has remained popular in heightened performance forms such as opera.

In other forms of theatre, a more intimate relationship between the audience and the actors than that which is offered by the proscenium configuration is desired. This was the case in Shakespeare's era, in which the worlds of the audience and of the stage were arranged in what is commonly referred to as the "thrust" configuration. The stage of the thrust configuration extends into the audience, and it is surrounded by spectators on three sides. An example of an early modern, English thrust stage is contained in a 1596 drawing of the Swan Theatre in London, which was sketched by a Dutch tourist named Johannes de Witt (Fig. 2; Brockett and Hildy 2008: 120). As the de Witt drawing shows, the relationship between the audience and the performance in the thrust configuration is closer and more intimate than the same relationship in the proscenium arrangement. This intimacy was maintained even in large-capacity public theatres in Shakespeare's era. The Globe, for instance, may have sat up to three thousand spectators, but the thrust arrangement made it impossible for any spectators to be more than 75 feet from the stage (Brockett and Hildy 2008: 122). This close proximity informed the manner in which Elizabethan and Jacobean playwrights wrote dramas. Shakespeare used soliloquies, for instance, which called for actors to speak directly to the audience; this form of the actor-audience relationship is particularly effective in the thrust formation.

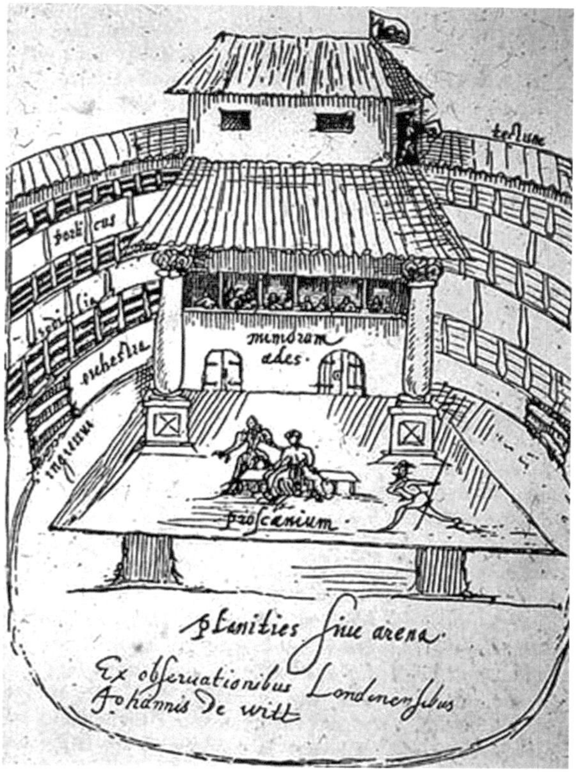

Fig. 2 Johannes deWitt's 1596 drawing of the Swan Theatre in London

By the early twentieth century, some experimental theatre artists began to resist the practice of separating the audience from the spectators at all. One of the best-known challenges of this division came from the highly-influential French theatre theorist, Antonin Artaud, who in his 1932 essay, "The Theatre of Cruelty (First Manifesto)" rejected the traditional practice of separating the world of the stage from that of the audience:

> THE STAGE.—THE AUDITORIUM. We are eliminating the stage and the auditorium and replacing them with a kind of single site, without partition or barrier of any kind, which will itself become the theater of the action. A direct communication will be reestablished between the spectator and the spectacle, between the actor and the spectator, because the spectator, by being placed in the middle of the action, is enveloped by it and caught in its cross fire. (Artaud 1988: 244)

By way of commingling the audience and the actors in this way, Artaud hoped to use the theatre to "call into question not only the objective and descriptive external world but the internal world, that is, man from a metaphysical point of view" (Artaud 1988: 244). Thus, Artaud undermined preconceived notions concerning the traditional borders between the worlds of the audience and of the actors. Artaud's ideas informed the creation of new approaches to scenography that would gain popularity in the late twentieth century through the work of radical theatre and performance groups who placed the audience within, rather than in front of, theatrical spectacles.

One of the best-known of such radical theatre companies who answered Artaud's call to unite spectator and actor within the same geographical space were The Living Theatre, the members of which removed their clothes and intermingled with audience members in their controversial production, *Paradise Now* (1968). Interestingly, Judith Malina, one of the founders of the Living Theatre, explained the aggressive audience "confrontations" that were part of *Paradise Now*, in terms of a journey taken through a terrain fraught with barriers and obstacles to be overcome:

> The Confrontations are guides to the relationship between the actor and the public at each stage of the trip. The Resistance to the Revolutionary Change is treated as the obstacle. The energy form designed is an appropriate strategy for the actor to use to transform the obstacle. (Malina and Beck 1971: 11–12)

Malina speaks about the events of *Paradise Now* as a map of movement from one condition (pre-Revolutionary Change) to a new condition (post-Revolutionary Change). In order to move from this condition to another condition, one must pass over obstacles via a "trip" or journey.

By speaking of *Paradise Now* in terms of a journey from one location ("resistance" and "confrontation") to another (acceptance of "revolutionary change"), Malina describes her performance in the terms of what performance theorist Richard Schechner describes as a "map of the action" (Schechner 1985: 186). Schechner, the artistic director of the Performance Group, which radically challenged the audience-actor divide in the 1968 production, *Dionysus in 69,* has used cartographic language and imagery to describe the theoretical basis of audience-actor interaction in Performance Group productions. In his 1973 book, *Environmental Theatre*, Schechner distinguishes the confrontational methods used by the Living Theatre from those of the Performance Group's harmonious and collaborative approach to audience interaction in *Dionysus in 69*. Schechner argues that the confrontational methods of the Living Theatre are rooted in "the need of the bourgeois to experience suffering as a relief of guilt," while the methods of The Performance Group are "collaborative" and have their basis in "the cooperation of performers and spectators" (Schechner 1973: 39). Schechner views what he refers to as "environmental theater" as a world in which the performers of the audience are two interdependent and essential parts of an organic planet (Fig. 3):

> Environmental theater encourages give-and-take throughout a globally organized space in which the areas occupied by the audience are a kind of sea through which the performers swim; and the performance areas are kinds of islands or continents. The audience does not sit in regularly arranged rows; there is one whole space rather than two opposing spaces. The environmental use of space is fundamentally *collaborative* ... (Schechner 1973: 39)

Schechner visually supports his collaborative and non-confrontational relationship between the performer and spectator with a simple drawing of a globe that includes continents that are labeled "performers" and seas that are labeled

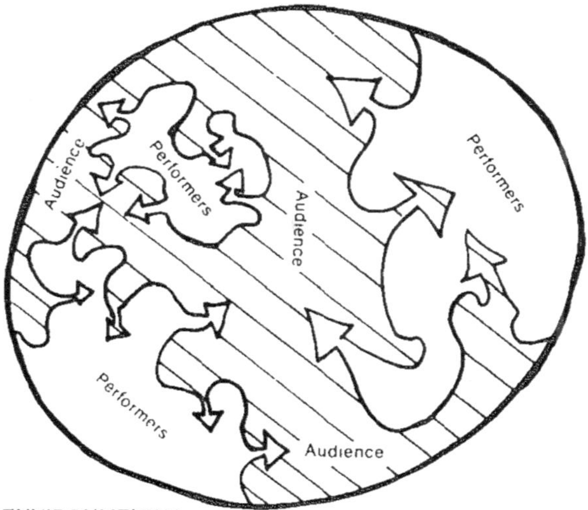

Fig. 3 Richard Schechner's visual representation of the audience-performer relationship as a world seas and continents. This image is contained in Schechner's book, *Environmental Theatre* (Hawthorn 1973; reprint, Applause Books 1994). Courtesy & copyright: Richard Schechner

"audience" (Schechner 1973: 39). Just as the seas and the continents co-exist in an organic relationship to one another in the physical world, so do audiences and actors co-exist in the moment of performance, according to Schechner.

4 Mapping Theatre: Physical Maps for Practical Purposes

Less flexible than the barrier between the actor and spectator is the one that separates the permanent architecture of the theatre from the temporary theatre set, which is created to depict the geography of a fictitious world upon the stage. In order to negotiate this boundary in the creation of an effective theatrical setting, scene designers must accurately map the performance space that will house the scenery. The map that scene designers create is called a floor plan, and this document gives a clear idea of the topography, height, width, depth of the performance space, as well as the dimensions of the theatrical set that will be placed within it. Although ground plans effectively represent the measurements of a theatrical space, as well as the boundaries between the architecture of the buildings, the performance space, and the audience seating areas, they do not provide all of the information that a director or designer needs in order to determine how and what kinds of scenery, lights, and sound to place within it. Set designer Lowell Detweiler, for instance, discusses aspects of a theatrical space that cannot be revealed by a ground plan:

> Theatres have a quality that a ground plan can only hint at. There are theatres that in ground plan look huge and in fact are incredibly fun and intimate to be in. And there are theatres that look more intimate, and you find yourself in Shea Stadium when you're in them. Some theatres are cold, like the Denver Theatre Center. Some are warm, like the Guthrie Theatre. (quoted in Condee 1995: 7)

Condee notes that some designers like to carry a ground plan with them while they explore a theatre for the first time, in order to "relate" the ground plan to their own "emotional, spatial, and instinctual response" to the space (Condee 1995: 7).

Interestingly, what Condee and Detweiler note about the ground plan's limited capacity to reveal the "emotional" nature of a performance resembles comments about the limitations of the map that William Cartright makes in his 2009 *Cartographic Journal* article, "Applying the Theatre Metaphor to Integrated Media for Depicting Geography." According to Cartright (2009: 24), "maps alone are not enough to appreciate the realities that one would experience" by "actually being" in a specific geographic space. To ameliorate this limitation, Cartright proposes that maps be augmented with a digital environment tool that operates according to what he describes as the "theatre metaphor." Such a digital environment would be "confined to the world of the representation" (in this case, the specific geographic area represented by a map) and depict characters who are all "situated in the same context," "have access to the same objects," and "speak the same language." With these elements in place, the characters in the digital environment would "move through a predetermined path" influenced by conceptualizations of "enactment," "pattern," "language," "thought," "character," and "action" that originated in various traditions of dramatic criticism (Cartright 2009: 28–29). The digital environments that Cartright proposes use theatrical elements to allow people to determine their emotional responses to an area of the world before they actually go to visit it. For the creators of a theatrical production, there is no choice but to inhabit the space, because no digital representation could ever replace the creative effect of actually visiting the theatre. Each theatrical space has its own unique qualities, and the actors, directors, and designers must interact with these qualities long before the curtain rises in front of an audience in order to realize a comprehensive work of art that harmonizes with the space in which it takes place.

5 The Border Between Architecture and Scenery

When Cartright writes that maps fail to capture the "emotional" quality of a space, he is asserting that neighborhoods, mountain ranges, cities, towns, etc. are rarely neutral: they usually have a personality that affects the moods and feelings of those who live in and visit them. The same is of course true in the case of theatrical spaces, because theatrical spaces and their permanent structures are not entirely neutral. On the contrary, they have an impact upon the nature of the productions that take place within them, and also upon the response of audiences to those productions. Some designers and directors seek a space that is as neutral as possible —one which does not exert what lighting designer and theatre consultant Jules Fisher describes as a "personality of its own" that must be fought in the process of staging a play (quoted in Condee 1995: 9). Other designers, on the other hand, seek to find a way to make what seems at first to be inconvenient architectural elements into a natural point of fusion that transcends the barrier between the permanent theatre building and the theatrical setting. Theatrical designer David Potts reveals how the second sort of designer thinks when he speaks of an "old steam pipe" that intrudes into the performance space. Rather than dismissing such permanent structures as a "nuisance," Potts thinks of them as objects that can be made into something that belongs in the visual world of the theatrical productions he designs (quoted in Condee 1995: 11).

The border between the theatre building and the fictitious world that occupies its stage can be made less distinct not only by disguising intrusive architectural elements, but also by designing theatrical settings that seem to fuse organically with the theatres in which they exist and, thereby, seem to expand into the permanent

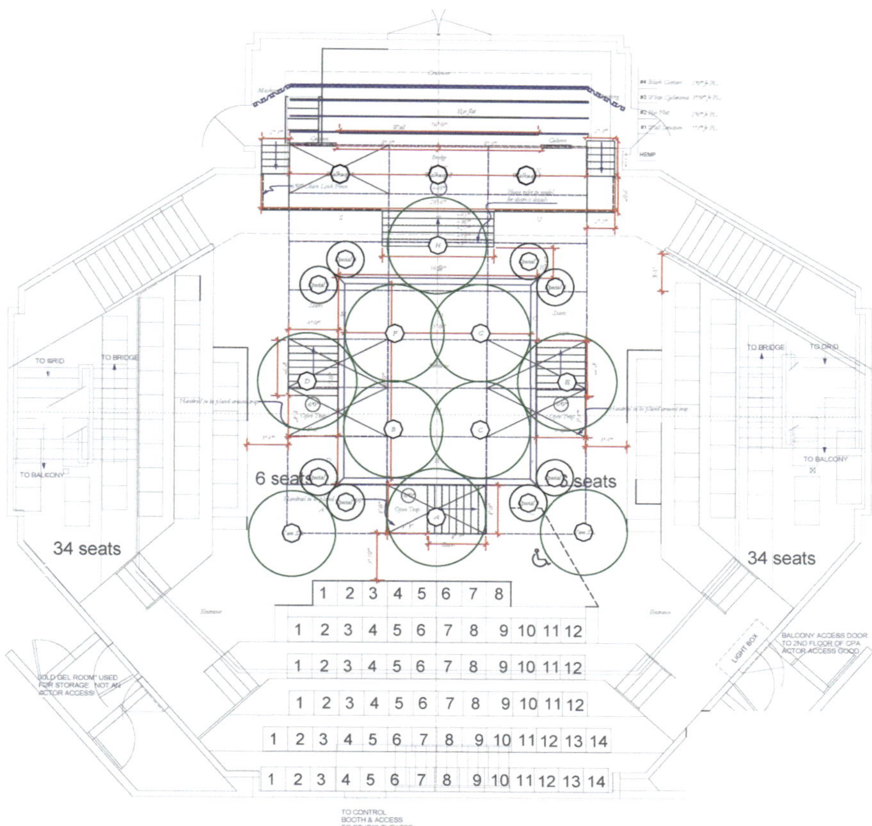

Fig. 4 Donald Robert Fox's ground plan for the 2012 production of *King Oedipus*, which was produced in the Center Theatre for the Performing Arts at the University of Toledo. Courtesy & copyright: Donald Robert Fox

architecture of the space. To demonstrate this point, I will speak briefly about a production of W.B. Yeats's version of *King Oedipus*, which I directed in February 2012 at the University of Toledo's Center Theatre for the Performing Arts for the university's Department of Theatre and Film (Fig. 4). The architectural style of the Center Theatre is industrial. Unlike large proscenium theatres with beautiful decorative elements, such as the Toledo, Ohio Peristyle Theatre, the Center Theatre has exposed brick walls, visible backdrop flying systems and lighting grids, and an unhidden system of wrought-iron piping for hanging sets and lights. Function, rather than decoration, was central to the design of the Center Theatre, and this style of design is what gives the Center Theatre its industrial appearance and atmosphere. For *King Oedipus*, designer Donald Robert Fox took a stylistic approach to the set that softened the barrier between the theatrical setting and the architecture of the Center Theater, and made it seem as if the set was an organic part of the space in which it took place.

Fox designed Oedipus's temple to look like an industrial complex, and the repeated use of industrial piping in the set mirrored the pipes that are visible in the Center Theatre. Although Fox painted the set a rust color to set it apart visually from the black paint of the theatre, its visual qualities seemed to connect organically with the space it inhabited (Fig. 5). This characteristic was noted by Professor Matthew Moore of Cedarville University, who adjudicated the production of *King Oedipus* as a representative of the Kennedy Center American College Theatre Festival. While responding to the actors, designers, and director after the

Fig. 5 Photograph of the set that Donald Robert Fox designed for the University of Toledo's production of *King Oedipus*. Note how the style of the set harmonizes with the industrial architecture style of the Center Theatre. Courtesy & copyright: Donald Robert Fox

production, Professor Moore humorously commented, "You guys really worked hard on this set!" The point of this comment was that the set harmonized so well with the architecture of the theatre that it almost seemed like an entire theatre was built to house *King Oedipus.* This effect was exactly what the Fox had in mind when he designed the set, for he seeks to transcend the boundary between scenery and architecture in his work as a scene designer (personal communication).

Fox's design for *King Oedipus* not only seemed like a natural extension of the theatre space, but it also seemed to be an organic part of the architectural landscape of urban Northwest Ohio and Southwest Michigan, which is the region in which it took place. Although the setting for *King Oedipus* represented an imaginary world, its appearance was based upon the abandoned industrial structures that can be found in places like Toledo and nearby Detroit, Michigan. Fox drew visual inspiration from photographs of decaying structures that were taken by "urban spelunkers," who daringly enter abandoned buildings to document the decay of once-great architectural structures that have now come to represent the decline of wealth and industry in what is known today as the "rust belt." (Examples of such photographs can be found at websites such as www.weburbanist.com.) The same fascination with the intersection of beauty and decay that drives urban spelunkers to photograph such buildings is what inspired the construction of what appeared to be an abandoned industrial complex upon the stage for this production of *King Oedipus.*

6 Conclusion

This essay has presented only a few examples of how cartography and geography inform the study and practice of theatre. In the most common usage, maps have served theatre historians in much the same way that they have served experts working other disciplines: they help the reader to visually imagine the geographic location of events and practices that are described in scholarly texts. The practical details of creating theatre, such as the size of the performance space and the relationship of that space to scene design and actor movement have necessitated the development of map-like documents, such as ground plans, which allow designers and directors to anticipate that performance space's spatial challenges and potential. On a more theoretical level, geographic and cartographic thinking, such as that exhibited by Richard Schechner, has enabled creators of innovative forms of theatre and performance to challenge preexisting notions about the necessity of separating actors from audience members. Even the imaginary aspects of performance—the fictitious actions that take place upon the stage—can be equated with mapping, for, as Agustin Zarzoza (2010: 237) phrases it, the arrangement of events within plays can be thought of as "strategies to map" human "experience." To be sure, the significance of maps, map-making, and geography to theatre and other performing arts has only been hinted at in this brief article. Despite its brevity, however, this essay does show that this relationship has been fruitful for centuries, and that it can be expected to continue flourishing well into the distant future.

References

Artaud, A. (1988). Selected writings. In Sontag, S. (Ed.) Berkeley: University of California Press.

Brockett, O. G., & Hildy, F. J. (2008). *History of the theatre* (10th ed.). Boston: Pearson.

Cartright, W. (2009). Applying the theatre metaphor to integrated media depicting geography. *Cartographic Journal, 46*(1), 24–35.

Condee, W. F. (1995). *Theatrical space: A guide for directors and designers*. Lanham, MD: Scarecrow Press.

Malina, J., & Beck, J. (1971). *Paradise now: A collective creation of the living theatre written down by Judith Malina and Julian Beck* (p. 1971). New York: Vintage Books.

Schechner, R. (1985). *Between theatre and anthropology*. Philadelphia: University of Pennsylvania Press.

Schechner, R. (1973). *Environmental theatre*. New York: Hawthorn Books.

Tillis, S. (2011). Conceptualizing space: The geographic dimension of world theatre. *Theatre Survey, 52*(2), 301–327.

Wiles, D. (2003). *A short history of western performance space*. Cambridge: Cambridge University Press.

Zarzosa, A. (2010). Melodrama and the modes of the world. *Discourse, 32*(2), 236–255.

Index

S.D. Brunn and M. Dodge (eds.), *Mapping Across Academia*,
DOI 10.1007/978-94-024-1011-2

MIX
Papier aus verantwortungsvollen Quellen
Paper from responsible sources
FSC® C105338

If you have any concerns about our products,
you can contact us on
ProductSafety@springernature.com

In case Publisher is established outside the EU,
the EU authorized representative is:
Springer Nature Customer Service Center GmbH
Europaplatz 3, 69115 Heidelberg, Germany

Printed by Libri Plureos GmbH
in Hamburg, Germany